Dear Student

Welcome to your new American Book Company Learning Program! This book has been created especially for you. Our writers have covered 100% of the standards and concepts as clearly and simply as possible.

As with every ABC Learning Program, this book now comes with an eBook to expand your educational experience! There are instructions on the following pages that show you how to access your eBook. We hope that having this digital copy will allow you to go deeper into your studies.

We look forward to hearing of your success!

E-Mail Address

Password

Forgot Your Password?

New to American Book Company? Sign Up!

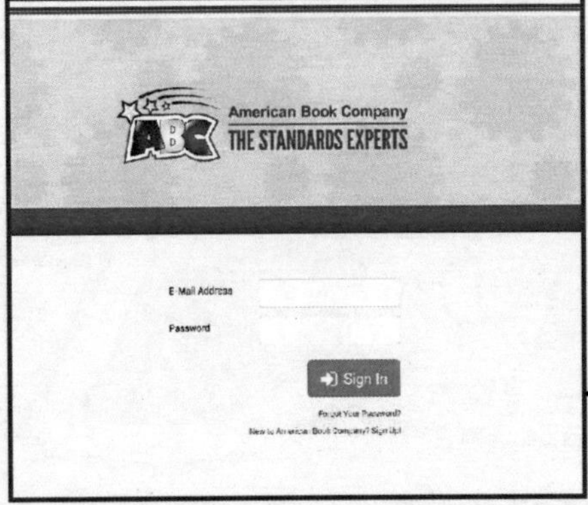

1 - Go to: http://americanbookcompany.com/ebooks

2 - Click on **"New to American Book Company? Sign Up!"**
Then Follow steps in the Registration Process using RedShelf.

3 - After registration you will be taken to your **"My Shelf"** page.
Here you will be able to redeem the access codes for you eBooks via the **"Redemption Code"** box. Each book will show up in **"My Shelf."**

Your eBook codes can be located on the front cover of your workbook.

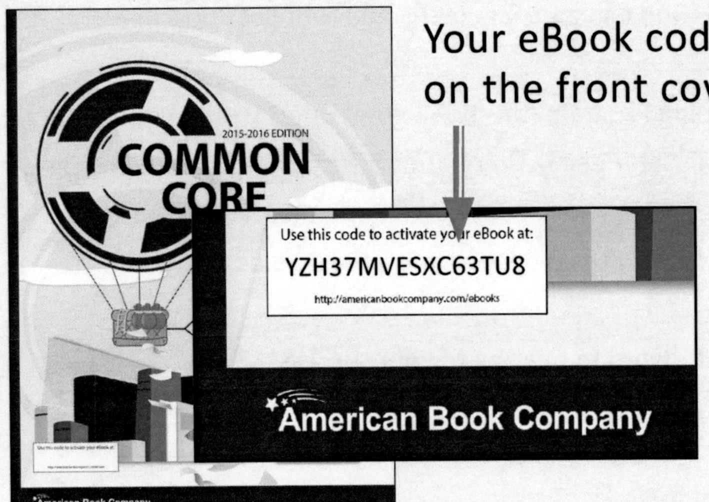

eBook Features

- Linked table of contents
- Searchable text for keywords
- Wikipedia and Google key term define
- Note-taking, highlighting, and free draw tool

PASSING THE NEVADA HIGH SCHOOL PROFICIENCY EXAMINATION

IN MATHEMATICS

REVISED 2010

COLLEEN PINTOZZI
ERICA DAY

AMERICAN BOOK COMPANY
P O BOX 2638
WOODSTOCK, GEORGIA 30188-1383
TOLL-FREE PHONE: 1-888-264-5877 TOLL-FREE FAX: 1-866-827-3240
Web site: www.americanbookcompany.com

P.O. Box 2638
Woodstock, GA 30188-1383

Printed in the United States of America
09/10

TABLE OF CONTENTS

PASSING THE NEVADA PROFICIENCY EXAMINATION IN MATHEMATICS

Preface

PASSING THE NEVADA HIGH SCHOOL PROFICIENCY EXAMINATION IN MATHEMATICS will help students preparing for the mathematics portion of the exam. This book will also assist students who have failed the mathematics exam and who want to review concepts, skills, and strategies before taking the exam again. **The materials in this book are based on the objectives and content descriptions for the Nevada High School Proficiency Exam in Mathematics published by the Nevada Department of Education.**

This book contains several sections. These sections are as follows: 1) General information about the exam; 2) A Diagnostic Exam; 3) Chapters that teach the concepts and skills emphasized on the exam; 4) Two Practice Exams. Answers to the exams and exercises are in a separate manual.

We welcome comments and suggestions about the book. Please contact the author at

American Book Company
PO Box 2638
Woodstock, GA 30188-1383

Toll Free: 1 (888) 264-5877
Phone: (770) 928-2834
Fax: (770) 928-7483
Web site: www.americanbookcompany.com

ABOUT THE AUTHORS

Colleen Pintozzi has taught mathematics at the middle school, junior high, senior high, and adult level for 22 years. She holds a B.S. degree from Wright State University in Dayton, Ohio and has done graduate work at Wright State University, Duke University, and the University of North Carolina at Chapel Hill. She is the author of numerous mathematics books including such best-sellers as ***Basics Made Easy: Mathematics Review*, *Passing the LEAP GEE in Math 2nd Edition*, *Passing the California Aglebra I State Exam*, *Passing the Maryland Algebra/Data Analysis HSA*,** and ***California Mathematics Review II.***

Erica Day has a Bachelor of Science degree in Mathematics with high honors and has completed all the courses for her Master of Science degree in Mathematics. She has tutored all levels of mathematics ranging from high school level algebra and geometry to university level statistics, calculus, and linear algebra. While working with American Book Company, she has authored numerous books such as *Passing the Georgia Algebra I End of Course*, *Targeting the AIMS in Mathematics, New SAT Math, ACT Mathematics Test Preparation Guide, and Passing the Indiana ISTEP+ GQE in Mathematics* to name a few.

PREPARING FOR THE NEVADA HIGH SCHOOL PROFICIENCY EXAMINATION IN MATHEMATICS

INTRODUCTION

If you are a student in a Nevada school district, you must pass the **Nevada High School Proficiency Examination in Mathematics** to receive a high school diploma.

In this book, you'll prepare for the **Nevada High School Proficiency Examination in Mathematics**. The questions and answers that follow will provide you with general information about this exam.

In this book, you'll take a **Diagnostic Exam** to determine your strengths and areas for improvement. In the chapters, you'll learn and practice the skills and strategies that are important in preparing for this exam. The last sections contain two practice exams that will provide further preparation for the actual **Nevada High School Proficiency Examination in Mathematics**.

What Is On the Nevada High School Proficiency Examination in Mathematics?

The mathematics portion of the **Nevada High School Proficiency Examination** is administered in two parts each having about 30 questions. All questions are multiple-choice.

Why Must I Pass the Nevada High School Proficiency Examination in Mathematics?

You are required to pass this exam for several reasons. First, the state of Nevada, your future employers, and your community need an educated workforce. Secondly, today's high school graduates will need to adapt to rapidly changing technology throughout their lives. Employees without basic mathematics skills in computation, measurement, and geometry will be unable to compete in the workplace. Without these skills, there's a great chance they will not only be unemployed but unemployable. Thirdly, by demonstrating your mathematics ability, you can show what you have learned in school and apply this knowledge to new situations and experiences.

When Do I Take the Nevada High School Proficiency Examination in Mathematics?

During the 10th grade, you must take the **Nevada High School Proficiency Examination in Mathematics** for the first time.

Is The Nevada High School Proficiency Examination in Mathematics timed?

You should have plenty of time to complete this test. Be sure to answer all of the questions even if you must make your best guess for some of the answers. You should complete each part of the test in approximately 45 minutes, but you may be given more if needed. A short break is permitted after the completion of part 1.

May I Use a Calculator on the Nevada High School Proficiency Examination in Mathematics?

No, you may not use a calculator on the Nevada High School Proficiency Examination in Mathematics.

What Happens If I Don't Pass the Exam?

You will have at least 3 opportunities to pass the exam before graduation.

If I Fail the Mathematics Exam, Where Can I Get Help To Pass It the Next Time?

In some school districts, you may be able to sign up for special classes. The instructors in these classes will teach you how to study and prepare for the **Nevada High School Proficiency Examination in Mathematics.** You can also seek extra help in mathematics classes during your years in high school. Finally, you may work with tutors or counselors who can help you pass the **Mathematics Exam.**

NEVADA STANDARDS CORRELATION CHART

Prioritized Standards	Chapter(s)	Diagnostic	Practice Exam 1	Practice Exam 2
1.12.6	1	57	2, 16,	7, 12, 28, 32
1.12.7	1,2	18, 60	1, 19,	16
1.12.8	1	2, 23, 45	51	42, 55
2.12.1	3	29, 31, 32, 49	5, 24, 40	25
2.12.2	6	10, 11, 19, 47, 54	38, 50, 59	5, 6, 46
2.12.3	4,5,6,7	8, 25, 33, 56	4, 11, 13, 14, 18, 45, 39	10, 27, 37, 49, 52
2.12.4	6,10	15, 24, 58	29, 30,	17, 26, 33, 54
2.12.5	9,	39	22, 28, 35,	22, 40, 47
2.12.6	7,8	3, 12, 35, 50, 51, 52, 53	20, 25, 60	14, 20, 50
3.12.2	12,	27	36, 44	58, 59
3.12.3	12	4, 16,	12, 17,	48
3.12.4	18	1, 46	21, 27, 57	1, 44
3.12.5	15,16	7, 30, 34, 38	55	9, 13, 18, 31, 43
4.12.1	15	5, 44	3, 46	21, 56
4.12.2	14	14, 20	52	15, 45
4.12.5	9,11,13	26, 40, 42, 59	31, 33	8, 35, 53
4.12.6	13		37, 47	11, 34, 41
4.12.7	14	22	34, 43	19, 30
4.12.9	17	21,	23, 48, 56	60
5.12.1	19		41	39
5.12.2	19	6, 13, 28, 37, 50, 51	7, 9, 10, 42	3, 36, 51
5.12.3	21	40	32	
5.12.4	20	9, 43, 48	15, 26, 58	24, 29, 57
5.12.5	21	17, 36, 55	6, 8, 49	2, 4, 23, 38

NOTES

Nevada High School Proficiency Diagnostic Exam in Math

The following Diagnostic Exam will assess your strengths as well as identify areas for improvement. The following directions are similar to the directions you will read on the actual Nevada High School Proficiency Exit Exam in Mathematics. The actual Math Exam is timed, and you will be given a bubble sheet on which to mark your answers. If you are using a timed format for this diagnostic test and/or you are using a bubble sheet for scoring, follow the directions below carefully.

DIRECTIONS

DO NOT WASTE TIME ON DIFFICULT QUESTIONS. If a question is particularly difficult and taking a lot of time, go on to the next one and come back to it later. ***Be sure to skip this answer on your answer sheet as well!*** Make sure that the question that you are filling in on the answer sheet is ***the same number as the problem you are working on!***

ANSWER AS MANY QUESTIONS AS YOU CAN IN THE TIME PROVIDED. You should have plenty of time to answer all of the questions on this test. If the test administrator says that time is running out, however, you may wish to make ***your best guess*** on the questions that you have not yet completed.

EACH PROBLEM HAS FOUR POSSIBLE ANSWERS, LABELED A, B, C, AND D. Be sure that the problem number on the answer sheet matches the number on the test and then mark your answer by filling in the space that contains the letter of the correct answer–either A, B, C, or D. ***Be sure to fill in only one answer on the answer sheet for each question, or the question will be marked wrong!***

IF YOU NEED TO CHANGE AN ANSWER ON YOUR ANSWER SHEET, BE SURE TO ERASE YOUR FIRST MARK COMPLETELY! There can be ***only one mark that shows for each question*** or the question will be counted as wrong. ***Do not make any stray marks*** on the answer sheet!

Nevada HSPE Mathematics Formula Sheet

Note to student: You may use these formulas throughout the entire test. Feel free to use this sheet as needed during your testing time.

Parallelogram

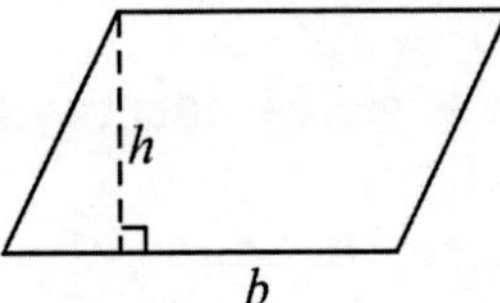

Area: $A = bh$

Trapezoid

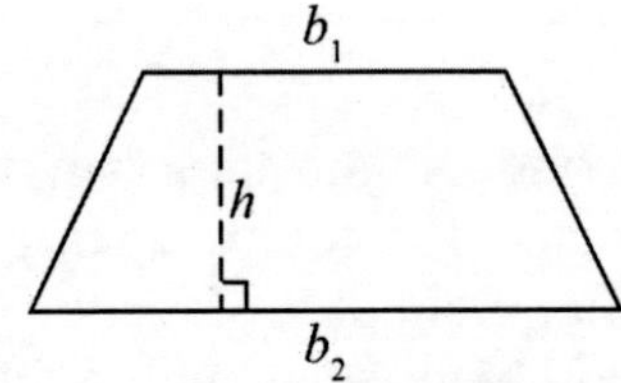

Area: $A = \frac{1}{2}(b_1 + b_2)h$

Circle

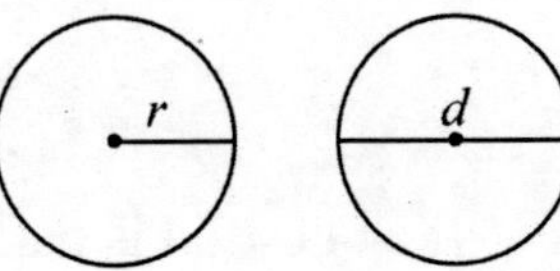

Circumference: $C = 2\pi r^2$

$C = \pi d$

Area: $A = \pi r^2$

Rectangular Solid

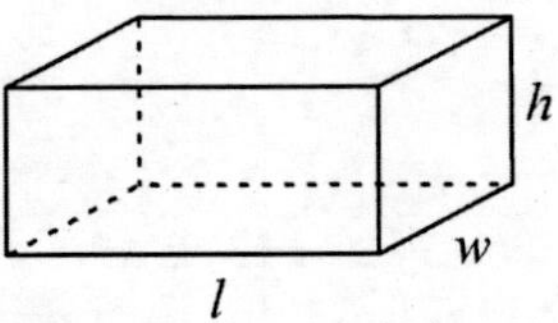

Volume: $V = lwh$

Surface Area: $SA = 2lw + 2lh + 2hw$

Pythogorean Theorem

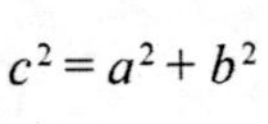

$c^2 = a^2 + b^2$

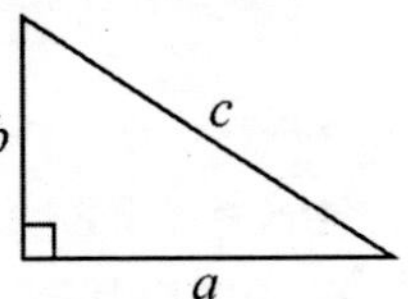

Cylinder

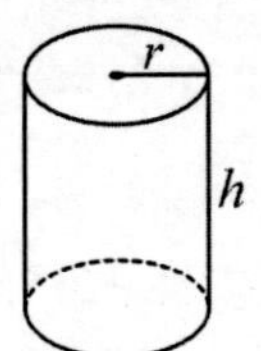

Cone

Volume: $V = \pi r^2 h$

Volume: $V = \frac{1}{3}\pi r^2 h$

Trigonomic Ratios

$\sin x = \frac{a}{c}$

$\cos x = \frac{b}{c}$

$\tan x = \frac{a}{b}$

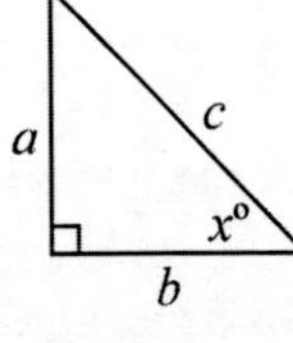

Permutations

$${}_nP_k = \frac{n!}{(n-k)!}$$

Combinations

$${}_nC_k = \frac{n!}{k!(n-k)!}$$

Special Right Triangles

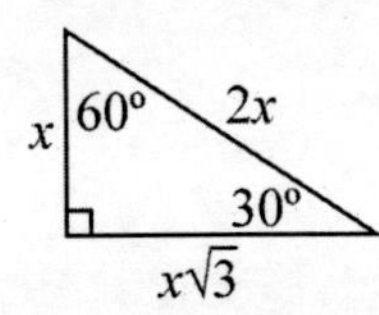

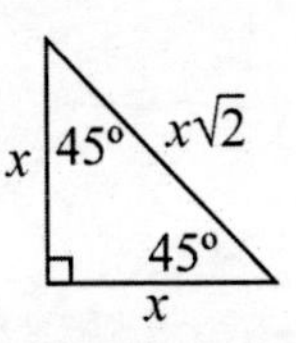

Temperature Formulas

$°F = \frac{9}{5}C + 32 \quad °C = \frac{5}{9}(F - 32)$

Math Diagnostic Exam: Part 1

1. Derek was shopping for tennis shoes that
3.12.4 normally sell for \$100. He went to 4 different stores. Which store had the best price?

Store A: The shoes were on sale for \$85.00
Store B: The shoes were selling for $\frac{1}{4}$ off regular price.
Store C: The shoes were reduced by 20%.
Store D: The shoes were reduced by \$27.25 off the original price.

A. Store A
B. Store B
C. Store C
D. Store D

2. Pat wants to divide 7.86 by 3.9, but he
1.12.8 forgets to enter the decimal points when he puts the numbers into the calculator. Using estimation, where should Pat put the decimal point?

A. 0.2015385
B. 2.015385
C. 20.15385
D. 201.5385

3. Factor: $x^2 - x - 12$
2.12.6

A. $(x-6)(x+2)$
B. $(x-3)(x-4)$
C. $(x-3)(x+4)$
D. $(x+3)(x-4)$

4. Boyle's Law is stated by the formula:
3.12.3

$$PV = pv$$

Find V when $P = 220$, $p = 25$, and $v = 440$.

A. 12.5
B. 50
C. 100
D. 200

5. In the diagram below, which arc is a
4.12.1 major arc?

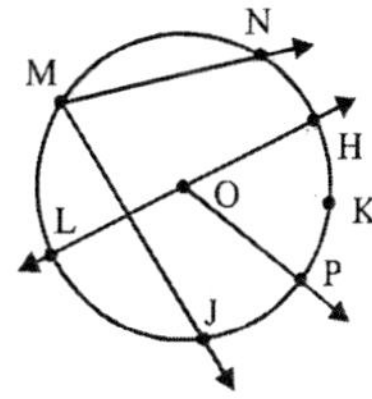

A. $\overset{\frown}{HKP}$
B. $\overset{\frown}{LMN}$
C. $\overset{\frown}{HMP}$
D. $\overset{\frown}{NKJ}$

6. Tom's math grades so far this semester
5.12.2 have been 94, 72, and 50. What is the mean of his grades?

A. 53
B. 72
C. 77
D. 86

7. If you double the radius of a circle, how
3.12.5 much does the area increase?

A. The area is four times greater.
B. The area doubles.
C. The area decreases by half.
D. The area is three times greater.

8. What is the perimeter of the rectangle
2.12.3 below?

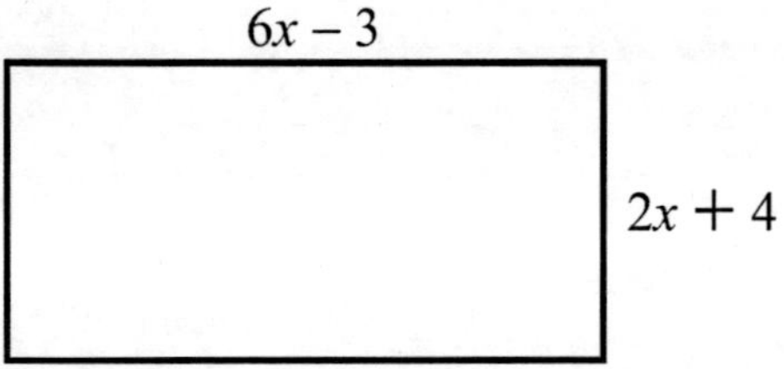

A. $8x - 12$
B. $16x + 2$
C. $24x^2 - 24$
D. $12x^2 - 12$

9. The Joyful Juice Co. sells 6 different
5.12.4 fruit juices. The sales manager has decided to conduct a taste test on juice blends. Each blend will consist of equal parts of one of the 6 fruit juices and one other juice, producing blends such as apple-grape, cherry-orange, etc. Using the 6 fruit juices, how many two-juice mixes can be made?

A. 12
B. 15
C. 30
D. 36

10.
2.12.2

Appliance	Ohms	Volts	Amps
Coffee maker	20	110	5.0
Toaster	22	110	5.5
Radio	220	110	0.5
Dryer	25	220	8.8
Dishwasher	70	220	3.1

Watts = Volts × Amps

Using this formula and the chart above, how many watts of electrical power are required to operate a toaster?

A. 115.5
B. 121
C. 550.5
D. 605

11. The number of feet any object will fall
2.12.2 in a number of seconds is found by the formula:

$$s = \frac{1}{2}gt^2$$

s = the number of feet an object will fall
t = the number of seconds
g = 32 feet per second per second

If an object is dropped from an airplane, how far will it fall in 10 seconds? (Use the formula in the box above.)

A. 160 feet
B. 1600 feet
C. 320 feet
D. 3200 feet

12. After t seconds, the height in feet of
2.12.6 an object with an initial upward velocity of v feet per second and an initial height of h feet is given by the equation:

$$h(t) = -16t^2 + vt + h$$

If Amanda hit a pop fly ball with an initial upward velocity of 50 feet per second and an initial height of 3 feet, what would be the height of the ball after 1.5 seconds (assuming no one catches the ball)?

A. 37 ft
B. 38 ft
C. 39 ft
D. 42 ft

13. A neighborhood surveyed the times of
5.12.2 day people water their lawns and tallied the data below.

Time	Tally
midnight - 3:59 a.m.	II
4:00 a.m. - 7:59 a.m.	卌 I
8:00 a.m. - 11:59 a.m.	卌 IIII
noon - 3:59 p.m.	卌
4:00 p.m. - 7:59 p.m.	卌 卌
8:00 p.m. - 11:59 p.m.	卌 III

If you wanted to find which was the most popular time of day to water the lawn, it would be best to find the ____________ of the data.

A. mean
B. median
C. mode
D. range

14. A 16-foot ramp is to be installed
4.12.2 between a parking area and a 5 foot high loading dock as shown in the figure below.

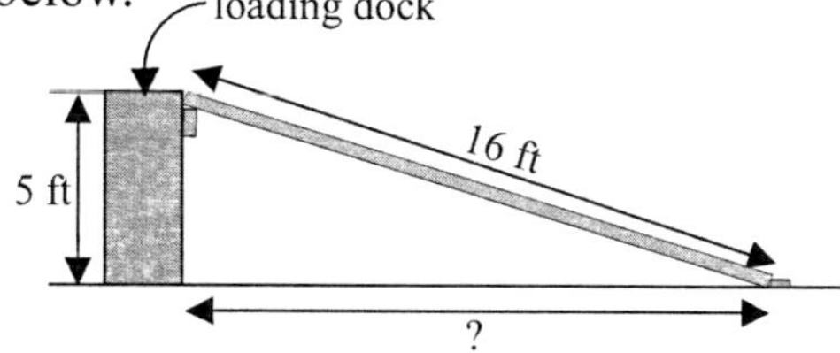

How far will the end of the ramp be from the base of the loading dock?

A. $\sqrt{256}$ ft
B. $\sqrt{231}$ ft
C. $\sqrt{121}$ ft
D. 121 ft

15. Solve: $|2n - 4| < 6$
2.12.4

A. $5 > n > 2$
B. $-2 < n < 5$
C. $-1 < n < 5$
D. $5 > n > 1$

16. Skip paints houses for a living. To
3.12.3 ensure that the paint dries properly, he only paints when the temperature is greater than 50°F. Below is a chart showing the projected temperatures for the next 4 days.

DAY	TEMP
Tues	15°C
Wed	10°C
Thurs	11°C
Fri	9°C

The formula for converting Celsius to Fahrenheit is $F = \frac{9}{5}C + 32$

Which of the following describes Skip's painting schedule for the next four days?

A. Skip can paint Tues, Wed, Thurs and Fri.
B. Skip can paint on Thurs and Fri only.
C. Skip can paint on Tues and Thurs only.
D. Skip can paint on Tues only.

17. Sarah deposits 50¢ into Miss Clucky, a
5.12.5 machine that makes chicken squawks and gives Sarah one plastic egg with a toy surprise. In the machine, 30 eggs contain a rubber frog, 43 eggs contain a plastic ring, 23 eggs contain a necklace, and 18 eggs contain a plastic car. What is the probability that Miss Clucky will give Sarah a necklace in her egg?

A. $\frac{1}{114}$

B. $\frac{23}{114}$

C. $\frac{23}{91}$

D. $\frac{1}{5}$

18. Solve the equation $x = \sqrt{4x - 3}$
1.12.7

A. $x = 1, 3$
B. $x = \pm\sqrt{3}$
C. $x = -1, -3$
D. $x = 2\sqrt{3}, -2\sqrt{3}$

19. Solve $I = PRT$ for R.
2.12.2

A. $R = IPT$
B. $R = I + PT$
C. $R = I - PT$
D. $R = \frac{I}{PT}$

20. The measures of the sides of triangles are given below. Which of the following triangles is similar to a triangle with sides 18, 21, and 27?
4.12.2

A. 12, 14, 18
B. 15, 18, 24
C. 20, 23, 29
D. 24, 36, 28

21. Four workers have been hired to assemble computer hard drive units. Their first weeks' production results are listed below.
4.12.9

	Number of defective units	Number of units assembled
Audra	11	208
Jessica	12	239
Krista	9	125
Megan	10	197

Krista claims that she is best at assembling the units because only 9 of the units she assembled were defective. Is this a valid conclusion?

A. Yes
B. No, because a higher percentage of her units were defective.
C. No, because a lower percentage of her units were defective.
D. No, because she didn't assemble as many units as the others.

22. For the triangle below, which expression could you use to determine the length of x?
4.12.7

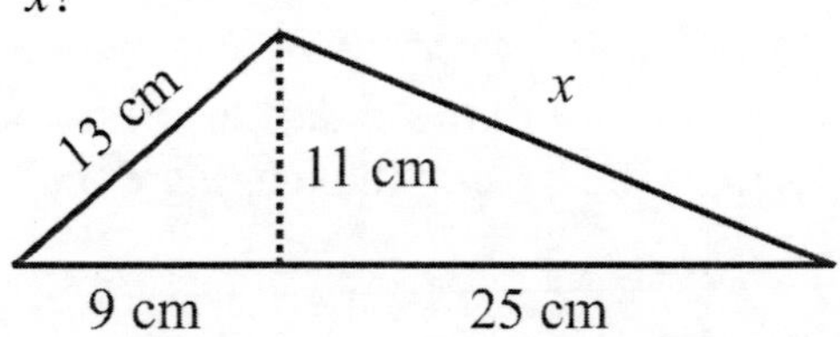

A. $\sqrt{34^2 + 11^2}$

B. $\sqrt{25^2} + \sqrt{11^2}$

C. $\sqrt{25^2 + 11^2}$

D. $\frac{25 \times 11}{2}$

23. Which of the following computations
1.12.8 will result in an irrational number?

A. 7π
B. $3\frac{1}{2} + 7\frac{1}{4}$
C. $6.8 - 3.9$
D. $5 - \frac{1}{2}$

24. The following graph depicts the height
2.12.4 of a projectile as a function of time.

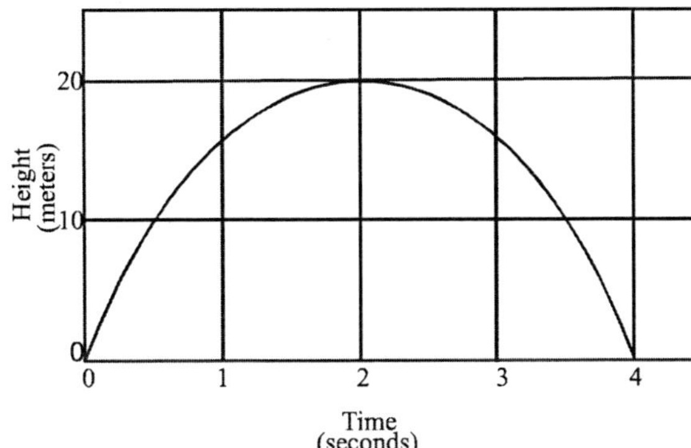

What is the domain (D) of this function?

A. 0 meters ≤ D ≤ 20 meters
B. 4 seconds ≤ D ≤ 20 meters
C. 20 meters ≤ D ≤ 4 seconds
D. 0 seconds ≤ D ≤ 4 seconds

25. Find the difference:
2.12.3 $(x^2 + 4x - 7) - (2x^2 + x - 1)$

A. $-3x^2 - 3x - 8$
B. $-x^2 - 3x - 6$
C. $-x^2 + 3x - 6$
D. $-2x^2 + 4x + 7$

26. Which graph represents the equation
4.12.5 $y = -2x - 1$?

A.
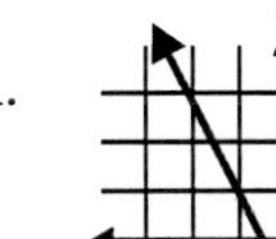
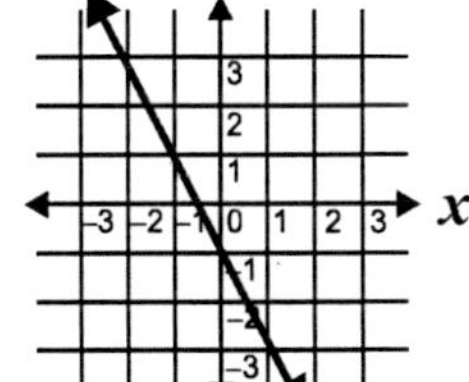

B.
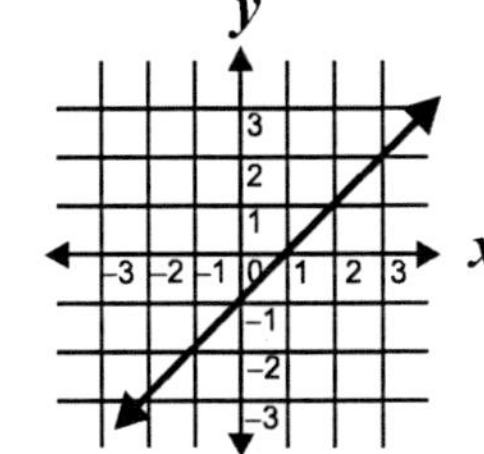

C.
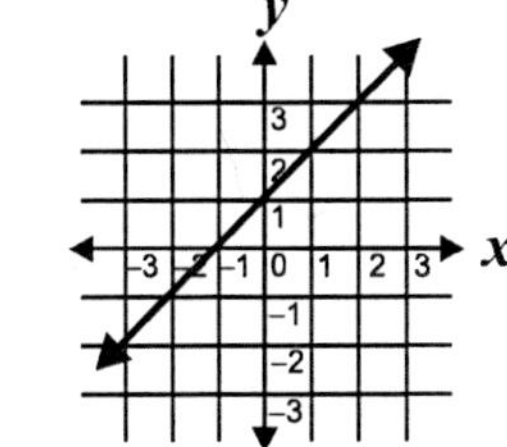

D.
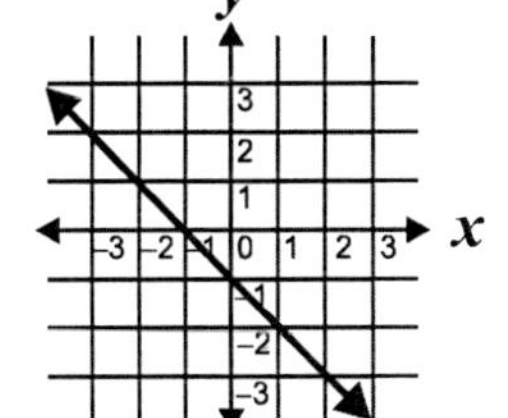

27. Choose which of the following is the
3.12.2 most precise measurement.

A. 45.26 m
B. 310 m
C. 4,200 m
D. 87.9 m

28. Justin measured the heights of 10
5.12.2 basketball players. Their heights, in inches, are given below.
71, 82, 72, 78, 73, 76, 72, 75, 73, 78
What is the median height of the 10 basketball players?

A. 73 inches

B. 74 inches

C. 75 inches

D. 76 inches

29. It was 61°F at 5:00 a.m. By 3:00 p.m.
2.12.1 the temperature rose to 106°F. If the temperature rose at a constant rate, what was the temperature at 11:00 a.m?

A. 83.5°F
B. 88°F
C. 86°F
D. 91°F

30. Find the area of the rectangle below.
3.12.5

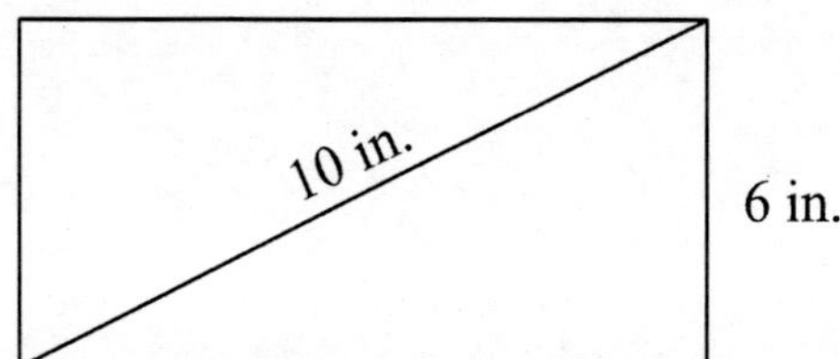

A. 42 sq. in.
B. 48 sq. in.
C. 54 sq. in.
D. 60 sq. in.

31. Nancy is watching her nephew, Drew,
2.12.1 arrange his marbles in rows on the kitchen floor. The figure below shows the progression of his arrangement.

Row 1
Row 2
Row 3
Row 4

Which expression below will show how many total marbles there are for n rows?

A. 2^{n-1}
B. $n^2 - n$
C. $2n^2 - n$
D. $n^2 - 1$

32. Handy and Andy's Electric Repair
2.12.1 Service charges a fixed amount for a service call plus a charge for each hour worked. The charge for each hour is the same. Use the chart below to determine the cost of a service call plus 8 hours of work.

Number of Hours	Cost
1	\$63
2	\$81
3	\$99
4	\$117

A. \$26
B. \$ 98
C. \$144
D. \$189

Math Diagnostic Exam: Part 2

33. Simplify $2^3 \times 2^5$
2.12.3

A. 2
B. 2^{-2}
C. 2^{15}
D. 2^8

34. Emily needs to make a glass case with
3.12.5 the following measurements:

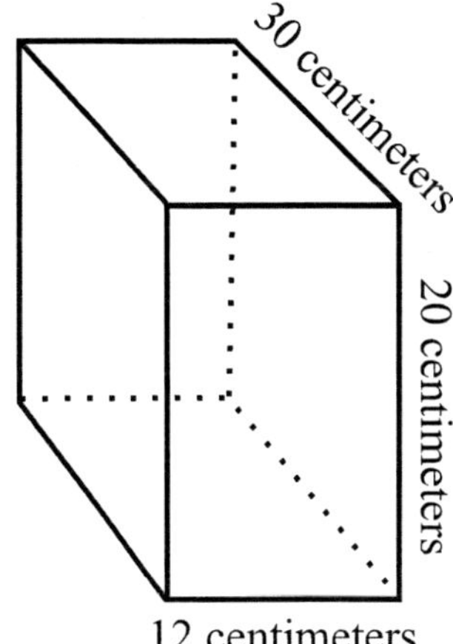

How many square centimeters of glass would it take to construct the case enclosed on all sides?

A. 60 square centimeters
B. 612 square centimeters
C. 2,400 square centimeters
D. 6,200 square centimeters

35. The regular price of a stereo (r) is \$560.
2.12.6 The stereo is on sale for 25% off. Which equation will help you find the sale price (s) of the stereo?

A. $s = r - 0.25$
B. $s = r - 0.25s$
C. $s = r - 0.25r$
D. $s = r - s$

36. David owns a dog named Wishes.
5.12.5 David reached into his box of 8 lamb-flavored, 12 liver-flavored, 6 chicken-flavored, and 20 milk-flavored dog biscuits and gave one to Wishes without looking. What is the probability of Wishes getting a liver treat?

A. $\frac{1}{12}$
B. $\frac{6}{17}$
C. $\frac{6}{23}$
D. $\frac{1}{23}$

37. Find $c : \frac{c}{2} > -6$
2.12.4
A. $c > -12$
B. $c > 12$
C. $c < 12$
D. $c < -12$

38. Two picture frames have the same area.
3.12.5 The first is a square with 12 inch sides. The second is a rectangle with a width of 9 inches. What is the length of the second frame?

A. 8 inches
B. 12 inches
C. 16 inches
D. 24 inches

39. According to this chart, what are the
2.12.5 solution(s) (x-intercepts) to the equation $f(x) = -9x^2 + 9$?

x	y
−2	−27
−1	0
0	9
1	0
2	−27

A. 9
B. −1, 1
C. −1, 1, 9
D. None of the above

40. Tom's school was considering making
5.12.3 uniforms mandatory starting with the next school year. Tom hated the idea and wanted to do his own survey to see if parents were really in favor of it. He considered 4 places to conduct his survey. Which would give the most valid results?

A. He would stop people at random walking through the mall.
B. He would survey parents in the car pool lanes picking up students after school.
C. He would survey the teachers after school.
D. He would survey the students in his biology class to ask what their parents thought.

41. In the triangle below, what is the length
4.12.5 of side x?

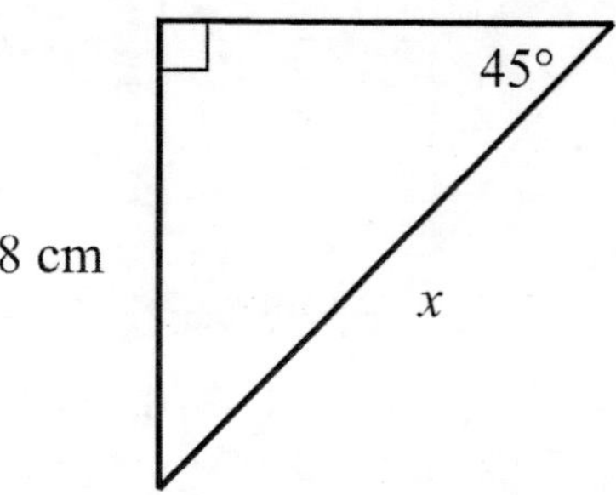

A. 16 cm
B. $8\sqrt{2}$ cm
C. 10 cm
D. $2\sqrt{8}$ cm

42. Which graph shows line m and n as a system of equations with an infinite number of solutions?
4.12.5

A.
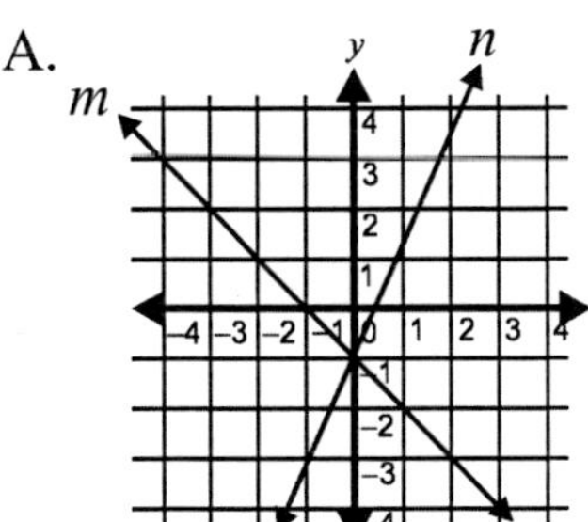

B.
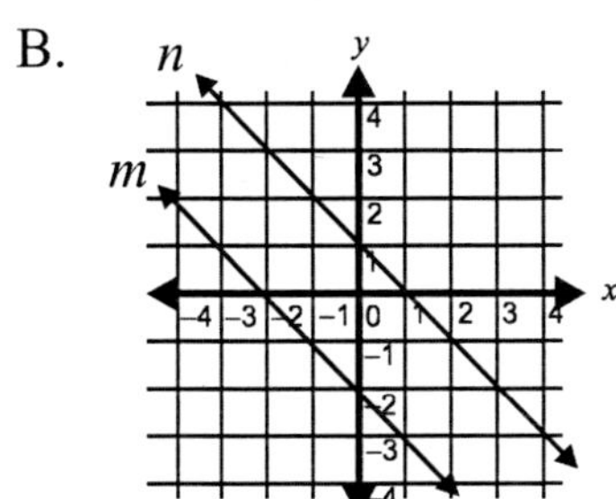

C.
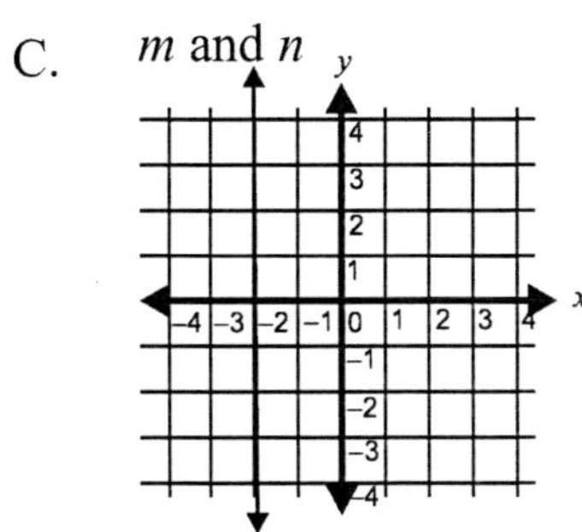

D.
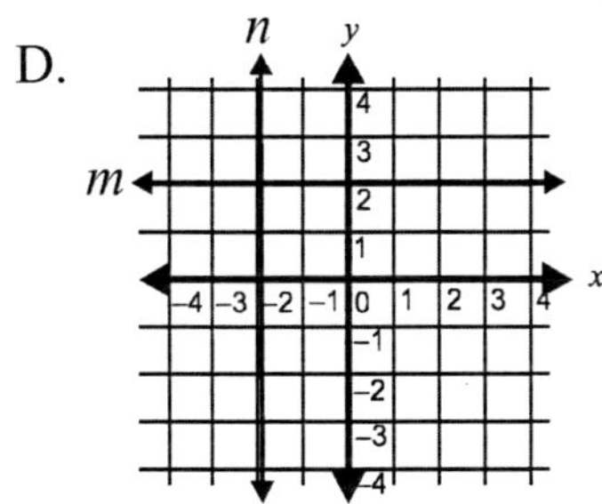

43. Terry is ordering a new truck. He has a choice of 6 exterior colors, 3 interior colors, 3 sound system,s and 2 types of seats. Which expression below represents the number of different colors, sound systems, and seat type combinations that are available for Terry to choose from?
5.12.4

A. $6 + 3 + 3 + 2$
B. $6 (3 + 3 + 2)$
C. $6 \times 3 \times 3 \times 2$
D. $6 \times 4 \times 3 \times 2$

44. Triangle ABC, shown in the diagram below is an isosceles triangle.
4.12.1

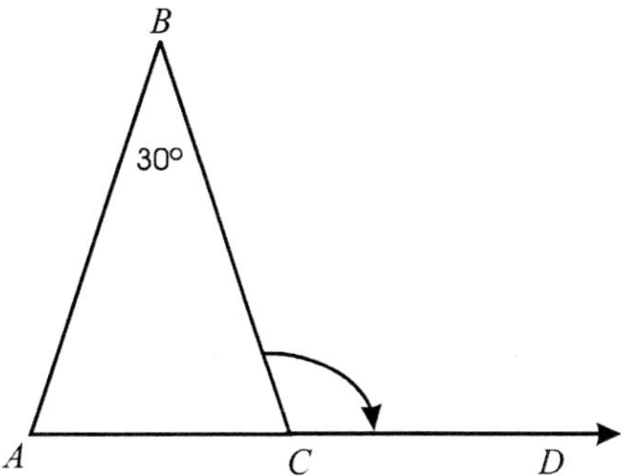

What is the measure, in degrees, of $\angle BCD$?

A. 75°
B. 105°
C. 150°
D. 165°

45. Which equation illustrates the distributive property?
1.12.8

A. $(4 + x) + 3x = 3x + (4 + x)$
B. $3x(4 + x) + 0 = 3x(4 + x)$
C. $(4 + x) + 3x = 4 + (x + 3x)$
D. $3x(4 + x) = 3x(4) + 3x(x)$

46. Billy bought the following items at the store:
3.12.4

1 gal milk	\$2.17
2 lb hamburger	\$3.98
3 donuts	\$0.99
1 tube of toothpaste	\$2.79
1 pair of socks	\$1.57

He paid 6% sales tax. What was the total cost of his purchases?

A. \$11.50
B. \$11.70
C. \$12.19
D. \$13.65

47. Ralph has borrowed \$600 for 2 years at an annual rate of 5%.
2.12.2
Use the formula $I = prt$ to find the amount of interest he will pay.

A. \$15
B. \$24
C. \$60
D. \$150

48. Zandra went into a candy store to buy jelly beans. She wanted to choose 3 flavors from the 12 flavors they sold. How many combinations of 3 flavors could she make?
5.12.4

A. 12
B. 36
C. 220
D. 1320

49. What is the missing number in this sequence?
2.12.1

0.04, 0.12, 0.36, ___?___, 3.24

A. 1.08
B. 0.48
C. 0.6
D. 12.96

50. Examine the following two data sets:
5.12.2
Set #1: 49, 55, 68, 72, 98
Set #2: 20, 36, 47, 68, 75, 82, 89
Which of the following statements is true?

A. They have the same mode.
B. They have the same median.
C. They have the same mean.
D. None of the above.

51. The information in the table below shows the number of games won by a number of baseball teams.
5.12.2

Team	**Games Won**
Raccoons	8
Armadillos	10
Cougars	3
Panthers	1
Bears	6
Lions	13

Which measure of central tendency is most representative of the data?

A. mean
B. median
C. range
D. mode

52. An arrow is shot upward with an initial velocity of 128 feet per second. The height (h) of the arrow is a function of time (t) in seconds since the arrow left the ground and can be expressed by the equation $h = 128t - 16t^2$. When will the arrow be at a height of 240 feet?
2.12.6

A. At 3 seconds and at 5 seconds
B. Only at 3 seconds
C. Only at 5 seconds
D. Only at 8 seconds

53. The table shows values of x and y for the equation $y = x^3 - x^2 - 2x$.
2.12.6
What is the value of y when $x = 5$?

x	y
0	0
1	−2
2	0
3	12
4	40
5	?

A. −90
B. 160
C. 90
D. −160

54. If $2a = 12$, then $a =$
2.12.4

A. 3

B. 6

C. 4

D. 2

55. Adam has a bag containing 6 white marbles and 4 black marbles. Bob's marble bag contains 2 white marbles, 4 red marbles, and 3 blue marbles. If Adam and Bob each randomly take one marble from their respective bags, what is the probability that both boys will choose white marbles?
5.12.5

A. $\frac{2}{19}$

B. $\frac{2}{15}$

C. $\frac{12}{15}$

D. $\frac{8}{19}$

56. Isabella is simplifying this expression:
2.12.3
$2(5a + 3b - c) - 5(4a - 2b - 3c)$
The expression above is equivalent to which of the following expressions?

A. $-10a + 16b + 13c$
B. $-10a - 4b + 4c$
C. $30a + b + 2c$
D. $30a - 4b + 17c$

57. Simplify: $\sqrt{45} \times \sqrt{27}$
1.12.6

A. $9\sqrt{15}$

B. $\sqrt{72}$

C. $\sqrt{1215}$

D. $\sqrt{9} \times \sqrt{15}$

58. George measured the height of 11
5.12.2 football players in inches.
73, 72, 75, 82, 75, 76, 77, 81, 76, 80, 77
Which of the box-and-whisker plots below correctly represents these heights?

A.

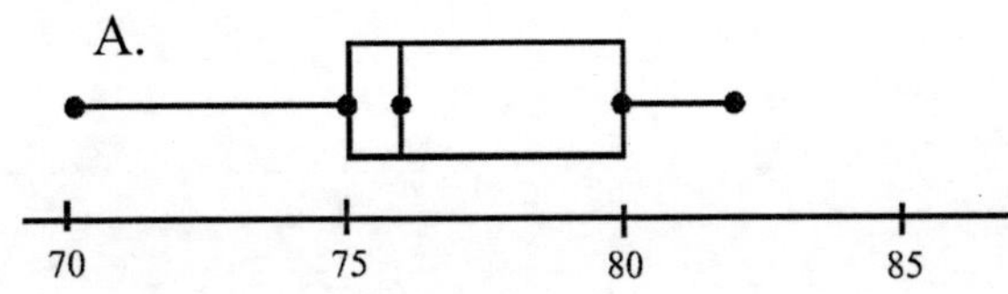

B.

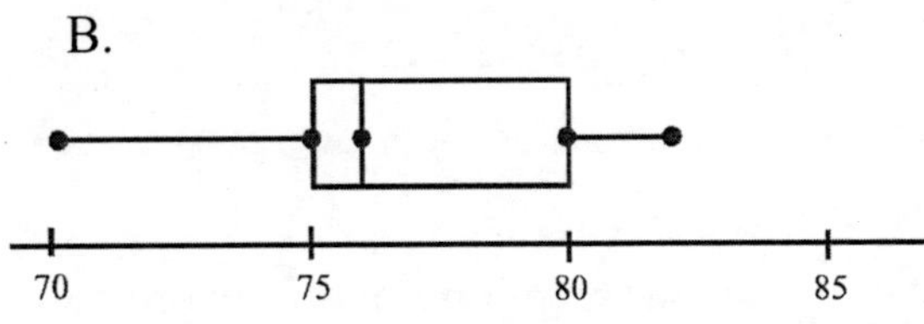

C.

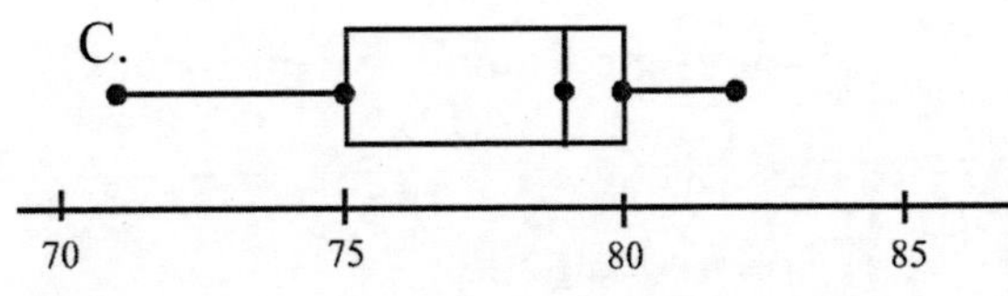

D.

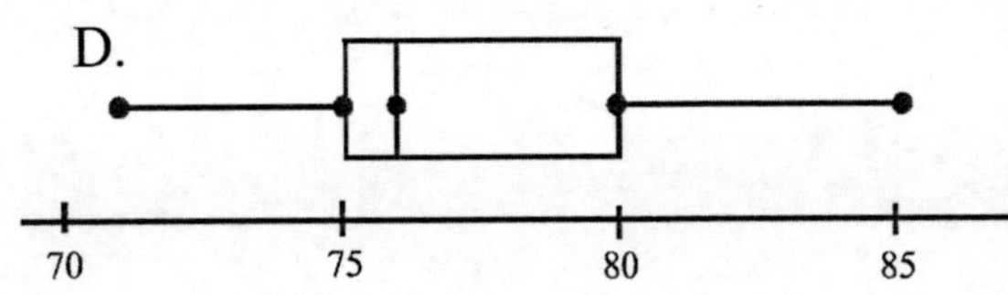

59. Which is the graph of $f(x) = |x|$?
4.12.5

A.

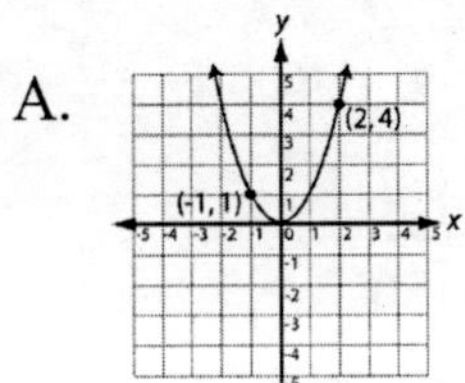

B.

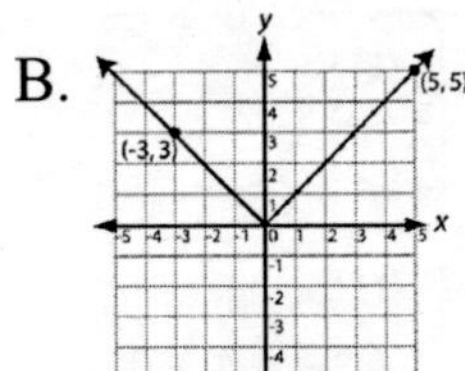

C.

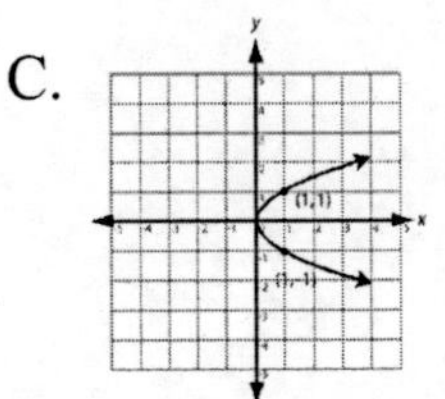

D.

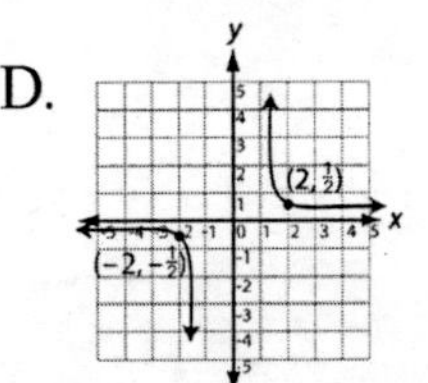

60. A grocery store sells two brands of
1.12.7 whole milk and skim milk in pint sized bottles. Brand A's prices are \$1.13 and \$0.97, respectively. Brand B's prices are \$1.57 and \$0.80. If the grocery store has a sale where everything is 20% off, which matrix below models the sale milk prices?

A. Brand A, Brand B $\begin{bmatrix} \$1.36 & \$1.16 \\ \$1.88 & \$0.96 \end{bmatrix}$

B. Brand A, Brand B $\begin{bmatrix} \$0.91 & \$1.15 \\ \$1.87 & \$0.65 \end{bmatrix}$

C. Brand A, Brand B $\begin{bmatrix} \$1.35 & \$0.79 \\ \$1.30 & \$\ 1.00 \end{bmatrix}$

D. Brand A, Brand B $\begin{bmatrix} \$0.90 & \$0.78 \\ \$1.26 & \$0.64 \end{bmatrix}$

EVALUATION CHART
DIAGNOSTIC MATHEMATICS EXAM

Directions: On the following chart, circle the question numbers that you answered incorrectly, and evaluate the results. Then turn to the appropriate topics (listed by chapters), read the explanations, and complete the exercises. Review the other chapters as needed. Finally, complete the **Nevada High School Proficiency Mathematics Practice Exams** to further prepare yourself for the **Nevada High School Proficiency in Mathematics.**

	QUESTIONS	PAGES
Chapter 1: Number Sense	2,23,45,57	16-22
Chapter 2: Matrices	18, 60	23-31
Chapter 3: Patterns and Problem Solving	29, 31, 32, 49	32-35
Chapter 4: Polynomials	8, 33, 56	36-51
Chapter 5: Solving Equations and Inequalities	37, 54	52-66
Chapter 6: Solving Multistep Equations and Inequalities	10, 11, 15, 19, 47, 53	67-78
Chapter 7: Factoring and Solving Quadratic Equations	3, 12, 25, 52	79-88
Chapter 8: Algebra Word Problems	35	89-97
Chapter 9: Systems of Equations	39, 59	98-108
Chapter 10: Relations and Functions	24	109-119
Chapter 11: Graphing & Writing Equations & Inequalities	26, 41, 42	120-138
Chapter 12: Measurement	4, 16, 27	139-151
Chapter 13: Angles	5, 34	152-172
Chapter 14: Triangles	14, 20, 22	173-196
Chapter 15: Plane Geometry	7, 38, 44	197-216
Chapter 16: Solid Geometry	30	217-225
Chapter 17: Sets and Logic	21	234-240
Chapter 18: Consumer Math	1, 46	233-245
Chapter 19: Data Analysis	6, 13, 28, 50, 51, 58	246-253
Chapter 20: Permutations and Combinations	9, 43, 48	257-268
Chapter 21: Probability	17, 36, 40, 55	269-286

CHAPTER 1: NUMBER SENSE

Standards 1.12.6, 1.12.7and 1.12.8

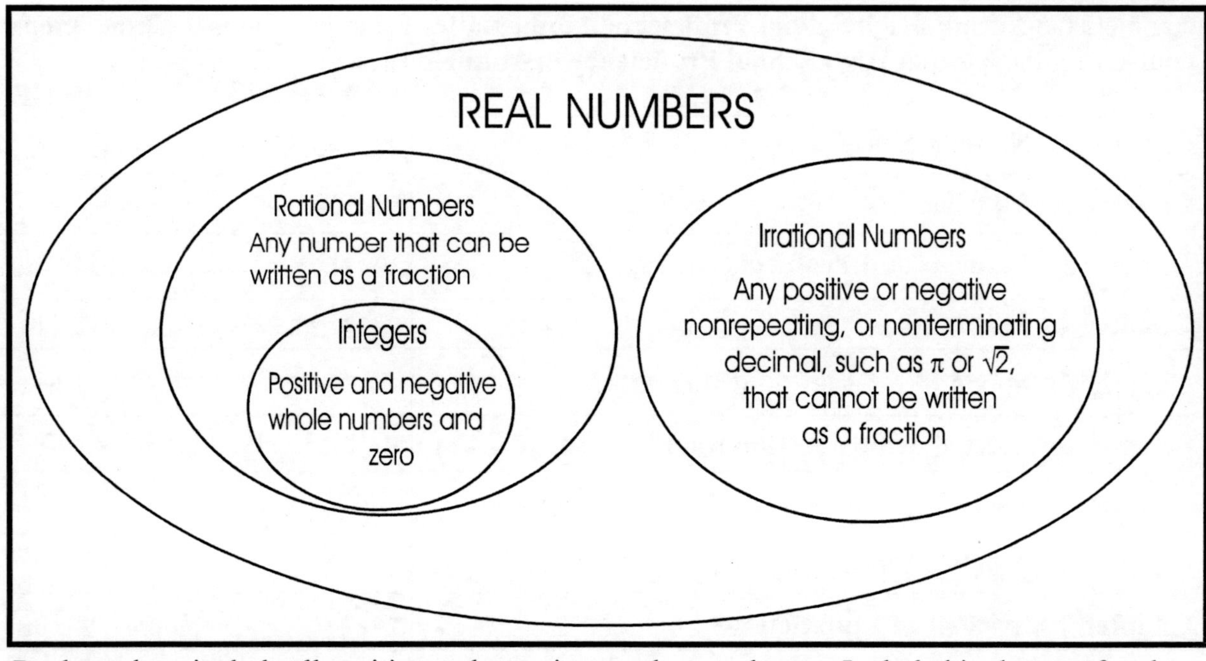

Real numbers include all positive and negative numbers and zero. Included in the set of real numbers are positive and negative fractions, decimals, and rational and irrational numbers.

Use the diagram above and your calculator to answer the following questions.

1. Using your calculator, find the square root of 7. Does it repeat? Does it end? Is it a rational or an irrational number?
2. Find $\sqrt{25}$. Is it rational or irrational? Is it an integer?
3. Is an integer an irrational number?
4. Is an integer a real number?
5. Is $\frac{1}{8}$ a real number? Is it rational or irrational?

Identify the following numbers as rational (R) or irrational (I).

6. 5π
7. $\sqrt{8}$
8. $\frac{1}{3}$
9. -7.2
10. $-\frac{3}{4}$
11. $\frac{\sqrt{2}}{2}$
12. $9 + \pi$
13. 1.0004
14. $-\frac{4}{5}$
15. $1.1\overline{8}$
16. $\sqrt{81}$
17. $\frac{\pi}{4}$
18. $-\sqrt{36}$
19. $17\frac{1}{2}$
20. $-\frac{5}{3}$

PROPERTIES OF ADDITION AND MULTIPLICATION

The Associative, Commutative, and Distributive Properties and the Identity Property of Addition and Multiplication are listed below by example as a quick refresher.

Property	Example
1. Associative Property of Addition	$(a + b) + c = a + (b + c)$
2. Associative Property of Multiplication	$(a \times b) \times c = a \times (b \times c)$
3. Commutative Property of Addition	$a + b = b + a$
4. Commutative Property of Multiplication	$a \times b = b \times a$
5. Distributive Property	$a \times (b + c) = (a \times b) + (a \times c)$
6. Identity Property of Addition	$0 + a = a$
7. Identity Property of Multiplication	$1 \times a = a$
8. Inverse Property of Addition	$a + (-a) = 0$
9. Inverse Property of Multiplication	$a \times \frac{1}{a} = \frac{a}{a} = 1 \quad a \neq 0$

In the blanks provided, write the number of the property listed above that describes each of the following statements.

1. $4 + 5 = 5 + 4$ ________
2. $4 + (2 + 8) = (4 + 2) + 8$ ________
3. $10 (4 + 7) = (10)(4) + (10)(7)$ ________
4. $(2 \times 3) \times 4 = 2 \times (3 \times 4)$ ________
5. $1 \times 12 = 12$ ________
6. $8 (\frac{1}{8}) = 1$ ________
7. $1c = c$ ________
8. $18 + 0 = 18$ ________
9. $9 + (-9) = 0$ ________
10. $p \times q = q \times p$ ________
11. $\mathrm{t} + 0 = \mathrm{t}$ ________
12. $x (y + z) = xy + xz$ ________
13. $(m)(n \cdot p) = (m \cdot n)(p)$ ________
14. $-y + y = 0$ ________

UNDERSTANDING EXPONENTS

Sometimes it is necessary to multiply a number by itself one or more times. For example, in a math problem, you may need to multiply 3×3 or $5 \times 5 \times 5 \times 5$. In these situations, mathematicians have come up with a shorter way of writing out this kind of multiplication. Instead of writing 3×3, you can write 3^2; also, instead of $5 \times 5 \times 5 \times 5$, 5^4 means the same thing. The first number is the **base**. The small, raised number is called the **exponent**. The exponent tells how many times the base should be multiplied by itself.

EXAMPLE 1: 6^3 **This means multiply 6 three times:** $6 \times 6 \times 6$.

You also need to know two special properties of exponents:

1. Any base number raised to the exponent of 1 equals the base number.
2. Any base number raised to the exponent of 0 equals 1.

EXAMPLE 2: $4^1 = 4$ $10^1 = 10$ $25^1 = 25$
$4^0 = 1$ $10^0 = 1$ $25^0 = 1$

Rewrite the following problems using <u>exponents</u>.

Example: $2 \times 2 \times 2 = 2^3$

1. $4 \times 4 \times 4 =$ ______
2. $5 \times 5 \times 5 \times 5 =$ ______
3. $11 \times 11 =$ ______
4. $8 \times 8 =$ ______
5. $17 \times 17 \times 17 =$ ______
6. $12 \times 12 =$ ______
7. $1 \times 1 \times 1 \times 1 \times 1 =$ ______
8. $7 \times 7 \times 7 \times 7 \times 7 \times 7 =$ ______
9. $100 \times 100 \times 100 =$ ______

Write the whole number equivalent of the following.

Example: $2^3 = 2 \times 2 \times 2 = 8$

10. $2^5 =$ ______
11. $10^2 =$ ______
12. $8^3 =$ ______
13. $3^4 =$ ______
14. $25^1 =$ ______
15. $10^5 =$ ______
16. $15^0 =$ ______
17. $9^2 =$ ______
18. $3^3 =$ ______

Express each of the following numbers as a number with an exponent.

Example: $4 = 2 \times 2 = 2^2$

19. $32 =$ ______
20. $64 =$ ______ or ______
21. $1000 =$ ______
22. $27 =$ ______
23. $81 =$ ______ or ______
24. $121 =$ ______
25. $16 =$ ______ or ______
26. $8 =$ ______
27. $49 =$ ______

SQUARE ROOT

Just as working with exponents is related to multiplication, so finding square roots is related to division. In fact, the sign for finding the square root of a number looks similar to a division sign. The best way to learn about square roots is to look at examples.

EXAMPLES: **This is a square root problem:** $\sqrt{64}$

It is asking, "What is the square root of 64?"

It means, "What number multiplied by itself equals 64?"

The answer is 8. $8 \times 8 = 64$.

Look at the square root of the following numbers.

$\sqrt{36}$ $6 \times 6 = 36$ so $\sqrt{36} = 6$ $\sqrt{144}$ $12 \times 12 = 144$ so $\sqrt{144} = 12$

Find the square roots of the following numbers.

1. $\sqrt{49}$ _______	5. $\sqrt{121}$ _______	9. $\sqrt{900}$ _______
2. $\sqrt{81}$ _______	6. $\sqrt{100}$ _______	10. $\sqrt{64}$ _______
3. $\sqrt{25}$ _______	7. $\sqrt{36}$ _______	11. $\sqrt{9}$ _______
4. $\sqrt{16}$ _______	8. $\sqrt{4}$ _______	12. $\sqrt{144}$ _______

SIMPLIFYING SQUARE ROOTS

Square roots can sometimes be simplified even if the number under the square root is not a perfect square. One of the rules of roots is that if a and b are two positive real numbers, then it is always true that $\sqrt{a \cdot b} = \sqrt{a} \cdot \sqrt{b}$. You can use this rule to simplify square roots.

EXAMPLE 1: $\sqrt{100} = \sqrt{4 \cdot 25} = \sqrt{4} \cdot \sqrt{25} = 2 \cdot 5 = 10$

EXAMPLE 2: $\sqrt{200} = \sqrt{100 \cdot 2} = 10\sqrt{2}$ ⟵ Means 10 multiplied by the square root of 2

EXAMPLE 3: $\sqrt{160} = \sqrt{10 \cdot 16} = 4\sqrt{10}$

Simplify.

1. $\sqrt{98}$ ______	5. $\sqrt{8}$ ______	9. $\sqrt{54}$ ______	13. $\sqrt{90}$ ______
2. $\sqrt{600}$ ______	6. $\sqrt{63}$ ______	10. $\sqrt{40}$ ______	14. $\sqrt{175}$ ______
3. $\sqrt{50}$ ______	7. $\sqrt{48}$ ______	11. $\sqrt{72}$ ______	15. $\sqrt{18}$ ______
4. $\sqrt{27}$ ______	8. $\sqrt{75}$ ______	12. $\sqrt{80}$ ______	16. $\sqrt{20}$ ______

ESTIMATING SQUARE ROOTS

EXAMPLE 1: Is $\sqrt{52}$ closer to 7 or 8? Look at the perfect squares above and below 52.

To answer this question, first look at 7^2 which is equal to 49 and 8^2 which is equal to 64 Then ask yourself whether 52 is closer to 49 or to 64? The answer is 49, of course. Therefore, $\sqrt{52}$ is closer to 7 than 8.

Follow the steps above to answer the following questions. Do not use a calculator.

1. Is $\sqrt{66}$ closer to 8 or 9? ____________
2. Is $\sqrt{27}$ closer to 5 or 6? ____________
3. Is $\sqrt{13}$ closer to 3 or 4? ____________
4. Is $\sqrt{78}$ closer to 8 or 9? ____________
5. Is $\sqrt{11}$ closer to 3 or 4? ____________
6. Is $\sqrt{8}$ closer to 2 or 3? ____________
7. Is $\sqrt{20}$ closer to 4 or 5? ____________
8. Is $\sqrt{53}$ closer to 7 or 8? ____________
9. Is $\sqrt{60}$ closer to 7 or 8? ____________

CUBE ROOTS

Cube roots look like square roots, except that there is a raised "3" in front of the root symbol:

Square root of 64: $\sqrt{64}$
Cube root of 64: $\sqrt[3]{64}$

In fact they function very much like square roots, with one important difference. Recall the question, "What is the square root of 64?" That is the same as asking, "What number multiplied by itself is 64?"

Well, asking "What is the cube root of 64?" means "What number multiplied by itself 3 times equals 64? The answer is $4 \times 4 \times 4 = 64$. So, we say the cube root of 64, $\sqrt[3]{64}$, equals 4.

Find the cube root of the following numbers.

1. $\sqrt[3]{1}$
2. $\sqrt[3]{8}$
3. $\sqrt[3]{64}$
4. $\sqrt[3]{27}$
5. $\sqrt[3]{125}$
6. $\sqrt[3]{1000}$
7. $\sqrt[3]{216}$
8. $\sqrt[3]{\frac{27}{64}}$
9. $\sqrt[3]{\frac{125}{1000}}$

ADDING AND SUBTRACTING ROOTS

You can add and subtract terms with square roots only if the number under the square root sign is the same.

EXAMPLE 1: $2\sqrt{2} + 3\sqrt{2} = 5\sqrt{2}$
EXAMPLE 2: $12\sqrt{7} - 3\sqrt{7} = 9\sqrt{7}$

Or, look at the following examples where you can simplify the square roots and then add or subtract.

EXAMPLE 3: $2\sqrt{25} + \sqrt{36}$

Step 1: Simplify. You know that $\sqrt{25} = 5$, and $\sqrt{36} = 6$ so the problem simplifies to $2(5) + 6$

Step 2: Solve: $2(5) + 6 = 10 + 6 = 16$

EXAMPLE 4: $2\sqrt{72} - 3\sqrt{2}$

Step 1: Simplify what you know. $\sqrt{72} = \sqrt{36 \cdot 2} = 6\sqrt{2}$
Step 2: Substitute $6\sqrt{2}$ for $\sqrt{72}$ and simplify
$2(6)\sqrt{2} - 3\sqrt{2} = 12\sqrt{2} - 3\sqrt{2} = 9\sqrt{2}$

Simplify the following addition and subtraction problems.

1. $3\sqrt{5} + 9\sqrt{5}$
2. $3\sqrt{25} + 4\sqrt{16}$
3. $4\sqrt{8} + 2\sqrt{2}$
4. $3\sqrt{32} - 2\sqrt{2}$
5. $\sqrt{25} - \sqrt{49}$
6. $2\sqrt{5} + 4\sqrt{20}$
7. $5\sqrt{8} - 3\sqrt{72}$
8. $\sqrt{27} + 3\sqrt{27}$
9. $3\sqrt{20} - 4\sqrt{45}$
10. $4\sqrt{45} - \sqrt{125}$
11. $2\sqrt{28} + 2\sqrt{7}$
12. $\sqrt{64} + \sqrt{81}$
13. $5\sqrt{54} - 2\sqrt{24}$
14. $\sqrt{32} + 2\sqrt{50}$
15. $2\sqrt{7} + 4\sqrt{63}$
16. $8\sqrt{2} + \sqrt{8}$
17. $2\sqrt{8} - 4\sqrt{32}$
18. $\sqrt{36} + \sqrt{100}$
19. $\sqrt{9} + \sqrt{25}$
20. $\sqrt{64} - \sqrt{36}$
21. $\sqrt{75} + \sqrt{108}$
22. $\sqrt{81} + \sqrt{100}$
23. $\sqrt{192} - \sqrt{75}$
24. $3\sqrt{5} + \sqrt{245}$

MULTIPLYING ROOTS

You can also multiply square roots. To multiply square roots, you just multiply the numbers under the square root sign and then simplify. Look at the examples below.

EXAMPLE 1: $\sqrt{2} \times \sqrt{6}$

Step 1: $\sqrt{2} \times \sqrt{6} = \sqrt{2 \times 6} = \sqrt{12}$ Multiply the numbers under the square root sign.

Step 2: $\sqrt{12} = \sqrt{4 \times 3} = 2\sqrt{3}$ Simplify.

EXAMPLE 2: $3\sqrt{3} \times 5\sqrt{6}$

Step 1: $(3 \times 5)\sqrt{3 \times 6} = 15\sqrt{18}$ Multiply the numbers in front of the square root, and multiply the numbers under the square root sign.

Step 2: $15\sqrt{18} = 15\sqrt{2 \times 9}$ Simplify.
$15 \times 3\sqrt{2} = 45\sqrt{2}$

EXAMPLE 3: $\sqrt{14} \times \sqrt{42}$ For this more complicated multiplication problem, use the rule of roots that you learned on page 97, $\sqrt{a \cdot b} = \sqrt{a} \cdot \sqrt{b}$.

Step 1: $\sqrt{14} = \sqrt{7} \times \sqrt{2}$ and $\sqrt{42} = \sqrt{2} \times \sqrt{3} \times \sqrt{7}$ Instead of multiplying 14 by 42, divide these numbers into their roots.

$\sqrt{14} \times \sqrt{42} = \sqrt{7} \times \sqrt{2} \times \sqrt{2} \times \sqrt{3} \times \sqrt{7}$

Step 2: Since you know that $\sqrt{7} \times \sqrt{7} = 7$, and $\sqrt{2} \times \sqrt{2} = 2$, the problem simplifies to $(7 \times 2)\sqrt{3} = 14\sqrt{3}$

Simplify the following multiplication problems.

1. $\sqrt{5} \times \sqrt{7}$
2. $\sqrt{32} \times \sqrt{2}$
3. $\sqrt{10} \times \sqrt{14}$
4. $2\sqrt{3} \times 3\sqrt{6}$
5. $4\sqrt{2} \times 2\sqrt{10}$
6. $\sqrt{5} \times 3\sqrt{15}$
7. $\sqrt{45} \times \sqrt{27}$
8. $5\sqrt{21} \times \sqrt{7}$
9. $\sqrt{42} \times \sqrt{21}$
10. $4\sqrt{3} \times 2\sqrt{12}$
11. $\sqrt{56} \times \sqrt{24}$
12. $\sqrt{11} \times 2\sqrt{33}$
13. $\sqrt{13} \times \sqrt{26}$
14. $2\sqrt{2} \times 5\sqrt{5}$
15. $\sqrt{6} \times \sqrt{12}$

DIVIDING ROOTS

When dividing a number or a square root by another square root, you cannot leave the square root sign in the denominator (the bottom number) of a fraction. You must simplify the problem so that the square root is not in the denominator. Look at the examples below.

EXAMPLE 1: $\frac{\sqrt{2}}{\sqrt{5}}$

Step 1: $\frac{\sqrt{2}}{\sqrt{5}} \times \frac{\sqrt{5}}{\sqrt{5}}$ The fraction $\frac{\sqrt{5}}{\sqrt{5}}$ is equal to 1, and multiplying by 1 does not change the value of a number

Step 2: $\frac{\sqrt{2 \times 5}}{5} = \frac{\sqrt{10}}{5}$ Multiply and simplify. Since $\sqrt{5} \times \sqrt{5}$ equals 5, you no longer have a square root in the denominator.

EXAMPLE 2: $\frac{6\sqrt{2}}{2\sqrt{10}}$ In this problem, the numbers outside of the square root will also simplify.

Step 1: $\frac{6}{2} = 3$ so you have $\frac{3\sqrt{2}}{\sqrt{10}}$

Step 2: $\frac{3\sqrt{2}}{\sqrt{10}} \times \frac{\sqrt{10}}{\sqrt{10}} = \frac{3\sqrt{2 \times 10}}{10} = \frac{3\sqrt{20}}{10}$

Step 3: $\frac{3\sqrt{20}}{10}$ will further simplify because $\sqrt{20} = 2\sqrt{5}$ so you then have $\frac{3 \times 2\sqrt{5}}{10}$ which reduces to $\frac{3 \times \overset{1}{\cancel{2}}\sqrt{5}}{\underset{5}{\cancel{10}}}$ or $\frac{3\sqrt{5}}{5}$

Simplify the following division problems.

1. $\frac{9\sqrt{3}}{\sqrt{5}}$
2. $\frac{16}{\sqrt{8}}$
3. $\frac{24\sqrt{10}}{12\sqrt{3}}$
4. $\frac{\sqrt{121}}{\sqrt{6}}$
5. $\frac{\sqrt{40}}{\sqrt{90}}$
6. $\frac{33\sqrt{15}}{11\sqrt{2}}$
7. $\frac{\sqrt{32}}{\sqrt{12}}$
8. $\frac{\sqrt{11}}{\sqrt{5}}$
9. $\frac{\sqrt{2}}{\sqrt{6}}$
10. $\frac{2\sqrt{7}}{\sqrt{14}}$
11. $\frac{5\sqrt{2}}{4\sqrt{8}}$
12. $\frac{4\sqrt{21}}{7\sqrt{7}}$
13. $\frac{9\sqrt{22}}{2\sqrt{2}}$
14. $\frac{\sqrt{35}}{2\sqrt{14}}$
15. $\frac{\sqrt{40}}{\sqrt{15}}$

CHAPTER 1 REVIEW

Rewrite the following problems using exponents.

1. $3 \times 3 \times 3 \times 3$ ______
2. $5 \times 5 \times 5$ ______
3. $10 \times 10 \times 10 \times 10 \times 10$ ______
4. 25×25 ______

Write the whole number equivalent.

5. 2^2 = _____
6. 5^3 = _____
7. 12^1 = _____
8. 15^0 = _____
9. 10^4 = _____
10. 7^2 = _____

Answer the following square root questions.

11. Is $\sqrt{6}$ closer to 2 or 3?
12. Is $\sqrt{22}$ closer to 4 or 5?
13. Is $\sqrt{12}$ closer to 3 or 4?
14. Is $\sqrt{28}$ closer to 5 or 6?
15. $2\sqrt{3} + 3\sqrt{3}$ is closest to

    a. 5 b. 7 c. 9 d. 11

16. $5\sqrt{10} + 4\sqrt{10}$ is closest to

    a. 26 b. 28 c. 24 d. 22

17. Add the following square roots.

    a. $2\sqrt{5} + 6\sqrt{5}$
    b. $5\sqrt{3} + \sqrt{12}$

18. Subtract the following square roots.

    a. $4\sqrt{7} - \sqrt{28}$
    b. $9\sqrt{5} - 5\sqrt{5}$

19. Multiply the following square roots.

    a. $4\sqrt{5} \times 2\sqrt{3}$
    b. $\sqrt{18} \times 2\sqrt{3}$

20. Divide the following square roots.

    a. $\frac{\sqrt{10}}{\sqrt{5}}$ b. $\frac{4\sqrt{7}}{7\sqrt{14}}$

Name the property that describes each equation below.

21. $1 + 0 = 1$
22. $5(3 + 2) = (5 \times 3)+(5\times 2)$
23. $6\times 4 = 4\times 6$
24. $8\times\frac{1}{8} = 1$
25. $-2 + 2 = 0$

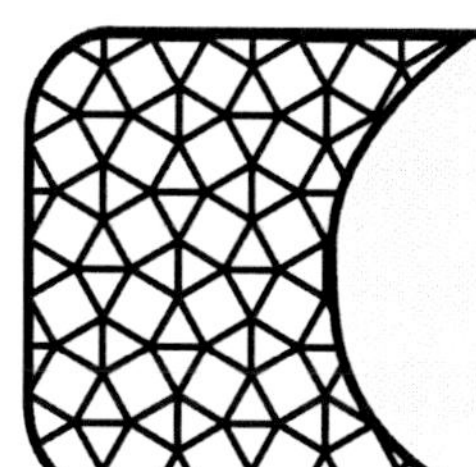

CHAPTER 2: MATRICES

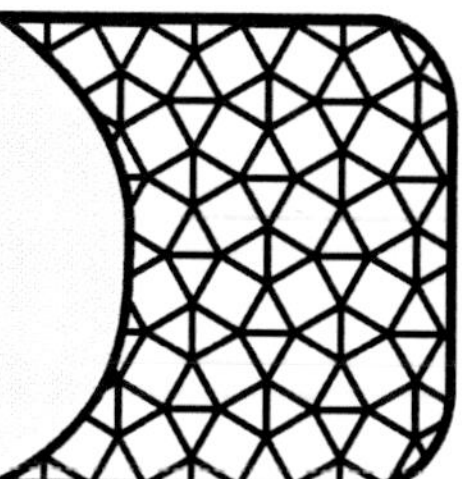

Standards 1.12.7 and 5.12.1

MATRICES

A **matrix** (plural: **matrices**) is an array of numbers, and each number in a matrix is called an **element**. The matrix shown below contains the elements 3, –1, 2, 4, 0, and 1. It is arranged in two rows and three columns and, therefore, is referred to as a 2×3 matrix. When describing a matrix, always give the number of rows and then the number of columns.

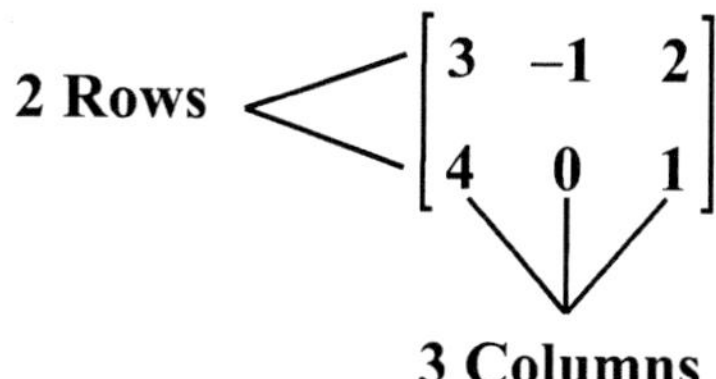

$$\begin{bmatrix} 3 & -1 & 2 \\ 4 & 0 & 1 \end{bmatrix}$$

Matrices can be used to solve systems of linear equations. This text will cover the use of basic operations on matrices and applications using matrices, but will not address solving systems of linear equations using matrices.

ADDITION OF MATRICES

To add matrices, they must be of the same size; they need to have the same number of rows and columns. A 2×2 matrix can only be added to another 2×2 matrix. A 2×3 matrix can be added to another 2×3 matrix, but **cannot** be added to a 3×2 matrix.

To add two matrices of the same size, add the corresponding elements of the two matrices. The resulting matrix is the same size as each of the two original matrices.

EXAMPLE 1: $\begin{bmatrix} 7 & -2 \\ -1 & 4 \end{bmatrix} + \begin{bmatrix} -6 & 2 \\ 4 & 0 \end{bmatrix} = \begin{bmatrix} 7+(-6) & (-2)+2 \\ (-1)+4 & 4+0 \end{bmatrix} = \begin{bmatrix} 1 & 0 \\ 3 & 4 \end{bmatrix}$

Note that the resulting matrix is a 2×2 matrix, as are the original matrices.

EXAMPLE 2: $\begin{bmatrix} 7 & -2 \\ -1 & 4 \\ 5 & -3 \end{bmatrix} + \begin{bmatrix} -6 & 2 & 2 \\ 4 & 0 & 0 \end{bmatrix} = \emptyset$ **(Not Possible)**

The matrices to be added are not of the same size. The first matrix is a 3×2 matrix, and the second matrix is a 2×3. Therefore, these two matrices cannot be added.

Note: Any number of matrices of the same size can be added together.

Add the matrices when possible. When the matrices cannot be added, write NP.

1. $\begin{bmatrix} 8 & 4 \\ 5 & -3 \end{bmatrix} + \begin{bmatrix} 0 & -7 \\ -4 & -9 \end{bmatrix} =$

2. $\begin{bmatrix} 6 \\ -4 \\ 5 \end{bmatrix} + \begin{bmatrix} 2 \\ -9 \\ -8 \end{bmatrix} + \begin{bmatrix} 9 \\ -1 \\ -3 \end{bmatrix} + \begin{bmatrix} 5 \\ -1 \\ -2 \end{bmatrix} =$

3. $\begin{bmatrix} 3 & -1 & 2 \\ 4 & 0 & 1 \end{bmatrix} + \begin{bmatrix} 6 & 0 & -5 \\ 4 & 0 & 1 \end{bmatrix} =$

4. $\begin{bmatrix} -6 \\ -2 \end{bmatrix} + \begin{bmatrix} -4 & -1 \end{bmatrix} =$

5. $\begin{bmatrix} -5 & -2 \\ 9 & -7 \\ 3 & 6 \end{bmatrix} + \begin{bmatrix} -1 & -2 \\ -8 & 7 \\ 9 & -4 \end{bmatrix} =$

6. $\begin{bmatrix} 8 & -1 & -6 & 3 \\ 0 & -7 & -5 & -4 \end{bmatrix} + \begin{bmatrix} -8 & 1 & 5 & -2 \\ 1 & 6 & 6 & 5 \end{bmatrix} =$

MULTIPLICATION OF A MATRIX BY A CONSTANT

A matrix can be multiplied by a constant. The constant is multiplied by each element in the matrix. The resulting matrix is the same size as the original matrix.

EXAMPLE 1: $4\begin{bmatrix} 1 & 0 \\ 3 & 4 \end{bmatrix} = \begin{bmatrix} (4\times1) & (4\times0) \\ (4\times3) & (4\times4) \end{bmatrix} = \begin{bmatrix} 4 & 0 \\ 12 & 16 \end{bmatrix}$

EXAMPLE 2: $-3\begin{bmatrix} -8 & 1 & 5 & -2 \\ 1 & 6 & 6 & 5 \end{bmatrix} = \begin{bmatrix} (-3\times-8) & (-3\times1) & (-3\times5) & (-3\times-2) \\ (-3\times1) & (-3\times6) & (-3\times6) & (-3\times5) \end{bmatrix} = \begin{bmatrix} 24 & -3 & -15 & 6 \\ -3 & -18 & -18 & -15 \end{bmatrix}$

Multiply each of the following.

1. $-2\begin{bmatrix} -8 & 1 \\ 6 & 7 \\ -5 & -4 \end{bmatrix} =$

2. $5\begin{bmatrix} 3 \\ 0 \\ -1 \end{bmatrix} =$

3. $\frac{1}{2}\begin{bmatrix} 2 & -4 & -9 & 7 \\ -1 & 6 & 3 & -8 \end{bmatrix} =$

4. $-\frac{1}{4}\begin{bmatrix} -1 & -6 & 4 & -2 \\ 8 & 0 & 3 & -8 \\ 4 & 6 & -3 & 12 \end{bmatrix} =$

5. $-9\begin{bmatrix} -10 & -1 & \frac{1}{3} & 7 & -\frac{3}{4} \end{bmatrix} =$

6. $6\begin{bmatrix} -4 & -5 \\ 1 & 0 \end{bmatrix} =$

SUBTRACTION OF MATRICES

Subtraction of matrices is similar to addition of matrices in that the matrices to be subtracted must be the same size. Suppose x and y represent two different matrices of the same size. $x - y$ can also be written $x + (-1)y$. Therefore, subtraction of matrices involves two steps: multiplying the second matrix by -1 and then adding it to the first matrix.

EXAMPLE : $\begin{bmatrix} 4 & -6 \\ 9 & -5 \end{bmatrix} - \begin{bmatrix} 3 & -1 \\ -4 & 7 \end{bmatrix}$ **can also be written** $\begin{bmatrix} 4 & -6 \\ 9 & -5 \end{bmatrix} + (-1)\begin{bmatrix} 3 & -1 \\ -4 & 7 \end{bmatrix}$

Step 1: **Multiply the second matrix by -1.**

$$(-1)\begin{bmatrix} 3 & -1 \\ -4 & 7 \end{bmatrix} = \begin{bmatrix} -3 & 1 \\ 4 & -7 \end{bmatrix}$$

Step 2: **Add the first matrix and the product from step 1.**

$$\begin{bmatrix} 4 & -6 \\ 9 & -5 \end{bmatrix} + \begin{bmatrix} -3 & 1 \\ 4 & -7 \end{bmatrix} = \begin{bmatrix} 1 & -5 \\ 13 & -12 \end{bmatrix}$$

Subtract the matrices when possible. When the matrices cannot be subtracted, write NP.

1. $\begin{bmatrix} 1 & 2 \\ 6 & 0 \\ -1 & 4 \end{bmatrix} - \begin{bmatrix} 3 & 2 \\ 1 & -3 \\ 5 & 1 \end{bmatrix} =$

2. $\begin{bmatrix} 1 & 2 & 3 \\ 4 & 5 & 6 \end{bmatrix} - \begin{bmatrix} 3 & 2 & 1 \\ 5 & 4 & 6 \end{bmatrix} =$

3. $\begin{bmatrix} 6 & 2 & 2 \\ -4 & -3 & -9 \\ 5 & 1 & -8 \end{bmatrix} - \begin{bmatrix} 6 & 9 & -4 \\ -4 & 0 & -8 \\ 2 & 3 & -7 \end{bmatrix} =$

4. $\begin{bmatrix} 2 & 3 \\ 1 & 5 \end{bmatrix} - \begin{bmatrix} 1 & 3 \\ -2 & 1 \end{bmatrix} =$

5. $\begin{bmatrix} 8 & -1 & -6 & 3 \\ 0 & -7 & -5 & -4 \end{bmatrix} - \begin{bmatrix} -8 & 1 & 5 & -2 \\ 1 & 6 & 6 & 5 \end{bmatrix} =$

6. $\begin{bmatrix} 1 & 2 & 5 & 5 \\ 7 & 9 & 1 & -8 \\ -3 & 2 & 5 & -7 \end{bmatrix} - \begin{bmatrix} 1 & 3 & 1 \\ -1 & 0 & -8 \\ 0 & 1 & -7 \end{bmatrix} =$

Use order of operations to solve the following matrices problem.

7. $3\begin{bmatrix} 3 & -2 \\ -4 & 1 \\ -1 & 0 \end{bmatrix} - \begin{bmatrix} 7 & -8 \\ -9 & 8 \\ -7 & 6 \end{bmatrix} - \begin{bmatrix} 0 & 7 \\ -1 & 1 \\ -3 & -2 \end{bmatrix} + 4\begin{bmatrix} 3 & 6 \\ -4 & 5 \\ 2 & -2 \end{bmatrix} + \begin{bmatrix} 2 & -3 \\ 1 & 1 \\ -3 & 4 \end{bmatrix} =$

APPLICATIONS WITH MATRICES

EXAMPLE 1: Find a, b, c, and d such that

$$\begin{bmatrix} 2 & -1 \\ 4 & 3 \end{bmatrix} + \begin{bmatrix} a & b \\ c & d \end{bmatrix} = \begin{bmatrix} 3 & -1 \\ 2 & 2 \end{bmatrix}$$

Step 1: Write the equations for the four sets of corresponding elements.

$2 + a = 3$ $\quad$ $-1 + b = -1$

$4 + c = 2$ $\quad$ $3 + d = 2$

Step 2: Solve each of the four equations. For review of solving one-step equations, see the chapter titled "Solving One-Step Equations and Inequalities."

$a = 1$ $\quad$ $b = 0$

$c = 2$ $\quad$ $d = -1$

EXAMPLE 2: Lucie and her friend Laura are shopping for a new cellular phone plan. Plan A offers 400 minutes per month for \$60 plus another 200 night and weekend minutes for an extra \$20. Plan B offers 500 monthly minutes for \$50 plus 150 night and weekend minutes for an extra \$20. The two plans can be represented in the following matrices:

$$\begin{bmatrix} \$60 & \$20 \\ 400 & 200 \end{bmatrix} = \text{A} \qquad \begin{bmatrix} \$50 & \$20 \\ 500 & 150 \end{bmatrix} = \text{B}$$

If Lucie chooses Plan A and Laura chooses Plan B, what will the combined total cost of their service be if they also select the night and weekend minutes, and how many total minutes will they receive? What is the average cost of each plan, and how many minutes does each include?

Step 1: To determine the total of the two plans, add the two matrices together, A + B.

$$\begin{bmatrix} \$60 + \$50 & \$20 + \$20 \\ 400 + 500 & 200 + 150 \end{bmatrix} = \begin{bmatrix} \$110 & \$40 \\ 900 & 350 \end{bmatrix}$$

Step 2: Calculate the average cost and minutes by multiplying the total matrix by $\frac{1}{2}$ (or dividing it by 2):

$$\frac{1}{2} \times \begin{bmatrix} \$110 & \$40 \\ 900 & 350 \end{bmatrix} = \begin{bmatrix} \$55 & \$20 \\ 450 & 175 \end{bmatrix}$$

Solve the following matrix problems.

1. $\begin{bmatrix} d & 3 \\ e & 1 \end{bmatrix} + \begin{bmatrix} 2 & f \\ 2 & g \end{bmatrix} = \begin{bmatrix} 5 & 3 \\ 1 & 2 \end{bmatrix}$

2. $\begin{bmatrix} 3d & 1 \\ 2 & 3g \\ 1 & f \end{bmatrix} - \begin{bmatrix} 2d & 0 \\ e & g \\ 1 & 2 \end{bmatrix} = \begin{bmatrix} -2 & 1 \\ 4 & 4 \\ 0 & 1 \end{bmatrix}$

3. $3\begin{bmatrix} 1 & 0 \\ 2 & 1 \end{bmatrix} + \frac{1}{2}\begin{bmatrix} 4 & 6 \\ 0 & 2 \end{bmatrix} = \begin{bmatrix} d & f \\ e & g \end{bmatrix}$

4. A computer company with one plant in the West and one in the East produces monitors and printers. The production for January and February are given as follows:

January

	West Plant	East Plant
Monitors	2,000	1,710
Printers	800	650

February

	West Plant	East Plant
Monitors	2,300	1,850
Printers	950	800

Record your answers as 2×2 matrices.

A. What is the average monthly production of monitors and of printers in the west plant and in the east plant?

B. What is the increase in production of monitors and of printers from January to February in the west and east plants?

C. What is the total production of monitors and of printers for January and February for the west and east plants?

5. The Yummy Candy Company produces a variety of candy products and packages them For various holidays. The Christmas package consists of three pieces of chocolate, two pecan candies, one peppermint twist, and four chocolate-covered cherries. The Valentine package consists of the same variety but containe three times as many pieceds of each candy. Write the number of candies in both the Christmas and Valentine packages in a 1×4 matrix form.

6. What are the total numbers of candies contained in one Christmas package and one Valentine package from Problem 5?

CHAPTER 2 REVIEW

Ray is a street vendor who sells on the 4th of July weekend in Las Vegas. The matrices below show the total of "USA" T-shits at Ray's Shirt Stand at the beginning and end of the 4th of July weekend.

Shirt Size

$$\text{Thursday, before weekend} = \begin{bmatrix} 40 & 60 & 80 \\ 50 & 80 & 90 \\ 30 & 40 & 60 \end{bmatrix} \begin{matrix} \text{White with Red Lettering} \\ \text{White with Blue Lettering} \\ \text{Blue with White Lettering} \end{matrix}$$

(columns: **S**, **M**, **L**)

$$\text{Wednesday, after weekend} = \begin{bmatrix} 34 & 37 & 12 \\ 42 & 65 & 0 \\ 2 & 32 & 52 \end{bmatrix} \begin{matrix} \text{White with Red Lettering} \\ \text{White with Blue Lettering} \\ \text{Blue with White Lettering} \end{matrix}$$

(columns: **S**, **M**, **L**)

1. How many Large shirts did Ray sell over the weekend?

2. How many White with Blue Lettering shirts did Ray sell over the weekend?

Solve each matrix. If not possible, write not possible.

3. $2\begin{bmatrix} -5 & -1 \\ 4 & -7 \\ 9 & 6 \end{bmatrix} + \begin{bmatrix} -1 & -5 \\ -8 & 7 \\ 1 & -4 \end{bmatrix} =$

4. $\begin{bmatrix} 4 \\ -1 \end{bmatrix} + \begin{bmatrix} -7 & -1 \end{bmatrix} =$

5. $\begin{bmatrix} 9 & -1 & -6 & 3 \\ 0 & -5 & -5 & -4 \end{bmatrix} + \begin{bmatrix} -8 & 1 & 2 & -3 \\ 1 & 6 & 11 & 5 \end{bmatrix} =$

6. $-1\begin{bmatrix} -1 & -6 & 4 & -8 \\ 12 & 0 & 3 & -2 \\ 4 & 6 & -4 & 16 \end{bmatrix} =$

Subtract the following matrices. If not possible, write NP.

7. $\begin{bmatrix} 2 & 4 \\ 0 & 5 \end{bmatrix} - \begin{bmatrix} 1 & 7 \\ -3 & 1 \end{bmatrix} =$

8. $\begin{bmatrix} 10 & -1 & -6 & 3 \\ 0 & -5 & -5 & -4 \end{bmatrix} - \begin{bmatrix} -8 & 9 & 1 & -2 \\ 1 & 0 & 6 & 4 \end{bmatrix} =$

9. $\begin{bmatrix} 1 & 2 & 5 & 4 \\ 7 & 0 & 1 & -8 \\ -3 & 2 & 5 & -2 \end{bmatrix} - \begin{bmatrix} 1 & 3 & 13 \\ -1 & 0 & -11 \\ 0 & 5 & -7 \end{bmatrix} =$

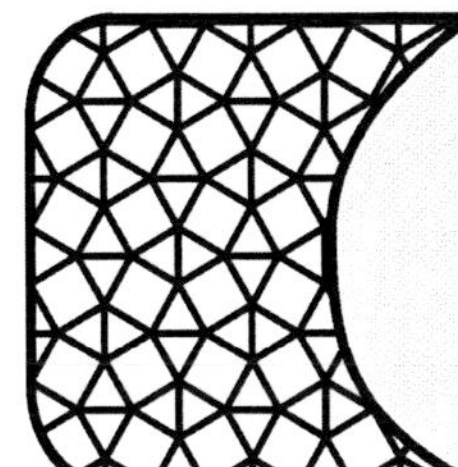

CHAPTER 3: PATTERNS & PROBLEM SOLVING

Standard 2.12.1

NUMBER PATTERNS

In each of the examples below, there is a sequence of data that follows a pattern. Think of the given sequence like the output for a function. You must find the pattern that holds true for each number in the data. Once you determine the pattern, you can write out a function that fits the data and figure out any other number in the sequence.

	Sequence	Pattern	Next Number	20th number in the sequence
EXAMPLE 1:	3, 4, 5, 6, 7	$n + 2$	$f(n) = n + 2 = 8$	$f(20) = 20 + 2 = 22$

In number patterns, the sequence is the output. The input can be the set of whole numbers starting with 1. But, you must determine the "rule" or pattern. Look at the table below.

input	sequence
1 →	3
2 →	4
3 →	5
4 →	6
5 →	7

What pattern or "rule" can you come up with that gives you the first number in the sequence, 3, when you input 1? $n + 2$ will work because when $n = 1$, the first number in the sequence = 3. Does this pattern hold true for the rest of the numbers in the sequence? Yes, it does. When $n = 2$, the second number in the sequence = 4. When $n = 3$, the third number in the sequence = 5, and so on. Therefore, $n + 2$ is the pattern. Even without knowing the algebraic form of the pattern, you could figure out that 8 is the next number in the sequence. The function describing this pattern would be $f(n) = n + 2$. To find the 20th number in the pattern, use $n = 20$ to get 22.

	Sequence	Pattern	Next Number	20th number in the sequence
EXAMPLE 2:	1, 4, 9, 16, 25	n^2	$f(n) = n^2 = 36$	400
EXAMPLE 3:	–2, –4, –6, –8, –10	$-2n$	$f(n) = -2n = -12$	–40

Find the pattern and the next number in each of the sequences below.

	Sequence	Pattern	Next Number	20th number in the sequence
1.	–2, –1, 0, 1, 2	__________	________	________
2.	5, 6, 7, 8, 9	__________	________	________
3.	3, 7, 11, 15, 19	__________	________	________
4.	–3, –6, –9, –12, –15	__________	________	________
5.	3, 5, 7, 9, 11	__________	________	________
6.	2, 4, 8, 16, 32	__________	________	________
7.	1, 8, 27, 64, 125	__________	________	________
8.	0, –1, –2, –3, –4	__________	________	________
9.	2, 5, 10, 17, 26	__________	________	________
10.	4, 6, 8, 10, 12	__________	________	________

MAKING PREDICTIONS

Use what you know about number patterns to answer the following questions.

Corn plants grow as tall as they will get in about 20 weeks. Study the chart of the rate of corn plant growth below, and answer the questions that follow.

Corn Growth	
Beginning Week	**Height (inches)**
2	9
7	39
11	63
14	??

1. If the growth pattern continues, how high will the corn plant be beginning week 14?

2. If the growth pattern was constant (at the same rate from week to week), how high was the corn in the beginning of the 8th week?

Peter Nichols is staining furniture for a furniture manufacturer. He stains large pieces of furniture that take longer to dry in the beginning of the day and smaller pieces of furniture as the day progresses.

Time	**# Pieces Completed per Hour**
Hour 1	3
Hour 3	5
~~Hour 6~~	~~8~~

3. How many pieces of furniture did Peter stain during his second hour of work?

~~4. How many pieces of furniture will Peter have stained by the end of an 8 hour day?~~

Brian Bailey is bass fishing down the Humbolt River. He has selected six locations to fish. Using his car, he drives to the first location near Golconda. His final location is near Valmy. As he travels south, he notices that the bass catches are getting larger.

Fishing Direction	Fishing Location	Number of bass caught
North	1	4
↓	2	unrecorded
	3	10
	4	unrecorded
South	5	16

5. How many bass would he likely catch in the sixth location?

6. If he fishes six locations, how many bass is he likely to catch altogether?

INDUCTIVE REASONING AND PATTERNS

Humans have always observed what happened in the past and used these observations to predict what would happen in the future. This is called **inductive reasoning**. Although mathematics is referred to as the "deductive science," it benefits from inductive reasoning. We observe patterns in the mathematical behavior of a phenomenon, then find a rule or formula for describing and predicting its future mathematical behavior. There are lots of different kinds of predictions that may be of interest.

EXAMPLE 1: Nancy is watching her nephew, Drew, arrange his marbles in rows on the kitchen floor. The figure below shows the progression of his arrangement.

Row 1 ○
Row 2 ○○
Row 3 ○○○○
Row 4 ○○○○○○○○

QUESTION 1: Assuming this pattern continues, how many marbles would Drew place in a fifth row?

ANSWER 1: It appears that Drew doubles the number of marbles in each successive row. In the 4th row he had 8 marbles, so in the 5th row we can predict 16 marbles.

QUESTION 2: How many marbles will Drew place in the *n*th row?

ANSWER 2: To find a rule for the number of marbles in the *n*th row, we look at the pattern suggested by the table below.

Which row	1st	2nd	3rd	4th	5th
Number of marbles	1	2	4	8	16

Observing closely, you will notice that the *n*th row contains 2^{n-1} marbles.

QUESTION 3: Suppose Nancy tells you that Drew now has 6 rows of marbles on the floor. What is the total number of marbles in his arrangement?

ANSWER 3: Again, organizing the data in a table could be helpful.

Number of rows	1	2	3	4	5
Total number of marbles	1	3	7	15	31

With careful observation, one will notice that the total number of marbles is always 1 less than a power of 2; indeed, for *n* rows there are $2^n - 1$ marbles total.

QUESTION 4: If Drew has 500 marbles, what is the maximum number of *complete* rows he can form?

ANSWER 4: With 8 complete rows, Drew will use $2^8 - 1 = 255$ marbles, and to form 9 complete rows he would need $2^9 - 1 = 511$ marbles; thus, the answer is 8 complete rows.

EXAMPLE 2: Manuel drops a golf ball from the roof of his high school while Carla videos the motion of the ball. Later, the video is analyzed and the results are recorded concerning the height of each bounce of the ball.

QUESTION 1: What height do you predict for the fifth bounce?

Initial height	1st bounce	2nd ounce	3rd bounce	4th bounce
30 ft	18 ft	10.8 ft	6.48 ft	3.888 ft

ANSWER 1: To answer this question, we need to be able to relate the height of each bounce to the bounce immediately preceding it. Perhaps the best way to do this is with **ratios** as follows:

$$\frac{\text{Height of 1st bounce}}{\text{Initial bounce}} = 0.6 \qquad \frac{\text{Height of 2nd bounce}}{\text{Height of 1st bounce}} = 0.6 \ldots. \qquad \frac{\text{Height of 4th bounce}}{\text{Height of 3rd bounce}} = 0.6$$

Since the ratio of the height of each bounce to the bounce before it appears constant, we have some basis for making predictions.

Using this, we can reason that the fifth bounce will be equal to 0.6 of the fourth bounce

Thus we predict the fifth bounce to have a height of **$0.6 \times 3.888 = 2.3328$ ft.**

QUESTION 2: Which bounce will be the last one with a height of one foot or greater?

ANSWER 2: For this question, keep looking at predicted bounce heights until a bounce less than 1 foot is reached.

The sixth bounce is predicted to be 1.39968 ft.
The seventh bounce is predicted to be 0.839808 ft.

Thus, the last bounce with a height greater than 1 foot is predicted to be the sixth one.

Read each of the following questions carefully. Use inductive reasoning to answer each question. You may wish to make a table or a diagram to help you visualize the pattern in some of the problems. Show your work.

George is stacking his coins as shown below.

1. How many coins do you predict he will place in the fourth stack?

2. How many coins in an *n*th row?

3. If George has exactly 6 "complete" stacks, how many coins does he have?

4. If George has 2,000 coins, how many complete stacks can he form?

Bob and Alice have designed and created a website for their high school. The first week they had 5 visitors to the site; during the second week, they had 10 visitors; and during the third week, they had 20 visitors.

5. If current trends continue, how many visitors can they expect in the fifth week?

6. How many in the *n*th week?

7. How many weeks will it be before they get more than 500 visitors in a single week?

---

8. In 1979 (the first year of classes), there were 500 students at Brookstone High. In 1989, there were 1000 students. In 1999, there were 2000 students. How many students would you predict at Brookstone in 2009 if this pattern continues (and no new schools are built)?

---

9. The number of new driver's licenses issued in the city of Boomtown, USA was 512 in 2002, 768 in 2004, 1,152 in 2006, and 1,728 in 2008. Estimate the number of new driver's licenses that will be issued in 2010.

10. The average combined (math and verbal) SAT scores for seniors at Brookstone High was 1,000 in 2002, 1,100 in 2003, 1,210 in 2004, and 1331 in 2005. Predict the combined SAT score for Brookstone seniors in 2010.

Juan wants to be a medical researcher, inspired in part by the story of how penicillin was discovered as a mold growing on a laboratory dish. One morning, Juan observes a mold on one of his lab dishes. Each morning thereafter, he observes and records the pattern of growth. The mold appeared to cover about 1/32 of the dish surface on the first day, 1/16 on the second day, and 1/8 on the third day.

11. If this rate of growth continues, on which day can Juan expect the entire dish to be covered with mold?

12. Suppose that whenever the original dish gets covered with mold Juan transfers half of the mold to another dish. How long will it be before *both* dishes are covered again?

13. Every year on the last day of school, the Brookstone High cafeteria serves the principal's favorite dish–Broccoli Surprise. In 1988, 1024 students chose to eat Broccoli Surprise on the last day of school, 512 students in 1992, and 256 students in 1996. Predict how many will choose Broccoli Surprise on the last day of school in 2000.

Part of testing a new drug is determining the rate at which it will break down (*decay*) in the blood. The decay results for a certain antibiotic after a 1000 milligram injection are given in the table below.

12:00 PM	1:00 PM	2:00 PM
1000 mg	800 mg	640 mg

14. Predict the number of milligrams that will be in the patient's bloodstream at 3:00 PM.

15. At which hour can the measurer expect to record a result of less than 300 mg?

16. Marie has a daylily in her mother's garden. Every Saturday morning in the spring, she measures and records its height in the table below. What height do you predict for Marie's daylily on April 29? (Hint: Look at the *change* in height each week when looking for the pattern.)

April 1	April 8	April 15	April 22
12 in	18 in	21 in	22.5 in

17. Bob puts a glass of water in the freezer and records the temperature every 15 minutes. The results are displayed in the table below. If this pattern of cooling continues, what will be the temperature at 2:15 PM? (Hint: Again, look at the *changes* in temperature in order to see the pattern.)

1:00 PM	1:15 PM	1:30 PM	1:45 PM
92°F	60°F	44°F	36°F

Suppose you cut your hand on a rusty nail that deposits 25 bacteria cells into the wound. Suppose also that each bacterium splits into two bacteria every 15 minutes.

18. How many bacteria will there be after two hours?

19. How many 15-minute intervals will pass before there are over a million bacteria?

20. Elias performed a psychology experiment at his school. He found that when someone is asked to pass information along to someone else, only about 70% of the original information is actually passed to the recipient. Suppose Elias gives the information to Brian, Brian passes it along to George, and George passes it to Montel. Using Elias's results from past experiments, what percentage of the original information does Montel actually receive?

FINDING A RULE FOR PATTERNS

EXAMPLE: Mr. Applegate wants to put desks together in his math class so that students can work in groups. The diagram below shows how he wishes to do it.

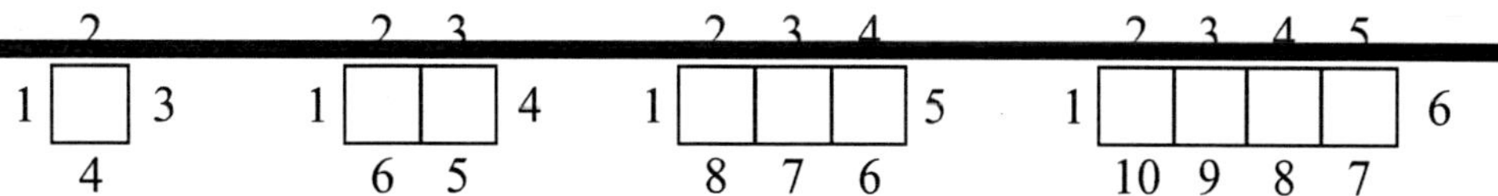

With 1 table he can seat 4 students, with 2 tables he can seat 6, with 3 tables 8, and with 4 tables 10.

QUESTION 1: How many students can he seat with 5 tables?

ANSWER 1: With 5 tables he could seat 5 students along the sides of the tables and 1 student on each end; thus, a total of 12 students could be seated.

QUESTION 2: Write a rule that Mr. Applegate could use to tell how many students could be seated at n tables. Explain how you got the rule.

ANSWER 2: For n tables, there would be n students along each of 2 sides and 2 students on the ends (1 on each end); thus, a total of $2n + 2$ students could be seated at n tables.

EXAMPLE 2: When he isn't playing football for the Brookstone Bears, Tim designs web pages. A car dealership paid Tim \$500 to start a site with photos of its cars. The dealer also agreed to pay Tim \$50 for each customer who buys a car first viewed on the web site.

QUESTION 1: Write and explain a rule that tells how much the dealership will pay Tim for the sale of n cars from his web site.

ANSWER 1: Tim's payment will be the initial \$500 plus \$50 for each sale. Translated into mathematical language, if Tim sells n cars he will be paid a total of $500 + 50n$ dollars.

QUESTION 2: How many cars have to be sold from his site in order for Tim to get \$1,000 from the dealership?

ANSWER 2: He earned \$500 just by establishing the site, so he only needs to earn an additional \$500, which at \$50 per car requires the sale of only 10 cars. (Note: Another way to solve this problem is to use the rule found in the first question. In that case, you simply solve the equation $500 + 50n = 1000$ for the variable n.)

EXAMPLE 3: Eric is baking muffins to raise money for the Homecoming dance. He makes 18 muffins with each batch of batter, but he must give one muffin each to his brother, his sister, and his dog, and himself (of course!) each time a batch is finished baking.

QUESTION 1: Write a rule for the number of muffins Eric produces for the fund raiser with n batches.

ANSWER 1: He bakes 18 with each batch, but only 14 are available for the fund raiser. Thus with n batches he will produce $14n$ muffins for the Homecoming.
The rule = 14*n*

QUESTION 2: Use your rule to determine how many muffins he will contribute if he makes 7 batches.

ANSWER 2: The number of batches, n, equals 7. Therefore, he will produce $14 \times 7 = 98$ muffins with 7 batches.

QUESTION 3: Determine how many batches he must bake in order to contribute at least 150 muffins.

ANSWER 3: Ten batches will produce $10 \times 14 = 140$ muffins. Eleven batches will produce $11 \times 14 = 154$ muffins. To produce at least 150 muffins, he must bake at least 11 batches.

QUESTION 4: Determine how many muffins he would actually bake in order to contribute 150 muffins.

ANSWER 4: Since Eric actually bakes 18 muffins per batch, 11 batches would result in Eric baking $11 \times 18 = 198$ muffins.

Carefully read and solve the problems below. Show your work.

Tito is building a picket fence along both sides of the driveway leading up to his house. He will have to place posts at both ends and at every 10 feet along the way because the rails come in prefabricated ten-foot sections.

1. How many posts will he need for a 180 foot driveway?

2. Write and explain a rule for determining the number of posts needed for n ten-foot sections.

3. How long of a driveway can he fence with 32 posts?

Linda is working as a bricklayer this summer. She lays the bricks for a walkway in *sections* according to the pattern depicted below.

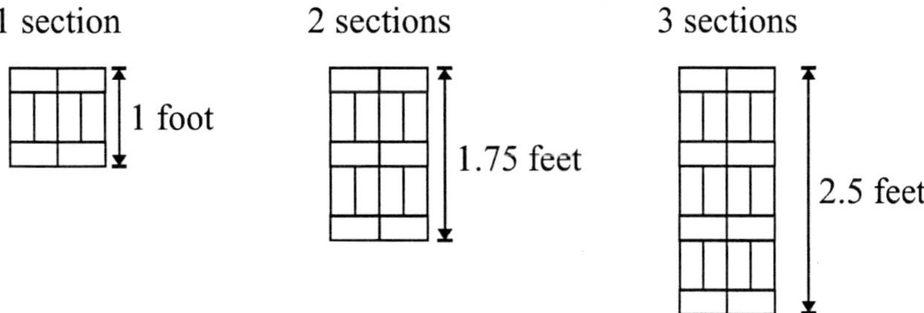

4. Write a formula for the number of bricks needed to lay n sections.

5. Write a formula for the number of feet covered by n sections.

6. How many bricks would it take to lay a walk that is 10 feet long?

Dakota's beginning pay at his new job is \$300 per week. For every three months, he continues to work there he will get a \$10 per week raise.

7. Write a formula for Dakota's weekly pay after n three-month periods.

8. After n years?

9. How long will he have to work before his pay gets to \$400 a week?

Amanda is selling shoes this summer. In addition to her hourly wages, Amanda got a \$100 bonus just for accepting the position, and she gets a \$2 bonus for each pair of shoes she sells.

10. Write and explain a rule that tells how much she will make in bonuses if she sells n pairs of shoes.

11. How many pairs of shoes must she sell in order to make \$200 in bonuses?

A certain teen telephone chat line, which sells itself as a benefit to teens but which is actually a money-making scheme, is a 900 telephone number that charges \$2.00 for the first minute and \$0.95 for each additional minute.

12. Write a formula for the cost of speaking n minutes on this line.

13. How many minutes does it take to accumulate charges of more than \$50.00?

CHAPTER 3 REVIEW

Find the pattern for the following number sequences, and then find the *n*th number requested.

1. 0, 1, 2, 3, 4 pattern ______

2. 0, 1, 2, 3, 4 20th number ______

3. 1, 3, 5, 7, 9 pattern ______

4. 1, 3, 5, 7, 9 25th number ______

5. 3, 6, 9, 12, 15 pattern ______

6. 3, 6, 9, 12, 15 30th number ______

7. The table below shows the estimated number of cases of a disease each year for 4 years.

Year	1	2	3	4	5	6
Total Cases	6,075	4,050	2,700	1,800	1,200	?

If the pattern continues, what will be the estimated number of cases by the 6th year?

A. 300 B. 600 C. 800 D. 900

8. Sandra is raising earthworms. Every week, she estimates the number of worms. The table below shows the population (number of worms) over a period of 6 weeks.

Week	1	2	3	4	5	6
Population	4,096	5,120	6,400	8,000	10,000	12,500

If this pattern continues, what will be the population of earthworms at 7 weeks?

A. 13,100 B. 14,500 C. 15,000 D. 15,625

Jessica Bloodsoe and Katie Turick are climbing Mt. Fuji in Japan which is 12,388 ft. High. The higher they go, the slower their climb due to lack of oxygen. The chart below shows their progress.

Days Ascending	Altitude
End of day 1	4,000 feet
End of day 2	7,200 feet
End of day 3	9,600 feet

9. If the weather holds, what will be their altitude at the end of day 4? __________

10. If they can keep the same rate, how many days would it take them to get to the top? __________

In a large city of 200,000, there was an outbreak of tuberculosis. Immediately, health care workers began an immunization campaign. The chart below records their results.

	No. of People Immunized	No. of TB Cases
Year 1	20,000	60
Year 2	60,000	45
Year 3	100,000	30

11. About how many cases of TB would you predict for year 4? ___________

Exotic goldfish are kept in different size containers of water. The larger the container, the bigger the size the goldfish can grow. The chart on the right shows how big one goldfish can grow in different size containers.

Fish Size	Container Size
1 inch	20 gallon or less
$2\frac{1}{2}$ inches	50 gallon
5 inches	100 gallon

12. Based on the chart, how large would you predict a goldfish could grow in a 140 gallon container? ___________

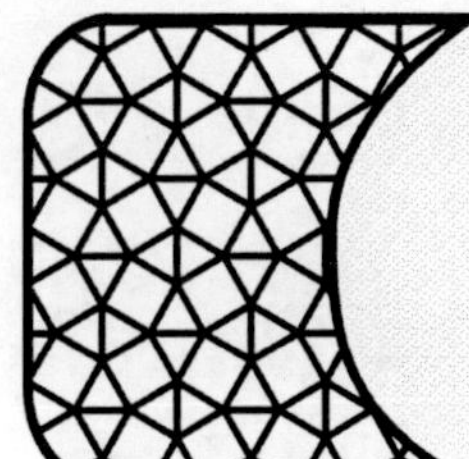

CHAPTER 4: POLYNOMIALS

Standard 2.12.3

Polynomials are algebraic expressions which include **monomials** containing one term, **binomials** which contain two terms, and **trinomials**, which contain three terms. Expressions with more than three terms are all called **polynomials**. **Terms** are separated by plus and minus signs.

EXAMPLES

Monomials:	**Binomials:**	**Trinomials:**	**Polynomials:**
$4f$	$4t + 9$	$x^2 + 2x + 3$	$x^3 - 3x^2 + 3x - 9$
$3x^3$	$9 - 7g$	$5x^2 - 6x - 1$	$p^4 + 2p^3 + p^2 - 5p + 9$
$4g^2$	$5x^2 + 7x$	$y^4 + 15y^2 + 100$	
2	$6x^3 - 8x$		

ADDING AND SUBTRACTING MONOMIALS

Two **monomials** can be added or subtracted as long as the **variable and its exponent** are the **same**. This is called combining like terms. Use the same rules you used for adding and subtracting integers.

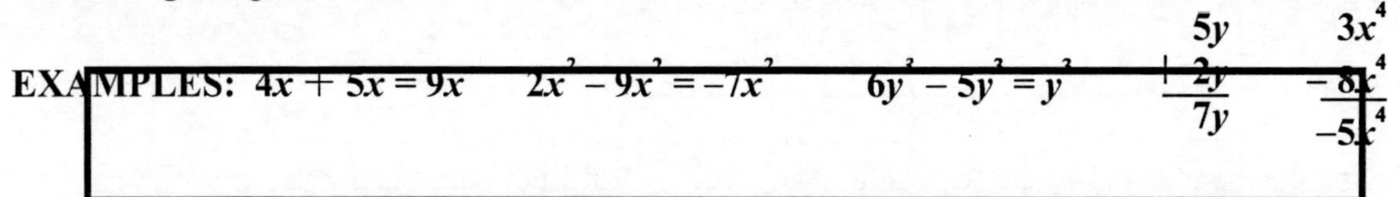

EXAMPLES: $4x + 5x = 9x$ $\quad 2x^2 - 9x^2 = -7x^2$ $\quad 6y^3 - 5y^3 = y^3$ $\quad \begin{array}{r} 5y \\ +\ 2y \\ \hline 7y \end{array}$ $\quad \begin{array}{r} 3x^4 \\ -\ 8x^4 \\ \hline -5x^4 \end{array}$

Remember: When the integer in front of the variable is "1", it is usually not written. $1x^2$ is the same as x^2, *and* $-1x$ is the same as $-x$.

Add or subtract the following monomials:

1.	$2x^2 + 5x^2 =$ _____	6.	$s^5 + s^5 =$ _____	11.	$3x^2 - 5x^2 =$ _____
2.	$5t + 8t =$ _____	7.	$-2x - 4x =$ _____	12.	$9t + 2t =$ _____
3.	$9y^3 - 2y^3 =$ _____	8.	$4w^2 - w^2 =$ _____	13.	$-7v^3 + 10v^3 =$ _____
4.	$6g - 8g =$ _____	9.	$z^4 + 9z^4 =$ _____	14.	$-2x^3 + x^3 =$ _____
5.	$7y^2 + 8y^2 =$ _____	10.	$-k + 2k =$ _____	15.	$10y^4 - 5y^4 =$ _____

16. $\begin{array}{r} y^4 \\ +\ 2y^4 \\ \hline \end{array}$ 18. $\begin{array}{r} 8t^2 \\ +\ 7t^2 \\ \hline \end{array}$ 20. $\begin{array}{r} 5w^2 \\ +\ 8w^2 \\ \hline \end{array}$ 22. $\begin{array}{r} -5z \\ +\ 9z \\ \hline \end{array}$ 24. $\begin{array}{r} 7t^3 \\ -\ 6t^3 \\ \hline \end{array}$

17. $\begin{array}{r} 4x^3 \\ -\ 9x^3 \\ \hline \end{array}$ 19. $\begin{array}{r} -2y \\ -\ 4y \\ \hline \end{array}$ 21. $\begin{array}{r} 11t^3 \\ -\ 4t^3 \\ \hline \end{array}$ 23. $\begin{array}{r} 4w^5 \\ +\ w^5 \\ \hline \end{array}$ 25. $\begin{array}{r} 3x \\ +\ 8x \\ \hline \end{array}$

ADDING POLYNOMIALS

When adding **polynomials**, make sure the exponents and variables are the same on the terms you are combining. The easiest way is to put the terms in columns with **like exponents** underneath each other. Each column is added as a separate problem. Fill in the blank spots with zeros if it helps you keep the columns straight. You never carry to the next column when adding polynomials.

EXAMPLE 1: Add $3x^2 + 14$ and $5x^2 + 2x$

$$\begin{array}{r} 3x^2 + 0x + 14 \\ (+)\ \underline{5x^2 + 2x + \ \ 0} \\ 8x^2 + 2x + 14 \end{array}$$

EXAMPLE 2: $(4x^3 - 2x) + (-x^3 - 4)$

$$\begin{array}{r} 4x^3 - 2x + 0 \\ (+)\ \underline{-x^3 + 0x - 4} \\ 3x^3 - 2x - 4 \end{array}$$

Add the following polynomials.

1. $y^2 + 3y + 2$ and $2y^2 + 4$

2. $(5y^2 + 4y - 6) + (2y^2 - 5y + 8)$

3. $5x^3 - 2x^2 + 4x - 1$ and $3x^2 - x + 2$

4. $-p + 4$ and $5p^2 - 2p + 2$

5. $(w - 2) + (w^2 + 2)$

6. $4t^2 - 5t - 7$ and $8t + 2$

7. $t^4 + t + 8$ and $2t^3 + 4t - 4$

8. $(3s^3 + s^2 - 2) + (-2s^3 + 4)$

9. $(-v^2 + 7v - 8) + (4v^3 - 6v + 4)$

10. $6m^2 - 2m + 10$ and $m^2 - m - 8$

11. $-x + 4$ and $3x^2 + x - 2$

12. $(8t^2 + 3t) + (-7t^2 - t + 4)$

13. $(3p^4 + 2p^2 - 1) + (-5p^2 - p + 8)$

14. $12s^3 + 9s^2 + 2s$ and $s^3 + s^2 + s$

15. $(-9b^2 + 7b + 2) + (-b^2 + 6b + 9)$

16. $15c^2 - 11c + 5$ and $-7c^2 + 3c - 9$

17. $5c^3 + 2c^2 + 3$ and $2c^3 + 4c^2 + 1$

18. $-14x^3 + 3x^2 + 15$ and $7x^3 - 12$

19. $(-x^2 + 2x - 4) + (3x^2 - 3)$

20. $(y^2 - 11y + 10) + (-13y^2 + 5y - 4)$

21. $3d^5 - 4d^3 + 7$ and $2d^4 - 2d^3 - 2$

22. $(6t^5 - t^3 + 17) + (4t^5 + 7t^3)$

23. $4p^2 - 8p + 9$ and $-p^2 - 3p - 5$

24. $20b^3 + 15b$ and $-4b^2 - 5b + 14$

25. $(-2w + 11) + (w^3 + w - 4)$

26. $(25z^2 + 13z + 8) + (z^2 - 2z - 10)$

SUBTRACTING POLYNOMIALS

When you subtract polynomials, it is important to remember to change all the signs in the subtracted polynomial (the subtrahend) and then add.

EXAMPLE: $(4y^2 + 8y + 9) - (2y^2 + 6y - 4)$

Step 1: Copy the subtraction problem into vertical form. Make sure you line up the terms with like exponents under each other just like you did for adding polynomials.

$$\begin{array}{rr} & 4y^2 + 8y + 9 \\ (-) & \underline{2y^2 + 6y - 4} \end{array}$$

Step 2: Change the subtraction sign to addition and all the signs of the subtracted polynomial to the opposite sign. The bottom polynomial in the problem becomes $-2y^2 - 6y + 4$.

Step 3: Add:

$$\begin{array}{rr} & 4y^2 + 8y + 9 \\ (+) & \underline{-2y^2 - 6y + 4} \\ & 2y^2 + 2y + 13 \end{array}$$

Subtract the following polynomials.

1. $(2x^2 + 5x + 2) - (x^2 + 3x + 1)$
2. $(8y - 4) - (4y + 3)$
3. $(11t^3 - 4t^2 + 3) - (-t^3 + 4t^2 - 5)$
4. $(-3w^2 + 9w - 5) - (-5w^2 - 5)$
5. $(6a^5 - a^3 + a) - (7a^5 + a^2 - 3a)$
6. $(14c^4 + 20c^2 + 10) - (7c^4 + 5c^2 + 12)$
7. $(5x^2 - 9x) - (-7x^2 + 4x + 8)$
8. $(12y^3 - 8y^2 - 10) - (3y^3 + y + 9)$

9. $(-3h^2 - 7h + 7) - (5h^2 + 4h + 10)$

10. $(10k^3 - 8) - (-4k^3 + k^2 + 5)$

11. $(x^2 - 5x + 9) - (6x^2 - 5x + 7)$

12. $(12p^2 + 4p) - (9p - 2)$

13. $(-2m - 8) - (6m + 2)$

14. $(13y^3 + 2y^2 - 8y) - (2y^3 + 4y^2 - 7y)$

15. $(7g + 3) - (g^2 + 4g - 5)$

16. $(-8w^3 + 4w) - (-10w^3 - 4w^2 - w)$

17. $(12x^3 + x^2 - 10) - (3x^3 + 2x^2 + 1)$

18. $(2a^2 + 2a + 2) - (-a^2 + 3a + 3)$

19. $(c + 19) - (3c^2 - 7c + 2)$

20. $(-6v^2 + 12v) - (3v^2 + 2v + 6)$

21. $(4b^3 + 3b^2 + 5) - (7b^3 - 8)$

22. $(15x^3 + 5x^2 - 4) - (4x^3 - 4x^2)$

23. $(8y^2 - 2) - (11y^2 - 2y - 3)$

24. $(-z^2 - 5z - 8) - (3z^2 - 5z + 5)$

A subtraction of polynomials problem may be stated in sentence form. Study the examples below.

EXAMPLE 1: Subtract $-5x^3 + 4x - 3$ from $3x^3 + 4x^2 - 6x$.

Step 1: Copy the problem in columns with terms with the same exponent and variable under each other. Notice the second polynomial in the sentence will be the top polynomial of the problem.

$$\begin{array}{rrrrr} & 3x^3 & + 4x^2 & - 6x & \\ (-) & -5x^3 & & + 4x & - 3 \\ \hline \end{array}$$

Step 2: Since this is a subtraction problem, change all the signs of the terms in the bottom polynomial. Then add.

$$\begin{array}{rrrrr} & 3x^3 & + 4x^2 & - 6x & \\ (+) & 5x^3 & & - 4x & + 3 \\ \hline & 8x^3 & + 4x^2 & -10x & + 3 \end{array}$$

EXAMPLE 2: From $6y^2 + 2$ subtract $4y^2 - 3y + 8$

In a problem phrased like this one, the first polynomial will be on top, and the second will be on bottom. Change the signs on the bottom polynomial and then add.

$$\begin{array}{rrrr} & 6y^2 & & + 2 \\ (-) & 4y^2 & - 3y & + 8 \\ \hline \end{array} \longrightarrow \begin{array}{rrrr} & 6y^2 & & + 2 \\ (+) & -4y^2 & + 3y & - 8 \\ \hline & 2y^2 & + 3y & - 6 \end{array}$$

Solve the following subtraction problems.

1. Subtract $3x^2 + 2x - 5$ from $5x^2 + 2$

2. From $5y^3 - 6y + 9$ subtract $8y^3 - 10$

3. From $4m^2 - 4m + 7$ subtract $2m - 3$

4. Subtract $8z^2 + 3z + 2$ from $4z^2 - 7z + 8$

5. Subtract $10t^3 + t^2 - 5$ from $-2t^3 - t^2 - 5$

6. Subtract $-7b^3 - 2b + 4$ from $-b^2 + b + 6$

7. From $10y^3 + 20$ subtract $5y^3 - 5$

8. From $14t^2 - 6t - 8$ subtract $4t^2 - 3t + 2$

9. Subtract $3p^2 + p - 2$ from $-7p^2 - 5p + 2$

10. Subtract $x^3 + 8$ from $3x^3 - 2x^2 + 9$

11. Subtract $12a^2 + 10$ from $a^3 - a^2 - 1$

12. From $6m^2 + 3m + 1$ subtract $-6m^2 - 3m$

13. From $-13z^3 - 3z^2 - 2$ subtract $-20z^3 + 20$

14. Subtract $9c^2 + 10$ from $8c^2 - 5c + 3$

15. Subtract $b^2 + b - 5$ from $5b^2 - 4b + 5$

16. Subtract $-3x - 4$ from $3x^2 + x + 9$

17. From $15y^2 + 2$ subtract $4y^2 + 3y + 7$

18. Subtract $3g^2 - 5g + 5$ from $9g^2 - 3g - 4$

19. From $-7m^2 - 8m$ subtract $3m^2 + 7$

20. Subtract $x + 1$ from $5x + 5$

21. Subtract $c^2 + c + 2$ from $-c^2 - c - 2$

22. From $8t^3 + 6t^2 - 4t + 2$ subtract $t^3 + 3t$

ADDING AND SUBTRACTING POLYNOMIALS REVIEW

Practice adding and subtracting the polynomials below.

1. Add $-3x^2 + x$ and $4x^2 - 2$

2. Subtract $(-2y^3 + 9y)$ from $(6y^3 - y + 4)$

3. $(8t^3 - 3t^2 - 9) + (-7t^3 + t - 4)$

4. $(7p^2 + 3p + 1) - (5p^2 + 4p + 6)$

5. From $4w^3 + 5w - 2$ subtract $6w^2 - 4$

6. Add $-8a^3 - 7a^2 + 10$ and $6a^3 + 4a^2$

7. $(-14b^2 + b + 2) + (6b^2 - b + 3)$

8. Subtract $(g^3 - 7g^2 - 5)$ from $(5g^3 - 10)$

9. $(4c - 6) - (2c^2 - 3c + 9)$

10. From $-m^3 + 2m^2 + m$ subtract $9m^3 + 2m$

11. $(-3v^2 + 9v - 6) + (3v^2 - 4v + 6)$

12. Add $10s^2 + 4$ and $5s - 6$

13. Subtract $-x^3 - 9x^2 - x$ from $3x^3 + 2x + 4$

14. $(-5y^2 - 4y - 1) - (5y^2 - 2y - 8)$

MULTIPLYING MONOMIALS

When two monomials have the **same variable**, they can be multiplied. The **exponents** are **added together**. If the variable has no exponent, it is understood that the exponent is 1.

Add

EXAMPLE: $4x^4 \times 3x^2 = 12x^6$

Multiply

EXAMPLE: $2y \times 5y^2 = 10y^3$

Multiply the following monomials

1. $6a \times 9a^5$ = _____
2. $2x^6 \times 5x^3$ = _____
3. $4y^{5} \times 3y^{2}$ = _____
4. $10t^2 \times 2t^2$ = _____
5. $2p^5 \times 4p^2$ = _____
6. $9b^2 \times 8b$ = _____
7. $3c^3 \times 3c^3$ = _____
8. $2d^8 \times 9d^2$ = _____
9. $6k^3 \times 5k^2$ = _____
10. $7m^{5} \times m$ = _____
11. $11z \times 2z^7$ = _____
12. $3w^4 \times 6w^5$ = _____
13. $4x^4 \times 5x^3$ = _____
14. $5n^2 \times 3n^3$ = _____
15. $8w^7 \times w$ = _____
16. $10s^6 \times 5s^3$ = _____
17. $4d^{5} \times 4d^{5}$ = _____
18. $5y^2 \times 8y^6$ = _____
19. $7t^{10} \times 3t^5$ = _____
20. $6p^8 \times 2p^3$ = _____
21. $x^3 \times 2x^3$ = _____

When problems include negative signs, follow the rules for multiplying integers.

22. $-7s^4 \times 5s^3 = -35s^7$
23. $-6a \times -9a^5$ = _____
24. $4x \times -x$ = _____
25. $-3y^2 \times -y^3$ = _____
26. $-5b^2 \times 3b^5$ = _____
27. $9c^4 \times -2c$ = _____
28. $-4t^3 \times 8t^3$ = _____
29. $10d \times -8d^7$ = _____
30. $-3g^6 \times -2g^3$ = _____
31. $-7s^4 \times 7s^3$ = _____
32. $-d^3 \times -2d$ = _____
33. $11p \times -2p^5$ = _____
34. $-5x^7 \times -3x^3$ = _____
35. $8z^4 \times 7z^4$ = _____
36. $-4w \times -5w^8$ = _____
37. $-5y^4 \times 6y^2$ = _____
38. $9x^3 \times -7x^5$ = _____
39. $-a^4 \times -a$ = _____
40. $-7k^2 \times 3k$ = _____
41. $-15t^2 \times -t^4$ = _____
42. $3x^8 \times 9x^2$ = _____

MULTIPLYING MONOMIALS WITH DIFFERENT VARIABLES

Warning: You cannot add the exponents of variables that are different.

EXAMPLE: $(-4wx)(6w^3x^2)$

To work this problem, first multiply the whole numbers, $-4 \times 6 = -24$. Then multiply the w's, $w \times w^3 = w^4$. Last, multiply the x's, $x \times x^2 = x^3$. The answer is $-24\,w^4x^3$.

Multiply the following monomials.

1. $(2x^2y^2)(-4xy^3)$ = ______	8. $(-3a^2b^2)(-4ab^3)$ = ______	15. $(2x^4z^2)(-9x^2z^4)$ = ______
2. $(9p^3q^4)(2p^2q)$ =	9. $(-5c^3d^2)(2c^4d^5)$ =	16. $(-4a^7c^9)(2a^2c)$ =
3. $(-3t^4v^2)(t^2v)$ = ______	10. $(10x^4y^2)(3x^3y)$ = ______	17. $(-bd^6)(-b^2d)$ = ______
4. $(7w^3z^2)(3wz)$ = ______	11. $(6f^3g^5)(-f^3g)$ = ______	18. $(3x^4y^2)(10x^3y^3)$ = ______
5. $(-2st^6)(-8s^2t)$ = ______	12. $(-4a^3v^4)(8a^4v)$ = ______	19. $(9p^2y)(3p^5y^3)$ = ______
6. $(xy^3)(4x^2y^2)$ = ______	13. $(5m^8n^5)(7m^2n^4)$ = ______	20. $(-2a^7x^2)(6ax^2)$ = ______
7. $(5y^2z)(3y^4z^2)$ = ______	14. $(7w^5y^3)(3wy)$ = ______	21. $(8c^4d^3)(-2c^2d^2)$ = ______

Multiplying three monomials works the same way.

22. $(3st)(4s^3t^2)(2\,s^2t^4)$ = $24s^6t^7$	32. $(8xy^2)(x^2y^3)(2x^2y)$ = ______
23. $(xy)(x^2y^2)(2x^3y^2)$ = ______	33. $(6p^3t)(2t^3)(p^2)$ = ______
24. $(2a^2b^2)(a^3b^3)(2ab)$ = ______	34. $(3bc)(b^2c)(4c^3)$ = ______
25. $(4y^2z^4)(2y^3)(2z^2)$ = ______	35. $(2y^4z^5)(2y^6)(y^2z^2)$ = ______
26. $(5cd^3)(3c^2d^2)(d^2)$ = ______	36. $(4p^3r^3)(4r^2)(p^2r)$ = ______
27. $(2w^2x^3)(3x^4)(2w^3)$ = ______	37. $(a^4z^6)(6a^2z^2)(3z^3)$ = ______
28. $(a^4d^2)(ad)(a^2d^3)$ = ______	38. $(5c^3)(6d^2)(2c^2d)$ = ______
29. $(7x^3t)(2t^4)(x^2t^2)$ = ______	39. $(9s^7t^2)(3st)(s^2t^3)$ = ______
30. $(p^2y^2)(4py)(p^3y^3)$ = ______	40. $(3a^3b^4)(2b^3)(3a^2)$ = ______
31. $(4x^3y)(5xy^3)(2y^4)$ = ______	41. $(5wz)(w^3z^3)(3w^2z^3)$ = ______

EXTRACTING MONOMIAL ROOTS

When finding the roots of monomial expressions, you must first divide the monomial expression into separate parts. Then, simplify each part of the expression.

Note: To find the square root of any variable raised to a positive exponent, simply divide the exponent by 2. For example, $\sqrt{y^8} = y^4$.

EXAMPLE: $\sqrt{16x^4y^2z^6}$

Step 1: Break each component apart.
$(\sqrt{16})(\sqrt{x^4})(\sqrt{y^2})(\sqrt{z^6})$

Step 2: Solve for each component.
$(\sqrt{16} = 4)(\sqrt{x^4} = x^2)(\sqrt{y^2} = y)(\sqrt{z^6} = z^3)$

Step 3: Recombine simplified expressions.
$(4)(x^2)(y)(z^3) = 4x^2yz^3$

Simplify the problems below.

1. $\sqrt{9a^4b^2c^8}$
2. $\sqrt{25h^{12}i^6j^8}$
3. $\sqrt{49p^{10}q^{12}r^6}$
4. $\sqrt{36a^{14}b^8c^4}$
5. $\sqrt{121t^{22}u^{18}v^4}$
6. $\sqrt{25k^6l^{16}m^{10}}$
7. $\sqrt{144s^4t^{14}u^{18}}$
8. $\sqrt{64x^6y^{18}z^{22}}$
9. $\sqrt{49a^6b^2c^4}$
10. $\sqrt{81u^8v^{12}w^{14}}$
11. $\sqrt{16x^{22}y^{14}z^6}$
12. $\sqrt{169d^{30}e^{42}f^8}$
13. $\sqrt{9f^2g^6h^{16}}$
14. $\sqrt{100l^{28}m^{16}n^2}$
15. $\sqrt{4g^{20}h^{18}i^{36}}$
16. $\sqrt{16a^{42}b^4c^{26}}$
17. $\sqrt{36j^{12}k^8l^{10}}$
18. $\sqrt{81q^2r^{10}s^{32}}$

MONOMIAL ROOTS WITH REMAINDERS

Monomial expressions which are not easily simplified under the square root symbol will also be covered under California's mathematics curriculum. Powers may be raised to odd numbers. In addition, the coefficients may not be perfect squares. Follow the example below to understand how to simplify these types of problems.

EXAMPLE: Simplify $\sqrt{20x^5y^9z^{13}}$

Step 1: Begin by simplifying the coefficient.

$\sqrt{20} = (\sqrt{5})(\sqrt{4})$,

$\sqrt{4} = 2$, so $\sqrt{20} = 2\sqrt{5}$

Step 2: Simplify the variable exponents.

$\sqrt{x^5} = (\sqrt{x^4})(\sqrt{x})$, $\sqrt{x^4} = x^2$, so $\sqrt{x^5} = x^2\sqrt{x}$

$\sqrt{y^9} = (\sqrt{y^8})(\sqrt{y})$, $\sqrt{y^8} = y^4$, so $\sqrt{y^9} = y^4\sqrt{y}$

$\sqrt{z^{13}} = (\sqrt{z^{12}})(\sqrt{z})$, $\sqrt{z^{12}} = z^6$, so $\sqrt{z^{13}} = z^6\sqrt{z}$

Step 3: Recombine simplified expressions.

$2x^2y^4z^6\sqrt{5xyz}$

Simplify the following square root expressions.

1. $\sqrt{45d^{14}e^{15}f^{11}}$
2. $\sqrt{50h^{16}i^{10}j^8}$
3. $\sqrt{30x^{22}y^{21}z^7}$
4. $\sqrt{84p^{11}q^7r^9}$
5. $\sqrt{48k^{25}l^{15}m^3}$
6. $\sqrt{54s^{13}t^5u^{17}}$
7. $\sqrt{60a^7b^{15}c^{21}}$
8. $\sqrt{18p^3q^{22}r^{18}}$
9. $\sqrt{72a^{19}b^9c^{15}}$
10. $\sqrt{68m^7n^3p^{11}}$
11. $\sqrt{75r^{15}s^{11}t^{13}}$
12. $\sqrt{20g^{21}h^{14}j^{17}}$
13. $\sqrt{80v^3w^9x^{12}}$
14. $\sqrt{44d^5e^8f^{13}}$
15. $\sqrt{24x^{17}y^6z^{11}}$
16. $\sqrt{32a^6b^{13}c^5}$
17. $\sqrt{52j^{12}k^{15}m^5}$
18. $\sqrt{96q^{12}r^{15}s^5}$

DIVIDING POLYNOMIALS BY MONOMIALS

EXAMPLE: $\dfrac{-8wx + 6x^2 - 16wx^2}{2wx}$

Step 1: Rewrite the problem. Divide each term from the top by the denominator, $2wx$.

$$\frac{-8wx}{2wx} + \frac{6x^2}{2wx} + \frac{-16wx^2}{2wx}$$

Step 2: Simplify each term in the problem. Then combine like terms.

$$-4 + \frac{3x}{w} - 8x$$

Simplify each of the following:

1. $\dfrac{bc^2 - 8bc - 2b^2c^2}{2bc}$

2. $\dfrac{3jk^2 + 12k + 9j^2k}{3jk}$

3. $\dfrac{5x^2y - 8xy^2 + 2y^3}{2xy}$

4. $\dfrac{16st^2 + st - 12s}{4st}$

5. $\dfrac{4wx^2 + 6wx - 12w^3}{2wx}$

6. $\dfrac{cd^2 + 10cd^3 + 16c^2}{2cd}$

7. $\dfrac{y^2z^3 - 2yz - 8z^2}{-2yz^2}$

8. $\dfrac{a^2b + 2ab^2 - 14ab^3}{2a^2}$

9. $\dfrac{pr^2 + 6pr + 8p^2r^2}{2pr^2}$

10. $\dfrac{6xy^2 - 3xy + 18x^2}{-3xy}$

11. $\dfrac{6x^2y + 12xy - 24y^2}{6xy}$

12. $\dfrac{5m^2n - 10mn - 25n^2}{5mn}$

13. $\dfrac{st^2 - 10st - 16s^2t^2}{2st}$

14. $\dfrac{7jk^2 - 14jk - 63j^2}{7jk}$

REMOVING PARENTHESES AND SIMPLIFYING

In the following problem, you must multiply each set of parentheses by the numbers and variables outside the parentheses, and then add the polynomials to simplify the expressions.

EXAMPLE: $8x\,(2x^2 - 5x + 7) - 3x\,(4x^2 + 3x - 8)$

Step 1: Multiply to remove the first set of parentheses.
$8x\,(2x^2 - 5x + 7) = 16x^3 - 40x^2 + 56x$

Step 2: Multiply to remove the second set of parentheses.
$-3x\,(4x^2 + 3x - 8) = -12x^3 - 9x^2 + 24x$

Step 3: Copy each polynomial in columns, making sure the terms with the same variable and exponent are under each other. Add to simplify.

$$\begin{array}{r} 16x^3 - 40x^2 + 56x \\ (+)\ -12x^3 - 9x^2 + 24x \\ \hline 4x^3 - 49x^2 + 80x \end{array}$$

Remove the parentheses, and simplify the following problems.

1. $4t\,(t + 7) + 5t\,(2t^2 - 4t + 1)$
2. $-5y\,(3y^2 - 5y + 3) - 6y\,(y^2 - 4y - 4)$
3. $-3\,(3x^3 + 4x) + 5x\,(x^2 + 3x + 2)$
4. $2b\,(5b^2 - 8b - 1) - 3b\,(4b + 3)$
5. $8d^2\,(3d + 4) - 7d\,(3d^2 + 4d + 5)$
6. $5a\,(3a^2 + 3a + 1) - (-2a^2 + 5a - 4)$
7. $3m\,(m + 7) + 8(4m^2 + m + 4)$
8. $4c^2\,(-6c^2 - 3c + 2) - 7c\,(5c^3 + 2c)$
9. $-8w\,(-w + 1) - 4w\,(3w - 5)$
10. $6p\,(2p^2 - 4p - 6) + 3p\,(p^2 + 6p + 9)$

MULTIPLYING TWO BINOMIALS

When you multiply two binomials such as $(x + 6)(x - 5)$, you must multiply each term in the first binomial by each term in the second binomial. The easiest way is to use the **FOIL** method. If you can remember the word **FOIL**, it can help you keep order when you multiply. The "F" stands for **first**, "O" stands for **outside**, "I" stands for **inside**, and "L" stands for **last**.

F	**O**	**I**	**L**
FIRST	**OUTSIDE**	**INSIDE**	**LAST**
Multiply the **first** terms in each binomial.	Next, multiply the **outside** terms.	Then, multiply the **inside** terms.	Last, multiply the **last** terms.
$(x + 6)(x - 5)$	$(x + 6)(x - 5)$	$(x + 6)(x - 5)$	$(x + 6)(x - 5)$
$x \times x = x^2$	$x \times -5 = -5x$	$6 \times x = 6x$	$6 \times -5 = -30$

$$x^2 \quad + \quad -5x \quad + \quad 6x \quad + \quad -30$$

Now just combine like terms, $6x - 5x = x$, and write your answer.

$(x + 6)(x - 5) = x^2 + x - 30$

Note: It is customary for mathematicians to write polynomials in descending order. That means that the term with the highest number exponent comes first in a polynomial. The next highest exponent is second and so on. When you use the **FOIL** method, the terms will always be in the customary order. You just need to combine like terms and write your answer.

Multiply the following binomials.

1. $(y - 7)(y + 3)$
2. $(2x + 4)(x + 9)$
3. $(4b - 3)(3b - 4)$
4. $(6g + 2)(g - 9)$
5. $(7k - 5)(-4k - 3)$
6. $(8v - 2)(3v + 4)$
7. $(10p + 2)(4p + 3)$
8. $(3h - 9)(-2h - 5)$
9. $(w - 4)(w - 7)$
10. $(6x + 1)(x - 2)$
11. $(5t + 3)(2t - 1)$
12. $(4y - 9)(4y + 9)$
13. $(a + 6)(3a + 5)$
14. $(3z - 8)(z - 4)$
15. $(5c + 2)(6c + 5)$

16. $(y + 3)(y - 3)$

17. $(2w - 5)(4w + 6)$

18. $(7x + 1)(x - 4)$

19. $(6t - 9)(4t - 4)$

20. $(5b + 6)(6b + 2)$

21. $(2z + 1)(10z + 4)$

22. $(11w - 8)(w + 3)$

23. $(5d - 9)(9d + 9)$

24. $(9g + 2)(g - 2)$

25. $(4p + 7)(2p + 3)$

26. $(m + 5)(m - 5)$

27. $(8b - 8)(2b - 1)$

28. $(z + 3)(3z + 5)$

29. $(7y - 5)(y - 3)$

30. $(9x + 5)(3x - 1)$

31. $(3t + 1)(t + 10)$

32. $(2w - 9)(8w + 7)$

33. $(8s - 2)(s + 4)$

34. $(4k - 1)(8k + 9)$

35. $(h + 12)(h - 2)$

36. $(3x + 7)(7x + 3)$

37. $(2v - 6)(2v + 6)$

38. $(2x + 8)(2x - 3)$

39. $(k - 1)(6k + 12)$

40. $(3w + 11)(2w + 2)$

41. $(8y - 10)(5y - 3)$

42. $(6d + 13)(d - 1)$

43. $(7h + 3)(2h + 4)$

44. $(5n + 9)(5n - 5)$

45. $(6z + 5)(z - 8)$

46. $(4p + 5)(2p - 9)$

47. $(b + 2)(5b + 7)$

48. $(9y - 3)(8y - 7)$

SIMPLIFYING EXPRESSIONS WITH EXPONENTS

EXAMPLE 1: Simplify $(2a + 5)^2$

When you simplify an expression such as $(2a + 5)^2$, it is best to write the expression as two binomials and use FOIL to simplify.

$(2a + 5)^2 = (2a + 5)(2a + 5)$

Using FOIL we have $4a^2 + 10a + 10a + 25 = 4a^2 + 20a + 25$

EXAMPLE 2: Simplify $4(3a + 2)^2$

Using order of operations, we must simplify the exponent first.

$= 4(3a + 2)(3a + 2)$

$= 4(9a^2 + 6a + 6a + 4)$

$= 4(9a^2 + 12a + 4)$ Now multiply by 4.

$= 4(9a^2 + 12a + 4) = 36a^2 + 48a + 16$

Note: It is customary for mathematicians to write polynomials in descending order. That means that the term with the highest number exponent comes first in a polynomial. The next highest exponent is second and so on. When you use the **FOIL** method, the terms will always be in the customary order. You just need to combine like terms, and write your answer.

Multiply the following binomials.

1. $(y + 3)^2$
2. $7(2x + 4)^2$
3. $6(4b - 3)^2$
4. $5(6g + 2)^2$
5. $(-4k - 3)^2$
6. $3(-2h - 5)^2$
7. $-2(8v - 2)^2$
8. $(10p + 2)^2$
9. $6(-2h - 5)^2$
10. $6(w - 7)^2$
11. $2(6x + 1)^2$
12. $(9x + 2)^2$
13. $(5t + 3)^2$
14. $3(4y - 9)^2$
15. $8(a + 6)^2$
16. $4(3z - 8)^2$
17. $3(5c + 2)^2$
18. $4(3t + 9)^2$

CHAPTER 4 REVIEW

Simplify:

1. $3a^2 + 9a^2$
2. $(7x^2y^4)(9xy^5)$
3. $-6z^2(z + 3)$
4. $(4b^2)(5b^3)$
5. $7x^2 - 9x^2$
6. $(5p - 4) - (3p + 2)$
7. $-5t(3t + 9)^2$
8. $(3w^3y^2)(4wy^5)$
9. $3(2g + 3)^2$
10. $14d^4 - 9d^4$
11. $(7w - 4)(w - 8)$
12. $15t^2 + 4t^2$
13. $(7c^4)(9c^2)$
14. $(9x + 2)(x + 5)$
15. $4y(4y^2 - 9y + 2)$
16. $(8a^4b)(2ab^3)(ab)$
17. $(5w^6)(9w^9)$
18. $8x^3 + 12x^3$
19. $15p^5 - 11p^5$
20. $(3s^4t^2)(4st^3)$
21. $(4d + 9)(2d + 7)$
22. $4w(-3w^2 + 7w - 5)$
23. $24z^6 - 10z^6$
24. $-7y^3 - 8y^3$
25. $(7x^4)(7x^5)$
26. $17p^2 + 9p^2$
27. $(a^2v)(2av)(a^3v^6)$
28. $4(6y - 5)^2$
29. $(3c^2)(6c^8)$
30. $(4x^5y^3)(2xy^3)$

31. Add $2x^2 + 9x$ and $5x^2 - 8x + 2$

32. $4t\,(6t^2 + 4t - 6) + 8t\,(3t + 3)$

33. Subtract $y^2 + 4y - 6$ from $3y^2 + 7$

34. $2x\,(4x^2 + 6x - 3) + 4x\,(x + 3)$

35. $(6t - 4) - (6t^2 + t - 2)$

36. $(4x + 6) + (7x^2 - 2x + 3)$

37. Subtract $5a - 2$ from $a + 9$

38. $(-2y + 4) + (4y - 6)$

39. $2t\,(t + 6) - 5t\,(2t + 7)$

40. Add $3c - 4$ and $c^2 - 3c - 2$

41. $2b\,(b - 4) - (b^2 + 2b + 1)$

42. $(6k^2 + 5k) + (k^2 + k + 9)$

43. $(q^2r^3)(3qr^2)(2q^4r)$

44. $(5df)(d^4f^2)(2df)$

45. $(7g^2h^3)(g^3h^6)(6gh^3)$

46. $(8v^2x^3)(3v^6x^2)(2v^4x^4)$

47. $(3n^2m^2)(9n^2m)(n^3m^7)$

48. $(11f^2g^2)(4f^3g^8)(2f^6g)$

49. $\sqrt{36r^6s^8t^2}$

50. $\sqrt{40g^3h^6j^7}$

51. $\sqrt{18m^5n^3p^7}$

52. $\sqrt{75a^7b^2c^9}$

53. $\sqrt{12x^3y^4z^7}$

54. $\sqrt{64f^6g^7h^5}$

ONE-STEP ALGEBRA PROBLEMS WITH ADDITION AND SUBTRACTION

You have been solving algebra problems since second grade by filling in blanks. For example, 5 + ____ = 8. The answer is 3. You can solve the same kind of problems using algebra. The problems only look a little different because the blank has been replaced with a letter. The letter is called a **variable**.

EXAMPLE: **Arithmetic** **5 + _____ = 14**
Algebra $\mathbf{5 + x = 14}$

The goal in any algebra problem is to move all the numbers to one side of the equal sign and have the letter (called a **variable**) on the other side. In this problem, the 5 and the "x" are on the same side. The 5 is added to x. To move it, do the **opposite** of **add**. The **opposite** of **add** is **subtract**, so subtract 5 from both sides of the equation. Now the problem looks like this:

$$\begin{array}{rcr} 5 + x & = & 14 \\ -5 & & -5 \\ \hline x & = & 9 \end{array}$$

To check your answer, put 9 in the place of x in the original problem. Does 5 + 9 = 14? Yes, it does.

EXAMPLE: $\mathbf{y - 16 = 27}$ Again, the 16 has to move. To move it to the other side of the equation, we do the **opposite** of **subtract**. We **add** 16 to both sides.

$$\begin{array}{rcr} y - 16 & = & 27 \\ +16 & & +16 \\ y & = & 43 \end{array}$$

Check by putting 43 in place of the y in the original problem. Does 43 – 16 = 27? Yes.

Solve the problems below.

1. $n + 9 = 27$
2. $12 + y = 55$
3. $51 + v = 67$
4. $f + 16 = 31$
5. $5 + x = 23$
6. $15 + x = 24$
7. $w - 14 = 89$
8. $t - 26 = 20$
9. $m - 12 = 17$
10. $c - 7 = 21$
11. $k - 5 = 29$
12. $a + 17 = 45$
13. $d + 26 = 56$
14. $15 + x = 56$
15. $y + 19 = 32$
16. $t - 16 = 28$
17. $m + 14 = 37$
18. $y - 21 = 29$
19. $f + 7 = 31$
20. $h - 12 = 18$
21. $r - 12 = 37$
22. $h - 17 = 22$
23. $x - 37 = 46$
24. $r - 11 = 28$
25. $t - 5 = 52$

MORE ONE-STEP ALGEBRA PROBLEMS

Sometimes the answer to the algebra problem is a negative number. The problems work the same. Study the **examples** below.

EXAMPLES:

$$\begin{array}{rcl} n + 8 & = & 6 \\ -8 & & -8 \\ \hline n & = & (-2) \end{array} \qquad \begin{array}{rcl} x - 10 & = & -14 \\ +10 & & +10 \\ \hline x & = & (-4) \end{array} \qquad \begin{array}{rcl} y - (-6) & = & (-2) \\ +(-6) & & +(-6) \\ \hline y & = & (-8) \end{array}$$

Solve the problems below.

1. $w + 4 = (-6)$
2. $q - 8 = (-9)$
3. $j + 7 = 1$
4. $y + 14 = 6$
5. $k - 5 = (-8)$
6. $h - 7 = (-2)$
7. $7 + d = (-3)$
8. $(-7) + k = (-4)$
9. $(-4) + h = 8$
10. $(-5) + g = -2$
11. $y - (-8) = (-5)$
12. $x + 21 = (-2)$
13. $w - (-3) = 8$
14. $q - (-6) = 12$
15. $z + 8 = 3$
16. $(-4) + m = 4$
17. $(-10) + r = (-2)$
18. $(-6) + b = 9$
19. $p - (-7) = (-2)$
20. $q - 5 = (-11)$
21. $x + 17 = (-4)$
22. $(-9) + x = 14$
23. $(-17) + r = -12$
24. $(-3) + y = 19$
25. $t - (-2) = (-8)$
26. $d - 9 = (-16)$
27. $x + 16 = -3$
28. $w + 14 = (-9)$
29. $q + 12 = (-5)$
30. $j + (-7) = (-1)$
31. $y - (-9) = 8$
32. $x + 12 = 6$

ONE-STEP ALGEBRA PROBLEMS WITH MULTIPLICATION AND DIVISION

Solving one-step algebra problems with multiplication and division is just as easy as solving addition and subtraction problems. Again, you perform the **opposite** operation. If the problem is a **multiplication** problem, you **divide** to find the answer. If it is a **division** problem, you **multiply** to find the answer. Carefully read the examples below, and you will see how easy they are.

EXAMPLE 1: $4x = 20$ (**$4x$** means **4 times *x*. 4** is the **coefficient** of ***x***.)

The goal is to get the numbers on one side of the equal sign and the variable ***x*** on the other side. In this problem, the **4** and ***x*** are on the same side of the equal sign. The **4** has to be moved over. **$4x$** means **4** times ***x***. The opposite of **multiply** is **divide**. If we divide both sides of the equation by **4**, we will find the answer.

$4x = 20$ **We need to divide both sides by 4.**

This means divide by 4. $\rightarrow \frac{\overset{1}{\cancel{4}}x}{\underset{1}{\cancel{4}}} = \frac{\overset{5}{\cancel{20}}}{\underset{1}{\cancel{4}}}$ **We see that $1x = 5$ so $x = 5$**

When you put 5 in place of *x* in the original problem, it is correct. $4 \times 5 = 20$

EXAMPLE 2: $\frac{y}{4} = 2$ This problem means *y* divided by 4 is equal to 2. In this case, the opposite of **divide** is **multiply**. We need to multiply both sides of the equation by 4.

$4 \times \frac{y}{4} = 2 \times 4$ so $y = 8$

When you put 8 in place of *y* in the original problem, it is correct. $\frac{8}{4} = 2$

Solve the problems below.

1. $2x = 14$
2. $\frac{w}{5} = 11$
3. $3h = 45$
4. $10y = 30$
5. $5a = 60$
6. $\frac{x}{3} = 9$
7. $6d = 66$
8. $\frac{w}{9} = 3$
9. $7r = 98$
10. $\frac{y}{3} = 2$
11. $\frac{x}{4} = 36$
12. $\frac{r}{4} = 7$
13. $8t = 96$
14. $\frac{z}{2} = 15$
15. $\frac{n}{9} = 5$
16. $4z = 24$
17. $6d = 84$
18. $\frac{t}{3} = 3$
19. $\frac{m}{6} = 9$
20. $9p = 72$

Sometimes the answer to the algebra problem is a **fraction**. Read the example below, and you will see how easy it is.

EXAMPLE

$4x = 5$ Problems like this are solved just like the problems on the previous page. The only difference is that the answer is a **fraction**.

In this problem, the 4 is **multiplied** by x. To solve, we need to divide both sides of the equation by 4.

$4x = 5$ Now **divide** by 4. $\frac{4x}{4} = \frac{5}{4}$ Now cancel. $\frac{\cancel{4}x}{\cancel{4}} = \frac{5}{4}$ so $x = \frac{5}{4}$

When you put $\frac{5}{4}$ in place of x in the original problem, it is correct.

$4 \times \frac{5}{4} = 5$ Now cancel. ⟶ $\cancel{4} \times \frac{5}{\cancel{4}} = 5$ so $5 = 5$

Solve the problems below. Some of the answers will be fractions. Some answers will be integers.

1. $2x = 3$	8. $4z = 64$	15. $3y = 8$	22. $7d = 12$
2. $4y = 5$	9. $7x = 126$	16. $2t = 10$	23. $2w = 13$
3. $5t = 2$	10. $6p = 10$	17. $3b = 2$	24. $9g = 81$
4. $12b = 144$	11. $2n = 9$	18. $5c = 14$	25. $6a = 18$
5. $9a = 72$	12. $5x = 11$	19. $4d = 3$	26. $2p = 16$
6. $8y = 16$	13. $15m = 180$	20. $5z = 75$	27. $15w = 3$
7. $7x = 21$	14. $5h = 21$	21. $9y = 4$	28. $5x = 13$

MULTIPLYING AND DIVIDING WITH NEGATIVE NUMBERS

EXAMPLE 1: $\mathbf{-3x = 15}$ In the problem, –3 is **multiplied** by x. To find the solution, we must do the opposite. The opposite of **multiply** is **divide**. We must **divide** both sides of the equation by –3.

$\frac{-3x}{-3} = \frac{15}{-3}$ Then cancel. $\frac{\not{-3}x}{\not{-3}} = \frac{\not{15}^{\,5}}{\not{-3}_{\,1}}$ $x = -5$

EXAMPLE 2: $\mathbf{\frac{y}{-4} = -20}$ In this problem, y is **divided** by –4. To find the answer, do the opposite. **Multiply** both sides by –4.

$-4 \times \frac{y}{-4} = (-20) \times (-4)$ so $y = 80$

EXAMPLE 3: $\mathbf{-6a = 2}$ The answer to an algebra problem can also be a negative fraction.

$\frac{-6a}{-6} = \frac{2}{-6}$ ← reduce to get $a = \frac{1}{-3}$ or $-\frac{1}{3}$

Note: A negative fraction can be written several different ways.

$\frac{1}{-3} = \frac{-1}{3} = -\frac{1}{3} = -\left(\frac{1}{3}\right)$

All mean the same thing.

Solve the problems below. Reduce any fractions to lowest terms.

1. $2z = -6$
2. $\frac{y}{-5} = 20$
3. $-6k = 54$
4. $4x = -24$
5. $\frac{t}{7} = -4$
6. $\frac{r}{-2} = -10$
7. $9x = -72$
8. $\frac{x}{-6} = 3$
9. $\frac{w}{-11} = 5$
10. $5y = -35$
11. $\frac{x}{-4} = -9$
12. $7t = -49$
13. $-14x = -28$
14. $\frac{m}{3} = -12$
15. $-8z = 32$
16. $-15w = -60$
17. $\frac{y}{-9} = -4$
18. $\frac{d}{8} = -7$
19. $-12v = 36$
20. $\frac{c}{-6} = -6$

21. $-4x = -3$

22. $-12y = 7$

23. $\frac{a}{-2} = 22$

24. $-18b = 6$

25. $13a = -36$

26. $\frac{b}{-2} = -14$

27. $-24x = -6$

28. $-6p = 42$

29. $\frac{x}{-23} = -1$

30. $\overline{7x} = -7$

31. $-9y = -1$

32. $\frac{d}{5} = -10$

33. $\frac{z}{-13} = -2$

34. $-5c = 45$

35. $2d = -3$

36. $-8d = -12$

37. $-24w = 9$

38. $\frac{y}{-9} = -6$

39. $-9a = -18$

40. $\frac{p}{-2} = 15$

VARIABLES WITH A COEFFICIENT OF NEGATIVE ONE

The answer to an algebra problem should not have a negative sign in front of the variable. For example, the problem $-x = 5$ is not completely simplified. Study the examples below to learn how to finish simplifying this problem.

EXAMPLE 1: $-x = 5$ $-x$ means the same thing as $-1x$ or -1 times x. To simplify this problem, **multiply** by -1 on both sides of the equation.

$(-1)(-1x) = (-1)(5)$ so $x = -5$

EXAMPLE 2: $-y = -3$ Solve the same way.

$(-1)(-y) = (-1)(-3)$ so $y = 3$

Simplify the following equations.

1. $-w = 14$
2. $-a = 20$
3. $-x = -15$
4. $-x = -25$
5. $-y = -16$
6. $-t = 62$
7. $-p = -34$
8. $-m = 81$
9. $-w = 17$
10. $-v = -9$
11. $-k = 13$
12. $-q = 7$

GRAPHING INEQUALITIES

An inequality is a sentence that contains a $\neq$, $<$, $>$, $\leq$, or $\geq$ sign. Look at the following graphs of inequalities on a number line.

NUMBER LINE

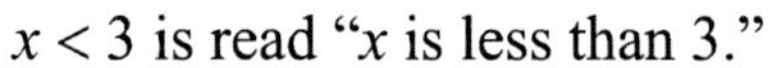

$x < 3$ is read "x is less than 3."

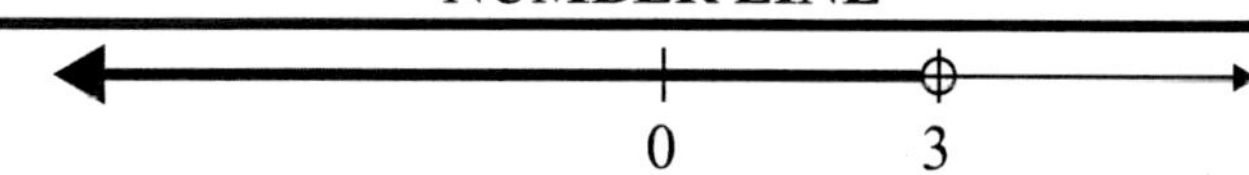

There is no line under the $<$ sign, so the graph uses an **open** endpoint to show x is less than 3 but does not include 3.

$x \leq 5$ is read "x is less than or equal to 5."

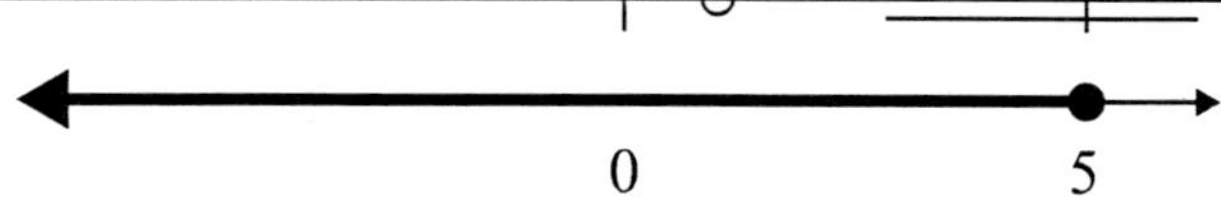

If you see a line under $<$ or $>$ ($\leq$ or $\geq$), the endpoint is filled in. The graph uses a **closed** circle because the number 5 **is** included in the graph.

$x > -2$ is read "x is greater than -2."

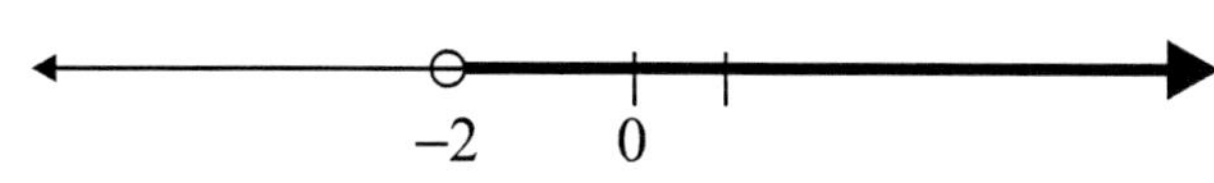

$x \geq 1$ is read "x is greater than or equal to 1."

There can be more than one inequality sign. For example:

$-2 \leq x < 4$ is read "-2 is less than or equal to x, and x is less than 4."

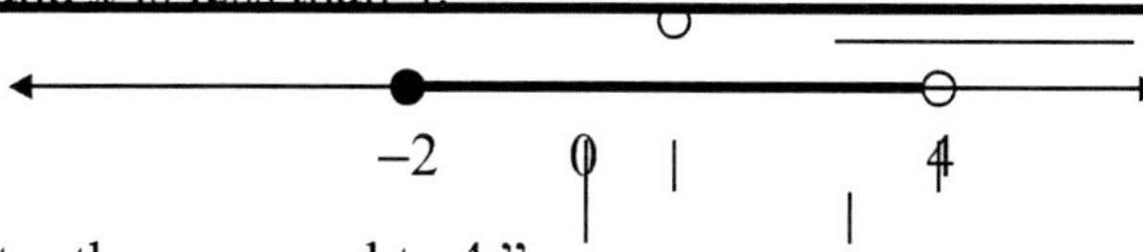

$x < 1$ or $x \geq 4$ is read "x is less than 1, or x is greater than or equal to 4."

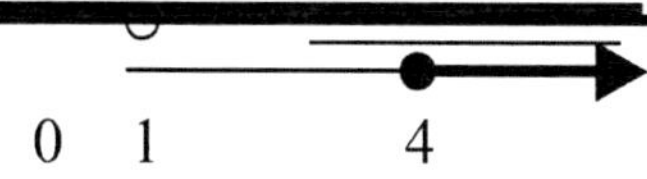

Graph the solution sets of the following inequalities.

1. $x > 8$
2. $x \leq 5$
3. $-5 < x < 1$
4. $x > 7$
5. $1 \leq x < 4$
6. $x < -2$ or $x > 1$
7. $x \geq 10$
8. $x < 4$
9. $x \leq 3$ or $x \geq 5$
10. $x < -1$ or $x > 1$

Give the inequality represented by each of the following number lines.

11. 0 ________
12. –4 10 ________
13. 2 4 ________
14. 8 ________
15. –10 –4 ________
16. –1 3 ________
17. –5 ________
18. 6 ________
19. –6 –4 ________
20. 4 7 ________

SOLVING INEQUALITIES BY ADDITION AND SUBTRACTION

If you add or subtract the same number to both sides of an inequality, the inequality remains the same. It works just like an equation.

EXAMPLE: Solve and graph the solution set for $x - 2 \leq 5$.

Step 1: Add 2 to both sides of the inequality.

$$\begin{array}{r} x - 2 \leq 5 \\ +2 \ \ +2 \\ \hline x \ \ \leq 7 \end{array}$$

Step 2: Graph the solution set for the inequality.

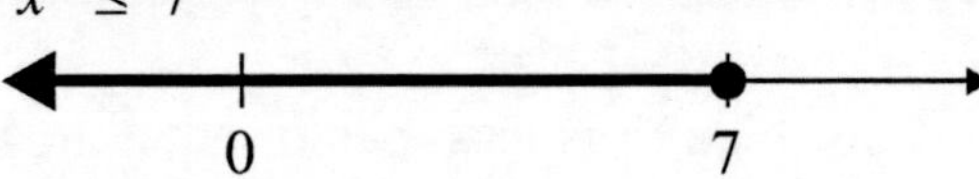

Solve and graph the solution set for the following inequalities.

1. $x + 5 > 3$
2. $x - 10 < 5$
3. $x - 2 \leq 1$
4. $9 + x \geq 7$
5. $x - 4 > -2$
6. $x + 11 \leq 20$
7. $x - 3 < -12$
8. $x + 6 \geq -3$
9. $x + 12 \leq 8$
10. $15 + x > 5$
11. $x - 6 < -2$
12. $x + 7 \geq 4$
13. $14 + x \leq 8$
14. $x - 8 > 24$
15. $x + 1 \leq 12$
16. $11 + x \geq 11$
17. $x - 3 < 17$
18. $x + 9 > -4$
19. $x + 6 \leq 14$
20. $x - 8 \geq 19$

SOLVING INEQUALITIES BY MULTIPLICATION AND DIVISION

If you multiply or divide both sides of an inequality by a **positive** number, the inequality symbol stays the same. However, if you multiply or divide both sides of an inequality by a **negative** number, **you must reverse the direction of the inequality symbol**.

EXAMPLE 1: Solve and graph the solution set for $4x \le 20$.

Step 1: Divide both sides of the inequality by 4.

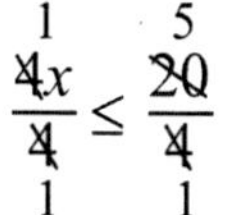

Step 2: Graph the solution. $x \le 5$

EXAMPLE 2: Solve and graph the solution set for $6 > -\frac{x}{3}$

Step 1: Multiply both sides of the inequality by −3, and **reverse the direction of the symbol**.

$$(-3) \times 6 < \frac{x}{-3} \times (-3)$$

Step 2: Graph the solution. $-18 < x$

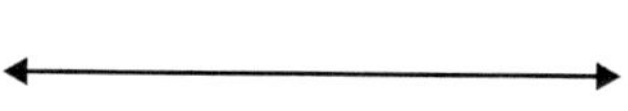

Solve and graph the following inequalities.

1. $\frac{x}{5} > 4$
2. $2x \le 24$
3. $-6x \ge 36$
4. $\frac{x}{10} > -2$
5. $-\frac{x}{4} > 8$
6. $-7x \le -49$
7. $-3x > 18$
8. $-\frac{x}{7} \ge 9$
9. $9x \le 54$
10. $\frac{x}{8} > 1$
11. $-\frac{x}{9} \le 3$
12. $-4x < -12$
13. $\frac{-x}{2} \ge -20$
14. $10x \le 30$
15. $\frac{x}{12} \ge -4$
16. $-6x < 24$

CHAPTER 5 REVIEW

Solve the following one-step algebra problems.

1. $5y = -25$
2. $x + 4 = 24$
3. $d - 11 = 14$
4. $\frac{a}{6} = -8$
5. $-t = 2$
6. $-14b = 12$
7. $\frac{c}{-10} = -3$
8. $z - 15 = -19$
9. $-13d = 4$
10. $\frac{x}{-14} = 2$
11. $-4k = -12$
12. $y + 13 = 27$
13. $15 + h = 4$
14. $14p = 2$
15. $\frac{b}{4} = 11$
16. $p - 26 = 12$
17. $x + (-2) = 5$
18. $m + 17 = 27$
19. $\frac{k}{-4} = 13$
20. $-18a = -7$
21. $21t = -7$
22. $z - (-9) = 14$
23. $23 + w = 28$
24. $n - 35 = -16$
25. $-a = 26$
26. $-19 + f = -9$
27. $\frac{w}{11} = 3$
28. $-7y = 28$
29. $x + 23 = 20$
30. $z - 12 = -7$
31. $-16 + g = 40$
32. $\frac{m}{-3} = -9$
33. $d + (-6) = 17$
34. $-p = 47$
35. $k - 16 = 5$
36. $9y = -3$
37. $-2z = -36$
38. $10h = 12$
39. $w - 16 = 4$
40. $y + 10 = -8$

Graph the solution sets of the following inequalities.

41. $x \leq -3$

42. $x > 6$

43. $x < -2$

44. $x \geq 4$

Give the inequality represented by each of the following number lines.

45. 

____

46. 

____

47. 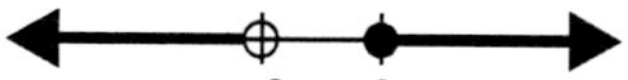

____

48. 

____

Solve and graph the solution set for the following inequalities.

49. $x - 2 > 8$

50. $4 + x < -1$

51. $6x \geq 54$

52. $-2x \leq 8$

53. $\frac{x}{2} > -1$

54. $-x < -9$

55. $-\frac{x}{3} \leq 5$

56. $x + 10 \leq 4$

57. $x - 6 \geq -2$

58. $7x < -14$

59. $-3x > -12$

60. $-\frac{x}{6} \leq -3$

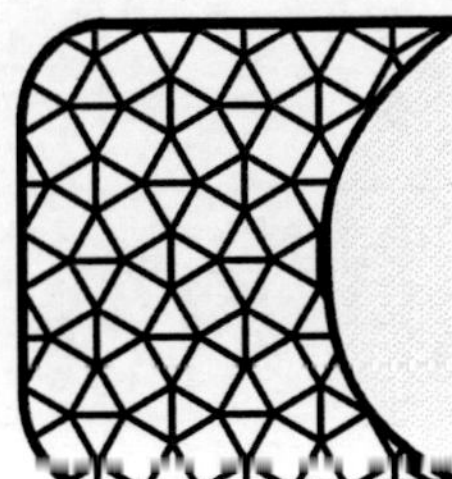

CHAPTER 6: SOLVING MULTI-STEP EQUATIONS & INEQUALITIES

Standards 2.12.2 & 2.12.3 & 2.12.4

TWO-STEP ALGEBRA PROBLEMS

In the following two-step algebra problems, **additions** or **subtractions** are performed **first** and *then* **divisions** or **multiplications**.

EXAMPLE 1: $-4x + 7 = 31$

Step 1: Subtract 7 from both sides.

$$\begin{array}{r} -4x + 7 = 31 \\ -7 \quad -7 \\ \hline -4x \quad = 24 \end{array}$$

Step 2: Divide both sides by –4.

$$\frac{-4x}{-4} = \frac{24}{-4} \quad \text{so} \quad x = -6$$

EXAMPLE 2: $-8 - y = 12$

Step 1: Add 8 to both sides.

$$\begin{array}{r} -8 - y = 12 \\ +8 \qquad +8 \\ \hline -y = 20 \end{array}$$

Step 2: **REMEMBER:** To finish solving an algebra problem with a negative sign in front of the variable, multiply both sides by –1. The variable needs to be positive in the answer.

$$(-1)(-y) = (-1)(20) \quad \text{so} \quad y = -20$$

Solve the two-step algebra problems below.

1. $6x - 4 = -34$
2. $5y - 3 = 32$
3. $8 - t = 1$
4. $10p - 6 = -36$
5. $11 - 9m = -70$
6. $4x - 12 = 24$
7. $3x - 17 = -41$
8. $9d - 5 = 49$
9. $10h + 8 = 78$
10. $-6b - 8 = 10$
11. $-g - 24 = -17$
12. $-7k - 12 = 30$
13. $9 - 5r = 64$
14. $6y - 14 = 34$
15. $12f + 15 = 51$
16. $21t + 17 = 80$
17. $20y + 9 = 149$
18. $15p - 27 = 33$
19. $22h + 9 = 97$
20. $-5 + 36w = 175$

TWO-STEP ALGEBRA PROBLEMS WITH FRACTIONS

An algebra problem may contain a fraction. Study the following example to understand how to solve algebra problems that contain a fraction.

EXAMPLE: $\frac{x}{2} + 4 = 3$

Step 1: $\frac{x}{2} + 4 = 3$ Subtract 4 from both sides.

$\phantom{\frac{x}{2}} - 4 \quad -4$

Step 2: $\frac{x}{2} = -1$ Now this looks like the one-step algebra problems you solved in Chapter 1. Multiply both sides by 2 to solve for x.

$\frac{x}{2} \times 2 = -1 \times 2 \quad x = -2$

Simplify the following algebra problems.

1. $4 + \frac{y}{3} = 7$
2. $\frac{a}{2} + 5 = 12$
3. $\frac{w}{5} - 3 = 6$
4. $\frac{x}{9} - 9 = -5$
5. $\frac{b}{6} + 2 = -4$
6. $7 + \frac{z}{2} = -13$
7. $\frac{x}{2} - 7 = 3$
8. $\frac{c}{5} + 6 = -2$
9. $3 + \frac{x}{11} = 7$
10. $16 + \frac{m}{6} = 14$
11. $\frac{p}{3} + 5 = -2$
12. $\frac{t}{8} + 9 = 3$
13. $\frac{v}{7} - 8 = -1$
14. $5 + \frac{h}{10} = 8$
15. $\frac{k}{7} - 9 = 1$
16. $\frac{y}{4} + 13 = 8$
17. $15 + \frac{z}{14} = 13$
18. $\frac{b}{6} - 9 = -14$
19. $\frac{d}{3} + 7 = 12$
20. $10 + \frac{v}{6} = 4$
21. $2 + \frac{p}{4} = -6$
22. $\frac{t}{7} - 9 = -5$
23. $\frac{a}{10} - 1 = 3$
24. $\frac{a}{8} + 16 = 9$

MORE TWO-STEP ALGEBRA PROBLEMS WITH FRACTIONS

Study the following example to understand how to solve algebra problems that contain a different type of fraction.

EXAMPLE: $\frac{x+2}{4} = 3$ — In this example, "$x + 2$" is divided by 4, and not *just* the x or the 2.

Step 1: $\frac{x+2}{\not{4}} \times \not{4} = 3 \times 4$ — First, multiply both sides by 4 to eliminate the fraction.

Step 2: $x + 2 = 12$ — Next, subtract 2 from both sides.

$$\begin{array}{r} x + 2 = 12 \\ -2 \quad -2 \\ \hline x = 10 \end{array}$$

Solve the following problems.

1. $\frac{x+1}{5} = 4$
2. $\frac{z-9}{2} = 7$
3. $\frac{b-4}{4} = -5$
4. $\frac{y-9}{3} = 7$
5. $\frac{d-10}{-2} = 12$
6. $\frac{w-10}{-8} = -4$
7. $\frac{x-1}{-2} = -5$
8. $\frac{c+40}{-5} = -7$
9. $\frac{13+h}{2} = 12$
10. $\frac{k-10}{3} = 9$
11. $\frac{a+11}{-4} = 4$
12. $\frac{x-20}{7} = 6$
13. $\frac{t+2}{6} = -5$
14. $\frac{b+1}{-7} = 2$
15. $\frac{f-9}{3} = 8$
16. $\frac{4+w}{6} = -6$
17. $\frac{3+t}{3} = 10$
18. $\frac{x+5}{5} = -3$
19. $\frac{g+3}{2} = 11$
20. $\frac{k+1}{-6} = 5$
21. $\frac{y-14}{2} = -8$
22. $\frac{z-4}{-2} = 13$
23. $\frac{w+2}{15} = -1$
24. $\frac{3+h}{3} = 6$

COMBINING LIKE TERMS

In an algebra problem, **terms** are separated by + and − signs. The expression $5x - 4 - 3x + 7$ has 4 terms: $5x$, 4, $3x$, and 7. Terms having the same variable can be combined (added or subtracted) to simplify the expression. In this case, $5x - 4 - 3x + 7$ simplifies to $2x + 3$.

$5x - 3x \quad -4 + 7$

Simplify the following expressions.

1. $7x + 12x$ = _____
2. $8y - 5y + 8$ = _____
3. $4 - 2c + 9$ = _____
4. $11a - 16 - a$ = _____
5. $9w + 3w + 3$ = _____
6. $-5x + x + 2x$ = _____
7. $w - 15 + 9w$ = _____
8. $21 - 10t + 9 - 2t$ = _____
9. $-3 + x - 4x + 9$ = _____
10. $7b + 12 + 4b$ = _____
11. $4h - h + 2 - 5$ = _____
12. $-6k + 10 - 4k$ = _____
13. $2a + 12a - 5 + a$ = _____
14. $5 + 9c - 10$ = _____
15. $-d + 1 + 2d - 4$ = _____
16. $-8 + 4h + 1 - \text{h}$ = _____
17. $12x - 4x + 7$ = _____
18. $10 + 3z + z - 5$ = _____
19. $14 + 3y - y - 2$ = _____
20. $11p - 4p + p$ = _____
21. $11m + 2 - m + 1$ = _____

SOLVING EQUATIONS WITH LIKE TERMS

When an equation has two or more like terms on the same side of the equation, like terms should be combined as the **first** step in solving the equation.

EXAMPLE: $7x + 2x - 7 = 21 + 8$

Step 1: Combine like terms on both sides of the equation.

Step 2: Solve the two-step algebra problem as explained previously.

$$7x + 2x - 7 = 21 + 8$$
$$9x - 7 = 29$$
$$+7 \quad +7$$
$$\frac{9x}{9} = \frac{36}{9}$$
$$x = 4$$

Solve the equations below, combining like terms first.

1. $3w - 2w + 4 = 6$
2. $7x + 3 + x = 16 + 3$
3. $5 - 6y + 9y = -15 + 5$
4. $-14 + 7a + 2a = -5$
5. $-2t + 4t - 7 = 9$
6. $9d + d - 3d = 14$
7. $-6c - 4 - 5c = 10 + 8$
8. $15m - 9 - 6m = 9$
9. $-4 - 3x - x = -16$
10. $9 - 12p + 5p = 14 + 2$
11. $10y + 4 - 7y = -17$
12. $-8a - 15 - 4a = 9$

If the equation has like terms on both sides of the equation, you must get all of the terms with a **variable** on one side of the equation and all of the **integers** on the other side of the equation.

EXAMPLE: $\mathbf{3x + 2 = 6x - 1}$

Step 1:	$3x + 2 = 6x - 1$ $\underline{-6x \qquad -6x}$	Subtract $6x$ from both sides to move all the **variables** to the left side.
Step 2:	$-3x + 2 = -1$ $\underline{\quad -2 \quad -2}$	Subtract 2 from both sides to move all the **integers** to the right side.
Step 3:	$\frac{-3x}{-3} = \frac{-3}{-3}$ $x = 1$	Divide by -3 to solve for x.

Solve the following problems.

1. $3a + 1 = a + 9$
2. $2d - 12 = d + 3$
3. $5x + 6 = 14 - 3x$
4. $15 - 4y = 2y - 3$
5. $9w - 7 = 12w - 13$
6. $10b + 19 = 4b - 5$
7. $-7m + 9 = 29 - 2m$
8. $5x - 26 = 13x - 2$
9. $19 - p = 3p - 9$
10. $-7p - 14 = -2p + 11$
11. $16y + 12 = 9y + 33$
12. $13 - 11w = 3 - w$
13. $-17b + 23 = -4 - 8b$
14. $k + 5 = 20 - 2k$
15. $12 + m = 4m + 21$
16. $7p - 30 = p + 6$
17. $19 - 13z = 9 - 12z$
18. $8y - 2 = 4y + 22$
19. $5 + 16w = 6w - 45$
20. $-27 - 7x = 2x + 18$
21. $-12x + 14 = 8x - 46$
22. $27 - 11h = 5 - 9h$
23. $5t + 36 = -6 - 2t$
24. $17y + 42 = 10y + 7$
25. $22x - 24 = 14x - 8$
26. $p - 1 = 4p + 17$
27. $4d + 14 = 3d - 1$
28. $7w - 5 = 8w + 12$
29. $-3y - 2 = 9y + 22$
30. $17 - 9m = m - 23$

SOLVING FOR A VARIABLE

Sometimes an equation has two variables and you may be asked to solve for one of the variables.

EXAMPLE 1: If $5x + y = 19$. then $y =$

Solution: The goal is to have only y on one side of the equation and the rest of the terms on the other side of the equation. Follow order of operations to solve.

$5x + y - 5x = 19 - 5x$ Subtract $5x$ from each side of the equation.

$y = 19 - 5x$

EXAMPLE 2: If $7m + n = 30$ then $m =$

Solution: The goal is to have only m on one side of the equation and the rest of the terms on the other side of the equation. Follow order of operations to solve.

$7m + n = 30$ Subtract n from both sides of the equation.

$7m + \cancel{n} - \cancel{n} = 30 - n$

$\frac{\cancel{7}m}{\cancel{7}} = \frac{30 - n}{7}$ Divide both sides of the equation by 7.

$m = \frac{30 - n}{7}$

Solve each of the equations below for the variable indicated. Be sure to follow order of operations.

1. If $5a + b = 14$, then $a =$
2. If $7c - d = 20$, then $d =$
3. If $4m - n = 10$. then $m =$
4. If $4r + 2s = 20$, then $r =$
5. If $15m - 9n - 6m = 9n$, then $m =$
6. If $-4y - 3x - x = -16y$, then $x =$
7. If $-2t + 4t - 7s = 9s$, then $s =$
8. If $5x - 4y + 9y = -15x + 5$, then $y =$
9. If $-14b + 7a + 2a = -5b$, then $a =$
10. If $7x + 3y + x = 16y + 3$, then $x =$

MANIPULATING FORMULAS AND EQUATIONS

Sometimes you are given a formula such as $A = l \times w$ (A = area, l = length, and w = width) and you need to solve for w. For example: The area of a playground is 4500 square feet. The length is 600 feet. What is the width of the playground? Starting with $A = l \times w$, you need to solve for w. You need to have w on one side of the equation and all the other variables on the other.

$A = l \times w \Rightarrow \frac{A}{l} = \frac{l \times w}{l} \Rightarrow \frac{A}{l} = w \Rightarrow w = \frac{A}{l}$ You have solved $A = l \times w$ for w.

Solve each of the following formulas and equations for the given variable.

1. $C = 2\pi r$ for r.
2. $I = PRT$ for R
3. $V = \pi r^2 h$ for h
4. $A = \frac{1}{2}bh$ for h
5. $d = 4a + 3c$ for c
6. $h = 6a + 9c^2$ for a
7. $y = 4xz$ for z
8. $5t = 9y + 22$ for y
9. $17 - 9m = n - 23$ for n
10. $7x + 4 = \frac{9y}{4}$ for y
11. $8 + 2a = 5b - 6$ for b
12. $A = s^2$ for s
13. $a^2 + b^2 = c^2$ for a
14. $I = PRT$ for P
15. $x = 4a + 7$ for a
16. $9 - 5y = 6x + 2$ for x
17. $D = rt$ for r
18. $A = lw$ for w
19. $a^2 + b^2 = c^2$ for b^2
20. $C = 2dr$ for r
21. $V = \pi r^2 h$ for r
22. $V = \frac{1}{3}Bh$ for B
23. $A = \pi r^2$ for r
24. $S = 4\pi r^2$ for r
25. $y = \frac{1}{4}x + 5$ for x
26. $x = -\frac{1}{5}y - 3$ for y
27. $a = \frac{b}{3}$ for b
28. $c = 3d + \frac{2}{5}$ for d
29. $g = \frac{2}{3}h - 2$ for h
30. $r = s^2$ for s
31. $F = \frac{9}{5}C + 32$ for C
32. $y = mx + b$ for m

USING DISTANCE = RATE × TIME

Remember, if you are given any two of the three measures: distance, rate, or time, you can easily find the third. Think of dirt to help you remember the formula. distance = rate · time.

$d = rt$. **Find *d***
Practice: If someone travels 55mph for 3 hours, find the distance, d.
d = 55mph(r) · 3 hours (t) = 165 miles

Using algebra, you can also find the rate and time using the same equation, dirt, $d = rt$.

Find *r*: Using $d = rt$, divide by t on both sides. $\frac{d}{t} = \frac{rt}{t}$, so $\frac{d}{t} = r$.

Using the formula we derived from dirt, we now know rate = $\frac{\text{distance}}{\text{time.}}$

Practice: Find the rate if someone travels 180 miles in three hours.

$r = \frac{\text{180 miles } (d)}{\text{3 hours } (t)}$ = 60 miles per hour.

Find *t*: Using $d = rt$, divide by r on both sides. $\frac{d}{r} = \frac{rt}{r}$, so $\frac{d}{r} = t$.

Using the formula we derived from dirt, we know time = $\frac{\text{distance}}{\text{rate}}$

Practice: If someone travels 600 miles at 50 miles per hour (mph), how many hours did it take to Travel that distance?

$t = \frac{\text{600 miles}}{\text{50 mph}}$ = 12 hours.

---

1. Jan drove for 165 miles. She set her speed control at 55 miles per hour. How many hours did she drive?

2. Tom drove 124 miles in 2 hours. What was his average speed?

3. A tour bus drove 4 hours averaging 58 mph. How many miles did it travel?

4. Anna drove 360 miles in 8 hours. What was her average speed?

5. How long will it take a bus averaging 54 mph to travel 378 miles?

6. Dustin raced for 3 hours averaging 176 mph. How many miles did he race?

7. Stacy drove 576 miles at a average speed of 48 mph. How many hours did she drive?

8. A commercial jet traveled 1572 miles in 3 hours. What was its average speed?

9. Oliver drove at an average of 93 mph for 3 hours. How far did he travel?

10. Rami drove 184 miles at an average of 46 mph. How many hours did he drive?

11. Joe drove 8 hours to a city 336 miles away. What was his average speed?

12. Javier flew his glider for 2 hours at 87 mph. How many miles did his glider fly?

REMOVING PARENTHESES

In this chapter, you will use the distributive principle to remove parentheses in problems with a variable (letter).

EXAMPLE 1: **$2(a + 6)$** You multiply 2 by each term inside the parentheses. $2 \times a = 2a$ and $2 \times 6 = 12$. The 12 is a positive number, so use a plus sign between the terms in the answer.

$2(a + 6) = 2a + 12$

EXAMPLE 2: **$7(2b - 5)$** You multiply 7 by each term inside the parentheses. $7 \times 2b = 14b$ and $7 \times -5 = -35$. The -35 is a negative number, so use a minus sign between the terms in the answer.

$7(2b - 5) = 14b - 35$

EXAMPLE 3: **$4(-5c + 2)$** The first term inside the parentheses could be negative. Multiply in exactly the same way as in the examples above. $4 \times (-5c) = -20c$ and $4 \times 2 = 8$

$4(-5c + 2) = -20c + 8$

Remove parentheses in the problems below.

1. $7(n + 6)$
2. $8(2g - 5)$
3. $11(5z - 2)$
4. $6(-y - 4)$
5. $3(-3k + 5)$
6. $4(d - 8)$
7. $2(-4x + 6)$
8. $7(4 + 6p)$
9. $5(-4w - 8)$
10. $6(11x + 2)$
11. $10(9 - y)$
12. $9(c - 9)$
13. $12(-3t + 1)$
14. $3(4y + 9)$
15. $8(b + 3)$
16. $5(8a + 7)$
17. $3(2b - 4)$
18. $2(-9x - 7)$
19. $4(8 - 7v)$
20. $10(3c + 5)$
21. $5(2x - 9)$
22. $11(y + 3)$
23. $9(7t + 4)$
24. $6(8 - g)$

The number in front of the parentheses can also be negative. Remove these parentheses the same way.

EXAMPLE: $-2(b - 4)$ First, multiply $-2 \times b$. $-2 \times b = -2b$
Second, multiply -2×-4. $-2 \times -4 = 8$

Copy the two products. The second product is a positive number, so put a plus sign between the terms in the answer.

$$-2(b - 4) = -2b + 8$$

Remove the parentheses in the following problems.

1. $-7(x + 2)$
2. $-5(4 - y)$
3. $-4(2b - 2)$
4. $-2(8c + 6)$
5. $-5(-w - 8)$
6. $-3(4x - 2)$
7. $-2(-z + 2)$
8. $-4(7p + 7)$
9. $-9(t - 6)$
10. $-10(2w + 4)$
11. $-3(9 - 7p)$
12. $-9(-k - 3)$
13. $-1(7b - 9)$
14. $-6(-5t - 2)$
15. $-7(-v + 4)$
16. $-3(-x - 5)$
17. $-11(4y + 2)$
18. $-1(-c + 100)$
19. $-5(-2t - 4)$
20. $-2(7z - 12)$
21. $-45(y - 1)$
22. $-100(a + 1)$
23. $-6(-x - 11)$
24. $-12(-2b + 1)$

MULTI-STEP ALGEBRA PROBLEMS

You can now use what you know about removing parentheses, combining like terms, and solving simple algebra problems to solve problems that involve three or more steps. Study the examples below to see how to solve multi-step problems.

EXAMPLE 1: $\mathbf{3(x + 6) = 5x - 2}$

Step 1: $3x + 18 = 5x - 2$ Use the distributive property to remove parentheses.

Step 2: $\underline{-5x \qquad -5x}$ Subtract $5x$ from each side to move the terms with variables to the left side of the equation.

$-2x + 18 = -2$

Step 3: $\underline{\qquad -18 \quad -18}$ Subtract 18 from each side to move the integers to the right side of the equation.

Step 4: $\dfrac{-2x}{-2} = \dfrac{-20}{-2}$ Divide both sides by -2 to solve for x.

$x = 10$

EXAMPLE 2: $\mathbf{\dfrac{3(x-3)}{2} = 9}$

Step 1: $\dfrac{3x-9}{2} = 9$ Use the distributive property to remove parentheses.

Step 2: $\dfrac{\cancel{2}(3x-9)}{\cancel{2}} = 2(9)$ Multiply both sides by 2 to eliminate the fraction.

Step 3: $3x - 9 = 18$ Add 9 to both sides, and combine like terms.

$\underline{\quad +9 \quad +9}$

Step 4: $\dfrac{3x}{3} = \dfrac{27}{3}$ Divide both sides by 3 to solve for x.

$x = 9$

Solve the following multi-step algebra problems.

1. $2(y - 3) = 4y + 6$
4. $\dfrac{12y - 18}{6} = 4y + 3$
7. $5(b - 4) = 3b - 6$
2. $\dfrac{2(a + 4)}{2} = 12$
5. $2x + 3x = 30 - x$
8. $-8(y + 4) = 10y + 4$
3. $\dfrac{10(x - 2)}{5} = 14$
6. $\dfrac{2a + 11}{3} = a + 5$
9. $\dfrac{x + 4}{-3} = 6 - x$

10. $\frac{4(n + 3)}{5} = n - 3$

11. $3(2x - 5) = 8x - 9$

12. $7 - 10a = 9 - 9a$

13. $7 - 5x = 10 - (6x + 7)$

14. $4(x - 3) - x = x - 6$

15. $4a + 4 = 3a - 4$

16. $-3(x - 4) + 5 = -2x - 2$

17. $5b - 11 = 13 - b$

18. $\frac{-4x + 3}{2x} = \frac{7}{2x}$

19. $-(x + 1) = -2(5 - x)$

20. $4(2c + 3) - 7 = 13$

21. $6 - 3a = 9 - 2(2a + 5)$

22. $-5x + 9 = -3x + 11$

23. $3y + 2 - 2y - 5 = 4y + 3$

24. $3y - 10 = 4 - 4y$

25. $-(a + 3) = -2(2a + 1) - 7$

26. $5m - 2(m + 1) = m - 10$

27. $\frac{1}{2}(b - 2) = 5$

28. $-3(b - 4) = -2b$

29. $4x + 12 = -2(x + 3)$

30. $\frac{7x + 4}{3} = 2x - 1$

31. $9x - 5 = 8x - 7$

32. $7x - 5 = 4x + 10$

33. $\frac{4x + 8}{2} = 6$

34. $2(c + 4) + 8 = 10$

35. $y - (y + 3) = y + 6$

36. $4 + x - 2(x - 6) = 8$

SOLVING RADICAL EQUATIONS

Some multi-step equations contain radicals. An example of a radical equation is a square root, $\sqrt{\ }$. Work these type of equations out similarly to the section on equations.

EXAMPLE: Solve the following radical equation for x. $\sqrt{4x-3} + 2 = 5$

Step 1: The first step is to get the constants that are not under the radical sign on one side of the equation. Subtract 2 from both sides of the equation.

$\sqrt{4x-3} + 2 - 2 = 5 - 2$

$\sqrt{4x-3} + 0 = 3$

$\sqrt{4x-3} = 3$

Step 2: Next, you must get rid of the radical sign by squaring both sides of the equation.

$(\sqrt{4x-3})^2 = 3^2$

$4x - 3 = 9$

Step 3: Add 3 to both sides of the equation to get the constants on just one side of the equation.

$4x - 3 + 3 = 9 + 3$

$4x + 0 = 12$

$4x = 12$

Step 4: Last, get x on one side of the equation by itself by dividing both sides by 4.

$\frac{4x}{4} = \frac{12}{4}$

$x = 3$

Solve the following radical equations.

1. $\sqrt{x+3} - 13 = -8$
2. $3 + \sqrt{7t-3} = 5$
3. $\sqrt{3q+12} - 4 = 5$
4. $\sqrt{11f+3} + 2 = 8$
5. $5 = \sqrt{6g-5} + (-2)$
6. $2 = \sqrt{x-3}$
7. $\sqrt{-8t} - 3 = 1$
8. $\sqrt{-d+1} - 9 = -6$
9. $10 - \sqrt{8x+2} = 9$
10. $\sqrt{15y+4} + 4 = 4$
11. $\sqrt{r+14} - 1 = 9$
12. $3 + \sqrt{2q-1} = 6$
13. $\sqrt{5t+16} + 4 = 13$
14. $17 = \sqrt{23-f} + 15$
15. $19 + \sqrt{7x-5} = 22$

MULTI-STEP INEQUALITIES

Remember that adding and subtracting with inequalities follow the same rules as with equations. When you multiply or divide both sides of an inequality by the same positive number, the rules are also the same as for equations. However, when you multiply or divide both sides of an inequality by a **negative** number, you must **reverse** the inequality symbol.

EXAMPLE 1: $-x > 4$

$(-1)(-x) < (-1)(4)$

$x < -4$

EXAMPLE 2: $-4x < 2$

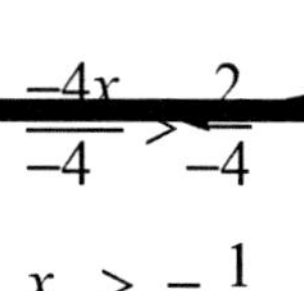

$\frac{-4x}{-4} > \frac{2}{-4}$

$x > -\frac{1}{2}$

Reverse the symbol when you multiply or divide by a negative number.

When solving multi-step inequalities, first add and subtract to isolate the term with the variable. Then, multiply and divide.

EXAMPLE 3: $2x - 8 > 4x + 1$

$-4x \quad -4x$ **Step 1:** Subtract $4x$ from both sides.

$-2x - 8 > 1$

$+8 \quad +8$ **Step 2:** Add 8 to both sides.

$\frac{-2x}{-2} < \frac{9}{-2}$ **Step 3:** Divide by -2. Remember to change the direction of the inequality sign.

$x < -\frac{9}{2}$

Solve each of the following inequalities.

1. $8 - 3x \leq 7x - 2$
2. $3(2x - 5) \geq 8x - 5$
3. $4 + 2(3 - 2y) \leq 6y - 20$
4. $7 + 3y > 2y - 5$
5. $3a + 5 < 2a - 6$
6. $3(a - 2) > -5a - 2(3 - a)$

Solve each of the following inequalities.

7. $2x - 7 \geq 4(x - 3) + 3x$

8. $6x - 2 \leq 5x + 5$

9. $-\frac{x}{4} > 12$

10. $-\frac{2x}{3} \leq 6$

11. $3b + 5 < 2b - 8$

12. $4x - 5 \leq 7x + 13$

13. $4x + 5 \leq -2$

14. $2y - 4 > 7$

15. $\frac{1}{3}b - 2 > 5$

16. $-4c + 6 \leq 8$

17. $-\frac{1}{2}x + 2 > 9$

18. $\frac{1}{4}y - 3 \leq 1$

19. $-3x + 4 > 5$

20. $\frac{y}{2} - 2 \geq 10$

21. $7 + 4c < -2$

22. $2 - \frac{a}{2} > 1$

23. $10 + 4b \leq -2$

24. $-\frac{1}{2}x + 3 > 4$

25. $12d - 8 > 28$

26. $\frac{3}{4}f + 21 \geq 18$

27. $-7m + 14 < 70$

28. $-15p + 19 \geq -6p + 13$

29. $\frac{z}{5} + 3 < 43$

30. $3t - 13 \leq 5t$

31. $12 - \frac{5}{6}g > 0$

32. $8s + 20 \geq 3s - 25$

33. $7p - 6 < 8p + 9$

SOLVING EQUATIONS AND INEQUALITIES WITH ABSOLUTE VALUES

When solving equations and inequalities which involve variables placed in absolute values (| |), remember that there will be two or more numbers that will work as correct answers. This is because the absolute value variable will signify both positive and negative numbers as answers.

EXAMPLE 1:	$5 + 3\lvert k\rvert = 8$	Solve as you would any equation.
Step 1:	$3\lvert k\rvert = 3$	Subtract 5 from each side.
Step 2:	$\lvert k\rvert = 1$	Divide by 3 on each side.
Step 3:	$k = 1$ or $-k = 1$	Because k is an absolute value, k can be positive or negative.
Answer:	$\mathbf{k = 1}$ **or** $\mathbf{k = -1}$	−1 0 1

EXAMPLE 2:	$2\lvert x\rvert - 3 < 7$	Solve as you normally would an inequality.
Step 1:	$2\lvert x\rvert < 10$	Add 3 to both sides.
Step 2:	$\lvert x\rvert < 5$	Divide by 2 on each side.
Step 3:	$\mathbf{x < 5}$ **or** $\mathbf{x > -5}$, which can also be written as $-5 < x < 5$	Because x is an absolute value, the answer is a set of both positive and negative numbers.

−5 5

Read each problem, and write the number or set of numbers which solves each equation or inequality. Graph the solution set on a number line.

1. $7 + 2\lvert y\rvert = 15$
2. $4\lvert x\rvert - 9 < 3$
3. $6\lvert k\rvert + 2 = 14$
4. $10 - 4\lvert n\rvert > -14$
5. $3 = 5\lvert z\rvert - 12$
6. $-4 + 7\lvert m\rvert < 10$
7. $5\lvert x\rvert - 12 > 13$
8. $21\lvert g\rvert + 7 = 49$
9. $-9 + 6\lvert x\rvert = 15$
10. $12 - 6\lvert w\rvert > -12$
11. $31 > 13 + 9\lvert r\rvert$
12. $-30 = 21 - 3\lvert t\rvert$
13. $9\lvert x\rvert - 19 < 35$
14. $-13\lvert c\rvert + 21 \geq -31$
15. $5 - 11\lvert k\rvert < -17$
16. $-42 + 14\lvert p\rvert = 14$
17. $15 < 3\lvert b\rvert + 6$
18. $9 + 5\lvert q\rvert = 29$
19. $-14\lvert y\rvert - 38 < -45$
20. $36 = 4\lvert s\rvert + 20$
21. $20 \leq -60 + 8\lvert e\rvert$

MORE SOLVING EQUATIONS AND INEQUALITIES WITH ABSOLUTE VALUE

Now, look at the following examples in which numbers and variables are added or subtracted in the absolute value symbols (| |).

EXAMPLE 1: $|3x - 5| = 10$

Remember, an equation with absolute value symbols has two solutions.

Step 1: To find the first solution, remove the absolute value symbol and solve the equation.

$$3x - 5 = 10$$
$$3x - 5 + 5 = 10 + 5$$
$$\frac{3x}{3} = \frac{15}{3}$$
$$x = 5$$

Step 2: To find the second solution, solve the equation for the negative of the expression in absolute value symbols.

$$-(3x - 5) = 10$$
$$-3x + 5 = 10$$
$$-3x + 5 - 5 = 10 - 5$$
$$-3x = 5$$
$$x = \frac{-5}{3}$$

Solutions: $x = 5$ and $x = \frac{-5}{3}$

EXAMPLE 2: $|5z - 10| < 20$

Step 1: Remove the absolute value symbols and solve the inequality.

$$5z - 10 < 20$$
$$5z - 10 + 10 < 20 + 10$$
$$\frac{5z}{5} < \frac{30}{5}$$
$$z < 6$$

Step 2: Next, solve the equation for the negative of the expression in absolute value symbols.

$$-(5z - 10) < 20$$
$$-5z + 10 < 20$$
$$-5z + 10 - 10 < 20 - 10$$
$$\frac{-5z}{5} < \frac{10}{5}$$
$$-z < 2$$
$$z > -2$$

Solution: $-2 < z < 6$

EXAMPLE 3: $|4y + 7| - 5 > 18$

Step 1:	$4y + 7 - 5 + 5 > 18 + 5$ $4y + 7 > 23$ $4y + 7 - 7 > 23 - 7$ $4y > 16$ $y > 4$	Remove the absolute value symbols and solve the inequality.
Step 2:	$-(4y + 7) - 5 > 18$ $-4y - 7 - 5 + 5 > 18 + 5$ $-4y - 7 + 7 > 23 + 7$ $-4y > 30$ $y < -7\frac{1}{2}$	Solve the equation for the negative of the expression in absolute value symbols.
Solutions:	$y > 4$ or $y < -7\frac{1}{2}$	

Solve the following equations and inequalities below.

1. $-4 + |2x + 4| = 14$
2. $|4b - 7| + 3 > 12$
3. $6 + |12e + 3| < 39$
4. $-15 + |8f - 14| > 35$
5. $70 + |5z + 12| = 18$
6. $|4g - 9| - 3 < 26$
7. $22 + |5a + 21| > 45$
8. $|11c - 25| + 13 = 21$
9. $|-9b + 13| - 12 = 10$
10. $-25 + |7b + 11| < 35$
11. $|7w + 2| - 60 > 30$
12. $63 + |3d - 12| = 21$
13. $|-23 + 8x| -12 > + 37$
14. $|61 + 20x| + 32 > 51$
15. $|21 + 11y| + 18 < 32$
16. $|4a + 13| + 31 = 50$
17. $4 + |4k - 32| < 51$
18. $8 + |4x + 3| = 21$
19. $|28 + 7v| - 28 < 77$
20. $35 + |5c - 12| > 48$
21. $89 > 62 + |9d - 8|$
22. $83 < 31 + |9e + 7|$
23. $|13x + 3| + 6 = 35$
24. $|62p + 31| + 43 = 136$
25. $18 - |6v + 22| < 22$
26. $12 = 4 + |42 + 10m|$
27. $53 < 18 + |12e + 31|$
28. $38 > -39 + |7j + 14|$
29. $9 = |14 + 15u| + 7$
30. $-73 + |24b - 16| > -61$
31. $11 - |2j + 50| > 45$
32. $|35 + 6i| - 3 = 14$
33. $|26 - 8r| - 9 > 41$
34. $|25 + 6z| - 21 = 28$
35. $|9u - 24| + 15 = 36$
36. $54 > 25 + |6f - 83|$
37. $|3g - 41| - 20 < 18$
38. $28 > |31 + 6k| - 15$
39. $12 < |2t + 6| - 14$
40. $50 > |9q - 10| + 6$
41. $12 + |8v - 18| > 26$
42. $-38 + |16i - 33| = 41$
43. $|-14 + 6p| - 9 < 7$
44. $28 > |25 - 5f| - 12$
45. $-41 + |10c - 32| = 67$

CHAPTER 6 REVIEW

Solve each of the following equations.

1. $4a - 8 = 28$
2. $-7 + 23w = 108$
3. $5 + \frac{x}{8} = -4$
4. $\frac{c}{3} - 13 = 5$
5. $\frac{y-8}{6} = 7$
6. $\frac{b+9}{12} = -3$

Simplify the following expressions by combining like terms.

7. $-4a + 8 + 3a - 9$
8. $14 + 3z - 8 - 5z$
9. $-7 - 7x - 2 - 9x$

Solve.

10. $19 - 8d = d - 17$
11. $6 + 16x = -2x - 12$
12. $7w - 8 = -4w - 30$

Simplify the following expression by removing parentheses.

13. $3(-4x + 7)$
14. $11(2y + 5)$
15. $6(8 - 9b)$
16. $-8(-2 + 3a)$
17. $-2(5c - 3)$
18. $-5(7y - 1)$

Solve each of the following equations and inequalities.

19. $6(b - 4) = 8b - 18$
20. $\frac{4x-16}{2} = 7x + 2$
21. $\frac{-11c-35}{4} = 4c - 2$
22. $5 + x - 3(x + 4) = -17$
23. $-9b - 3 = -3(b + 2)$
24. $7a - 5 = 2(2a - 13)$
25. $4(2x + 3) \geq 2x$
26. $3x - 5 + 3(x + 3) = 10$
27. $7(2x + 6) < -28$

Write the number or set of numbers which solves each equation or inequality.

28. $-11|k| < -22$
29. $3(x + 2) < 7x - 10$
30. $-\frac{3}{4}x \leq 6$
31. $2(3x - 1) \geq 3x - 7$
32. $\frac{t}{5} + 2 > 8$
33. $21 = -4 + |5x + 5|$

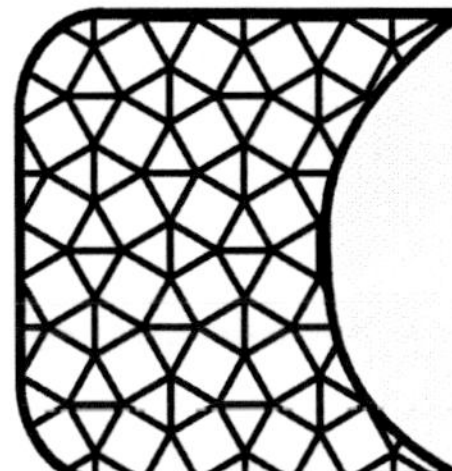

CHAPTER 7: FACTORING & SOLVING QUADRATIC EQUATIONS

Standards 2.12.3 & 2.1.6

In a multiplication problem, the numbers multiplied together are called **factors**. The answer to a multiplication problem is called the **product**.

$$5 \times 4 = 20$$

factors (5, 4) product (20)

If we reverse the problem, $20 = 5 \times 4$, we say we have **factored** 20 into 5×4.

In this chapter, we will factor **polynomials**.

EXAMPLE: Find the greatest common factor of $2y^3 + 6y^2$.

Step 1: Look at the whole numbers. The greatest common factor of 2 and 6 is 2. Factor the 2 out of each term.

$2(y^3 + 3y^2)$

Step 2: Look at the remaining terms, $y^3 + 3y^2$. What are the common factors of each term?

$y^3 = y \times \boxed{y \times y}$ ← common factors $= y^2$
$3y^2 = 3 \times \boxed{y \times y}$

Step 3: Factor 2 and y^2 out of each term: $2y^2(y + 3)$

Check: $2y^2(y + 3) = 2y^3 + 6y^2$

Factor each of the following by first finding greatest common factor.

1. $6x^3 + 18x^2$
2. $14y^3 + 7y$
3. $4b^4 + 12b^3$
4. $10a^3 + 5$
5. $2y^3 + 8y^2$
6. $6x^4 - 12x^2$
7. $18y^2 - 12y$
8. $15a^3 - 25a^2$
9. $4x^3 + 16x^2$
10. $6b^2 + 21b^3$
11. $27m^3 + 18m^2$
12. $100x^3 - 25x^2$
13. $4b^3 - 12b^4$
14. $18c^2 + 24c$
15. $20y^3 + 30y^2$
16. $16x^2 - 24x^3$
17. $15a^3 - 25a^2$
18. $24b^3 + 16b^2$
19. $36y^3 + 9y^2$
20. $42x^3 + 49x$

Factoring larger polynomials with 3 or 4 terms works the same way.

EXAMPLE: $4x^5 + 16x^4 + 12x^3 + 8x^2$

Step 1: Find the greatest common factor of the whole numbers. 4 can be divided evenly into 4, 16, 12, and 8; therefore, 4 is the greatest common factor.

$$4x^5 + 16x^4 + 12x^3 + 8x^2 = 4(x^5 + 4x^4 + 3x^3 + 2x^2)$$

Step 2: Find the greatest common factor of the variables. x^5, x^4, x^3, and x^2 can each be divided by x^2, the lowest power of x in each term.

$$4x^5 + 16x^4 + 12x^3 + 8x^2 = 4x^2(x^3 + 4x^2 + 3x + 2)$$

Factor each of the following polynomials.

1. $5a^3 + 15a^2 + 20a$
2. $18y^4 + 6y^3 + 24y^2$
3. $12x^5 + 21x^3 + x^2$
4. $6b^4 + 3b^3 + 15b^2$
5. $14c^3 + 28c^2 + 7c$
6. $15b^4 - 5b^2 + 20b$
7. $t^3 + 3t^2 - 5t$
8. $8a^3 - 4a^2 + 12a$
9. $16b^5 - 12b^4 - 20b^2$
10. $20x^4 + 16x^3 - 24x^2 + 28x$
11. $40b^7 + 30b^5 - 50b^3$
12. $20y^4 - 15y^3 + 30y^2$
13. $4m^5 + 8m^4 + 12m^3 + 6m^2$
14. $16x^5 + 20x^4 - 12x^3 + 24x^2$
15. $18y^4 + 21y^3 - 9y^2$
16. $3n^5 + 9n^3 + 12n^2 + 15n$
17. $4d^6 - 8d^2 + 2d$
18. $10w^2 + 4w + 2$
19. $6t^3 - 3t^2 + 9t$
20. $25p^5 - 10p^3 - 5p^2$
21. $18x^4 + 9x^2 - 36x$
22. $6b^4 - 12b^2 - 6b$
23. $y^3 + 3y^2 - 9y$
24. $10x^5 - 2x^4 + 4x^2$

FINDING THE NUMBERS

The next kind of factoring we will do requires thinking of two numbers with a certain sum and a certain product.

EXAMPLE: Which two numbers have a sum of 8 and a product of 12? In other words, what pair of numbers would answer both equations?

_____ + _____ = 8 and _____ × _____ = 12

You may think 4 + 4 = 8, but 4 × 4 does not equal 12.
Or you may think 7 + 1 = 8, but 7 × 1 does not equal 12.

6 + 2 = 8 and 6 × 2 = 12, so 6 and 2 are the pair of numbers that will work in both equations.

For each problem below, find one pair of numbers that will solve both equations.

1. _____ + _____ = 14 and _____ × _____ = 40
2. _____ + _____ = 10 and _____ × _____ = 21
3. _____ + _____ = 18 and _____ × _____ = 81
4. _____ + _____ = 12 and _____ × _____ = 20
5. _____ + _____ = 7 and _____ × _____ = 12
6. _____ + _____ = 8 and _____ × _____ = 15
7. _____ + _____ = 10 and _____ × _____ = 25
8. _____ + _____ = 14 and _____ × _____ = 48
9. _____ + _____ = 12 and _____ × _____ = 36
10. _____ + _____ = 17 and _____ × _____ = 72
11. _____ + _____ = 15 and _____ × _____ = 56
12. _____ + _____ = 9 and _____ × _____ = 18
13. _____ + _____ = 13 and _____ × _____ = 40
14. _____ + _____ = 16 and _____ × _____ = 63
15. _____ + _____ = 10 and _____ × _____ = 16
16. _____ + _____ = 8 and _____ × _____ = 16
17. _____ + _____ = 9 and _____ × _____ = 20
18. _____ + _____ = 13 and _____ × _____ = 36
19. _____ + _____ = 15 and _____ × _____ = 50
20. _____ + _____ = 11 and _____ × _____ = 30

MORE FINDING THE NUMBERS

Now that you have mastered positive numbers, take up the challenge of finding pairs of negative numbers or pairs where one number is negative and one is positive.

EXAMPLE: Which two numbers have a sum of –3 and a product of –40? In other words, what pair of numbers would answer both equations?

____ + ____ = –3 and ____ × ____ = –40

It is faster to look at the factors of 40 first. 8 and 5 and 10 and 4 are possibilities. 8 and 5 have a difference of 3, and in fact, 5 + (–8) = –3 and 5 (–8) = –40. This pair of numbers, 5 and –8, will satisfy both equations.

For each problem below, find one pair of numbers that will solve both equations.

1. ____ + ____ = –2 and ____ × ____ = –35
2. ____ + ____ = 4 and ____ × ____ = –5
3. ____ + ____ = 4 and ____ × ____ = –12
4. ____ + ____ = –6 and ____ × ____ = 8
5. ____ + ____ = 3 and ____ × ____ = –40
6. ____ + ____ = 10 and ____ × ____ = –11
7. ____ + ____ = 6 and ____ × ____ = –27
8. ____ + ____ = 8 and ____ × ____ = –20
9. ____ + ____ = –5 and ____ × ____ = –24
10. ____ + ____ = –3 and ____ × ____ = –28
11. ____ + ____ = –2 and ____ × ____ = –48
12. ____ + ____ = –1 and ____ × ____ = –20
13. ____ + ____ = –3 and ____ × ____ = 2
14. ____ + ____ = 1 and ____ × ____ = –30
15. ____ + ____ = –7 and ____ × ____ = 12
16. ____ + ____ = 6 and ____ × ____ = –16
17. ____ + ____ = 5 and ____ × ____ = –24
18. ____ + ____ = –4 and ____ × ____ = 4
19. ____ + ____ = –1 and ____ × ____ = –42
20. ____ + ____ = –6 and ____ × ____ = 8

FACTORING TRINOMIALS

A trinomial is a quadratic expression with three terms.

For example, $x^2 + x - 30$

A trinomial can be factored into two binomials (expressions with two terms).

EXAMPLE 1: Factor $x^2 + 6x + 8$

Step 1: When the trinomial is in descending order, as in the example above, you need to find a pair of numbers in which the sum of the two numbers equals the number in the second term, while the product of the two numbers equals the third term. In the above example, find the pair of numbers that has a sum of 6 and a product of 8.

_____ + _____ = 6 and _____ × _____ = 8

The pair of numbers that satisfy both equations is 4 and 2.

Step 2: Use the pair of numbers in the binomials.

The factors of $x^2 + 6x + 8$ are $(x + 4)(x + 2)$

Check: To check, use the FOIL method.

$(x + 4)(x + 2) = x^2 + 4x + 2x + 8 = x^2 + 6x + 8$

Notice, when the second term and the third term of the trinomial are both positive, both numbers in the solution pair are positive.

EXAMPLE 2: Factor $x^2 - x - 6$ Find the pair of numbers where ...

the sum is -1 and the product is -6

_____ + ____ = -1 and _____ × _____ = -6

The pair of numbers that satisfies both equations is 2 and -3.

The factors of $x^2 - x - 6$ are $(x + 2)(x - 3)$

Notice, if the third term is negative, one number in the solution pair is positive, and the other number is negative.

EXAMPLE 3: Factor $x^2 - 7x + 12$ Find the pair of numbers where ...

the sum is -7 and the product is 12

_____ + ____ = -7 and _____ × _____ = 12

The pair of numbers that satisfies both equations is -3 and -4.

The factors of $x^2 - 7x + 12$ are $(x - 3)(x - 4)$.

Notice, if the second term of a trinomial is negative and the third term is positive, both numbers in the solution pair are negative.

Find the factors of the following trinomials.

1. $x^2 - x - 2$
2. $y^2 + y - 6$
3. $w^2 + 3w - 4$
4. $t^2 + 5t + 6$
5. $x^2 + 2x - 8$
6. $k^2 - 4k + 3$
7. $t^2 + 3t - 10$
8. $x^2 - 3x - 4$
9. $y^2 - 5y + 6$
10. $y^2 + y - 20$
11. $a^2 - a - 6$
12. $b^2 - 4b - 5$
13. $c^2 - 5c - 14$
14. $c^2 - c - 12$
15. $d^2 + d - 6$
16. $x^2 - 3x - 28$
17. $y^2 + 3y - 18$
18. $a^2 - 9a + 20$
19. $b^2 - 2b - 15$
20. $c^2 + 7c - 8$
21. $t^2 - 11t + 30$
22. $w^2 + 13w + 36$
23. $m^2 - 2m - 48$
24. $y^2 + 14y + 49$
25. $x^2 + 7x + 10$
26. $a^2 - 7a + 6$
27. $d^2 - 6d - 27$

MORE FACTORING TRINOMIALS

Sometimes a trinomial has a greatest common factor which must be factored out first.

EXAMPLE : Factor $2x^2 + 8x - 32$

Step 1: Begin by factoring out the greatest common factor, 2.

$2(x^2 + 4x - 16)$

Step 2: Factor by finding a pair of numbers whose sum is 2 and product is –8.
4 and –2 will work, so

$4(x^2 + 2x - 8) = 4\,(x + 4)\,(x - 2)$

Check: Multiply to check. $4\,(x + 4)\,(x - 2) = 4x^2 + 8x - 32$

Factor the following trinomials. Be sure to factor out the greatest common factor first.

1. $2x^2 + 6x + 4$
2. $3y^2 - 9y + 6$
3. $2a^2 + 2a - 12$
4. $2b^2 + 28b + 80$
5. $3y^2 - 6y - 9$
6. $2x^2 - 200$
7. $2c^2 - 10c - 48$
8. $3d^2 + 12d - 36$
9. $2x^2 + 8x + 8$
10. $3a^2 - 6a - 24$
11. $2b^2 - 28b + 80$
12. $3c^2 - 6c - 24$
13. $2x^2 - 18x + 28$
14. $2y^2 - 20y + 32$
15. $2a^2 - 20a + 42$
16. $3b^2 - 27b + 60$
17. $2d^2 + 48d + 88$
18. $3x^2 - 24x + 45$

FACTORING MORE TRINOMIALS

Some trinomials have a whole number in front of the first term that cannot be factored out of the trinomial. The trinomial can still be factored.

EXAMPLE : Factor $2x^2 + 5x - 3$

Step 1: To get a product of $2x^2$, one factor must begin with $2x$ and the other with x.

$(2x \quad)(x \quad)$

Step 2: Now think: What two numbers give a product of −3? The two possibilities are 3 and −1 or −3 and 1. We know they could be in any order, so there are 4 possible arrangements.

$(2x + 3)(x - 1)$
$(2x - 3)(x + 1)$
$(2x + 1)(x - 3)$
$(2x - 1)(x + 3)$

Step 3: Multiply each possible answer until you find the arrangement of the numbers that works. Multiply the outside terms and the inside terms and add them together to see which one will equal $5x$.

$(2x + 3)(x - 1) = 2x^2 + x - 3$

$(2x - 3)(x + 1) = 2x^2 - x - 3$

$(2x + 1)(x - 3) = 2x^2 - 5x - 3$

$(2x - 1)(x + 3) = 2x^2 + 5x - 3$ ← This arrangement works so:

The factors of $2x^2 + 5x - 3$ are $(2x - 1)(x + 3)$

Alternative: You can do some of the multiplying in your head. For the above example, ask yourself the following question: What two numbers give a product of −3 and give a sum of 5 (the whole number in the second term) when one of the numbers is first multiplied by 2 (the whole number in front of the first term)? The pair of numbers, −1 and 3, have a product of −3 and a sum of 5 when the 3 is first multiplied by 2. Therefore, the 3 will go opposite the factor with the $2x$ so that when the terms are multiplied, you get −5.

You can use this method to narrow down the possible pairs of numbers when you have several to choose from.

Factor the following trinomials.

1. $3y^2 + 14y + 8$
2. $y^2 - 4y - 21$
3. $2a^2 + 22a + 36$
4. $2c^2 - 9c + 9$
5. $2y^2 - 7y - 15$
6. $3x^2 + 4x + 1$
7. $w^2 - 11w + 28$
8. $3y^2 + 23y + 30$
9. $3x^2 - 20x + 32$
10. $3a^2 + 4a - 7$
11. $2a^2 + 3a - 20$
12. $2m^2 + 4m + 2$
13. $3y^2 - 4y - 32$
14. $2x^2 - 17x + 36$
15. $3y^2 + 4y - 4$
16. $2y^2 + 5y - 12$
17. $d^2 + 10d + 16$
18. $2x^2 - 3x - 20$
19. $x^2 - 6x + 8$
20. $2x^2 - 2x - 84$
21. $3x^2 - 20x + 25$
22. $2a^2 - 7a - 4$
23. $x^2 + 7x + 12$
24. $x^2 + 17x + 30$
25. $2b^2 - 13b + 18$
26. $x^2 + 8x - 9$
27. $3c^2 - 2c - 21$

FACTORING THE DIFFERENCE OF TWO SQUARES

The product of a term and itself is called a **perfect square**.

25 is a perfect square because $5 \times 5 = 25$
49 is a perfect square because $7 \times 7 = 49$

Any variable with an even exponent is a perfect square.

y^2 is a perfect square because $y \times y = y^2$
y^4 is a perfect square because $y^2 \times y^2 = y^4$

When two terms that are both perfect squares are subtracted, factoring those terms is very easy. To factor the difference of perfect squares, you use the square root of each term, a plus sign in the first factor, and a minus sign in the second factor.

EXAMPLE 4 : Factor $4x^2 - 9$

The example has two terms which are both perfect squares, and the terms are subtracted.

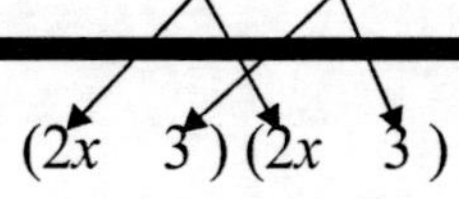

Step 1: $(2x \quad 3)\ (2x \quad 3)$

Find the square root of each term.
Use the square roots in each of the factors.

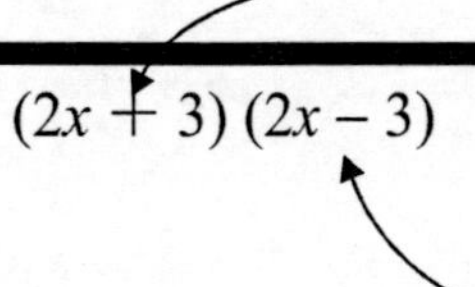

Step 2: $(2x + 3)\ (2x - 3)$

Use a plus sign in one factor
and a minus sign in the other factor.

Check: Multiply to check. $(2x + 3)\ (2x - 3) = 4x^2 - 6x + 6x - 9 = 4x^2 - 9$

The inner and outer terms add to zero.

Factor.

1. $d^2 - 144$	4. $a^2 - 25$	7. $a^2 - 16$	10. $x^2 - 4$
2. $c^2 - 36$	5. $x^2 - 49$	8. $y^2 - 9$	11. $x^2 - 100$
3. $b^2 - 25$	6. $x^2 - 81$	9. $b^2 - 36$	12. $y^2 - 121$

SOLVING QUADRATIC EQUATIONS

In this section, we learn that any equation that can be put in the form $ax^2 + bx + c = 0$ is a quadratic equation if a, b, and c are real numbers and $a \neq 0$. $ax^2 + bx + c = 0$ is the standard form of a quadratic equation. To solve these equations, follow the steps below.

EXAMPLE: Solve $y^2 - 4y - 5 = 0$

Step 1: Factor the left side of the equation.

$$y^2 - 4y - 5 = 0$$
$$(y + 1)(y - 5) = 0$$

Step 2: If the product of these two factors equals zero, then the two factors individually must be equal to zero. Therefore, to solve, we set each factor equal to zero.

$$\begin{array}{rcl} (y+1) & = & 0 \\ -1 & & -1 \\ \hline y & = & -1 \end{array} \qquad \begin{array}{rcl} (y-5) & = & 0 \\ +5 & & +5 \\ \hline y & = & 5 \end{array}$$

The equation has two solutions: $y = -1$ and $y = 5$.

Check: To check, substitute each solution into the original equation.

When $y = -1$, the equation becomes:

$$(-1)^2 - (4)(-1) - 5 = 0$$
$$1 + 4 - 5 = 0$$
$$0 = 0$$

When $y = 5$, the equation becomes:

$$5^2 - (4)(5) - 5 = 0$$
$$25 - 20 - 5 = 0$$
$$0 = 0$$

Both solutions produce true statements.
The solution set for the equation is $\{-1, 5\}$.

Solve each of the following quadratic equations by factoring and setting each factor equal to zero. Check by substituting answers back in the original equation.

1. $x^2 + x - 6 = 0$
2. $y^2 - 2y - 8 = 0$
3. $a^2 + 2a - 15 = 0$
4. $y^2 - 5y + 4 = 0$
5. $b^2 - 9b + 14 = 0$
6. $x^2 - 3x - 4 = 0$
7. $y^2 + y - 20 = 0$
8. $d^2 + 6d + 8 = 0$
9. $y^2 - 7y + 12 = 0$
10. $x^2 - 3x - 28 = 0$
11. $a^2 - 5a + 6 = 0$
12. $b^2 + 3b - 10 = 0$
13. $a^2 + 7a - 8 = 0$
14. $c^2 + 3c + 2 = 0$
15. $x^2 - x - 42 = 0$
16. $a^2 + a - 6 = 0$
17. $b^2 + 7b + 12 = 0$
18. $y^2 + 2y - 15 = 0$
19. $a^2 - 3a - 10 = 0$
20. $d^2 + 10d + 16 = 0$
21. $x^2 - 4x - 12 = 0$

MORE SOLVING QUADRATIC EQUATIONS

Sometimes quadratic equations are not in standard form. They are not already set equal to zero. These equations must first be put in standard form in order to solve.

EXAMPLE 1: $y^2 - 11y + 10 = -14$

Step 1: Add 14 to both sides so the equation will be set equal to 0.

$$\begin{array}{r} y^2 - 11y + 10 = -14 \\ +14 \quad +14 \\ \hline y^2 - 11y + 24 = 0 \end{array}$$

Step 2: Factor: $(y - 8)(y - 3) = 0$

Step 3: Set each factor equal to 0 and solve for y.

$y - 8 = 0$ $\qquad$ $y - 3 = 0$

$y = 8$ $\qquad$ $y = 3$

The solution set is $\{8, 3\}$.

Put in standard form and then solve.

1. $x^2 - x = 12$
2. $y^2 + 2y = 15$
3. $b^2 - 4b - 2 = 10$
4. $c^2 - 11c = -28$
5. $a^2 - 5a + 8 = 14$
6. $b^2 - b = 20$
7. $c^2 + 9c = -14$
8. $y^2 - 6y - 3 = 13$
9. $x^2 - 4x - 10 = 35$
10. $a^2 + 2a = 8$
11. $b^2 - 7b + 2 = 10$
12. $y^2 + y + 7 = 13$

SOLVING THE DIFFERENCE OF TWO SQUARES

To solve the difference of two squares, first factor. Then set each factor equal to zero.

EXAMPLE: $x^2 - 36 = 0$

Step 1: Factor the left hand side of the equation.

$x^2 - 36 = 0$

$(x + 6)(x - 6) = 0$

Step 2: Set each factor equal to zero and solve.

$$\begin{array}{rcl} x + 6 &=& 0 \\ -6 && -6 \\ \hline x &=& -6 \end{array} \qquad \begin{array}{rcl} x - 6 &=& 0 \\ +6 && +6 \\ \hline x &=& 6 \end{array}$$

$x = -6, 6$

Check: Substitute each solution in the equation to check.

for $x = -6$

$x^2 - 36 = 0$

$(-6)(-6) - 36 = 0$ ← Substitute -6 for x.

$36 - 36 = 0$ ← A true statement. $x = -6$ is a solution.

for $x = 6$:

$x^2 - 36 = 0$

$(6)(6) - 36 = 0$ ← Substitute 6 for x.

$36 - 36 = 0$ ← A true statement. $x = 6$ is a solution.

The solution set is $\{-6, 6\}$.

Find the solution sets for the following.

1. $a^2 - 16 = 0$
2. $c^2 - 36 = 0$
3. $x^2 - 64 = 0$
4. $y^2 - 49 = 0$
5. $y^2 - 1 = 0$
6. $y^2 - 25 = 0$
7. $d^2 - 121 = 0$
8. $b^2 - 9 = 0$
9. $x^2 - 4 = 0$
10. $a^2 - 100 = 0$
11. $x^2 - 144 = 0$
12. $y^2 - 81 = 0$

SOLVING PERFECT SQUARES

When the square root of a constant, variable, or polynomial results in a constant, variable, or polynomial without irrational numbers, the expression is a **perfect square**. Some examples are 49, x^2, and $(x-2)^2$.

EXAMPLE 1: Solve the perfect square for x. $(x-5)^2 = 0$

Step 1: Find the square root of both sides.
$\sqrt{(x-5)^2} = \sqrt{0}$
$(x-5) = 0$

Step 2: Solve the equation.
$(x-5) = 0$
$x - 5 + 5 = 0 + 5$
$\boldsymbol{x = 5}$

EXAMPLE 2: Solve the perfect square for x. $(x-5)^2 = 64$

Step 1: Find the square root of both sides.
$\sqrt{(x-5)^2} = \sqrt{64}$
$(x-5) = \pm 8$
$(x-5) = 8$ and $(x-5) = -8$

Step 2: Solve the two equations.
$(x-5) = 8$ and $(x-5) = -8$
$x - 5 + 5 = 8 + 5$ and $x - 5 + 5 = -8 + 5$
$x = 13$ and $x = -3$

Solve the perfect square for *x*.

1. $(x-5)^2 = 0$
2. $(x+1)^2 = 0$
3. $(x+11)^2 = 0$
4. $(x-4)^2 = 0$
5. $(x-1)^2 = 0$
6. $(x+8)^2 = 0$
7. $(x+3)^2 = 4$
8. $(x-5)^2 = 16$
9. $(x-10)^2 = 100$
10. $(x+9)^2 = 9$
11. $(x-4.5)^2 = 25$
12. $(x+7)^2 = 36$
13. $(x+2)^2 = 49$
14. $(x-1)^2 = 4$
15. $(x+8.9)^2 = 49$
16. $(x-6)^2 = 81$
17. $(x-12)^2 = 121$
18. $(x+2.5)^2 = 64$

COMPLETING THE SQUARE

"Completing the square" is another way of factoring a quadratic equation. To complete the square, the equation must be converted into a perfect square.

EXAMPLE 1: Solve $x^2 - 10x + 9 = 0$ by completing the square.

Completing the square:

Step 1: The first step is to get the constant on the other side of the equation. Subtract 9 from both sides:

$x^2 - 10x + 9 - 9 = 0 - 9$

$x^2 - 10x = -9$

Step 2: Determine the coefficient of the x. The coefficient in this example is 10. Divide the coefficient by 2 and square the result.

$(10 \div 2)^2 = 5^2 = 25$

Step 3: Add the resulting value, 25, to both sides:

$x^2 - 10x + 25 = -9 + 25$

$x^2 - 10x + 25 = 16$

Step 4: Now factor the $x^2 - 10x + 25$ into a perfect square:

$(x - 5)^2 = 16$

Solving the perfect square:

Step 5: Take the square root of both sides.

$\sqrt{(x - 5)^2} = \sqrt{16}$

$(x - 5) = \pm 4$

$(x - 5) = 4$ and $(x - 5) = -4$

Step 6: Solve the two equations.

$(x - 5) = 4$ and $(x - 5) = -4$

$x - 5 + 5 = 4 + 5$ and $x - 5 + 5 = -4 + 5$

$\mathbf{x = 9}$ and $\mathbf{x = 1}$

Solve for x by completing the square.

1. $x^2 + 2x - 3 = 0$
2. $x^2 - 8x + 7 = 0$
3. $x^2 + 6x - 7 = 0$
4. $x^2 - 16x - 36 = 0$
5. $x^2 - 14x + 49 = 0$
6. $x^2 - 4x = 0$
7. $x^2 + 12x + 27 = 0$
8. $x^2 + 2x - 24 = 0$
9. $x^2 + 12x - 85 = 0$
10. $x^2 - 8x + 15 = 0$
11. $x^2 - 16x + 60 = 0$
12. $x^2 - 8x - 48 = 0$
13. $x^2 + 24x + 44 = 0$
14. $x^2 + 6x + 5 = 0$
15. $x^2 - 11x + 5.25 = 0$

USING THE QUADRATIC FORMULA

On the Nevada High School Proficiency Exam in Mathematics, you may be asked to use the quadratic formula to solve a **quadratic equations**. The equation should be in the form

$$ax^2 + bx + c = 0.$$

EXAMPLE: Using the quadratic formula, find x in the following equation: $x^2 - 8x = -7$.

Step 1: Make sure the equation is set equal to 0.

$$\begin{array}{r} x^2 - 8x = -7 \\ \underline{+7 = +7} \\ x^2 - 8x + 7 = 0 \end{array}$$

The quadratic formula is: $\dfrac{-b \pm \sqrt{b^2 - 4ac}}{2a}$

Step 2: In the formula, a is the number x^2 is multiplied by, b is the number x is multiplied by, and c is the last term of the equation. For the equation in the example, $x^2 - 8x + 7$, $a = 1$, $b = -8$, and $c = 7$. When we look at the formula, we notice a $\pm$ sign. This means there will be two solutions to the equation, one when we use the plus sign and one when we use the minus sign. Substituting the numbers from the problem into the formula, we have:

$\dfrac{8 + \sqrt{8^2 - (4)(1)(7)}}{2(1)} = 7$ and $\dfrac{8 - \sqrt{8^2 - (4)(1)(7)}}{2(1)} = 1$ The solutions are (7,1).

For each of the following equations, use the quadratic formula to find two solutions.

1. $x^2 + x - 6 = 0$
2. $y^2 - 2y - 8 = 0$
3. $a^2 + 2a - 15 = 0$
4. $y^2 - 5y + 4 = 0$
5. $b^2 - 9b + 14 = 0$
6. $x^2 - 3x - 4 = 0$
7. $y^2 + y - 20 = 0$
8. $d^2 + 6d + 8 = 0$
9. $y^2 - 7y + 12 = 0$
10. $x^2 - 3x - 28 = 0$
11. $a^2 - 5a + 6 = 0$
12. $b^2 + 3b - 10 = 0$
13. $a^2 + 7a - 8 = 0$
14. $c^2 + 3c + 2 = 0$
15. $x^2 - x - 42 = 0$
16. $a^2 + a - 6 = 0$
17. $b^2 + 7b + 12 = 0$
18. $y^2 + 2y - 15 = 0$
19. $a^2 - 3a - 10 = 0$
20. $d^2 + 10d + 16 = 0$
21. $x^2 - 4x - 12 = 0$

REAL WORLD QUADRATIC EQUATIONS

The most common real life situation that would use a quadratic equation is the motion of an object under the force of gravity. Two examples are a ball being kicked into the air or a rocket being shot into the air.

EXAMPLE 1: A high school football player is practicing his field goal kicks. The equation below represents the height of the ball at a specific time.

$s = -9t^2 + 45t$ t = amount of time in seconds s = height in feet

Question 1: Where will the ball be at 4 seconds?

Solution 1: Since there are only two variables, you will only need the value of one variable to solve the problem. Simply plug in the number 4 in place of the variable t and solve the equation as shown below.

$s = -9(4)^2 + 45(4)$

$s = -9(16) + 180$

$s = -144 + 180$ $s = 36$

At 4 seconds the ball will be 36 ft in the air.

Question 2: If the ball is 54 ft in the air, how much time has gone by?

Solution 2: This question is similar to the previous one, except that the given variable is different. This time you would be replacing s with 54 and then solve the equation.

$54 = -9t^2 + 45t$ Subtract 54 on both sides.

$0 = -9t^2 + 45t - 54$ Divide the entire equation by -9.

$0 = t^2 - 5t + 6$ Factor the equation.

$0 = (t-3)(t-2)$ Solve for t. $t = 3$ $t = 2$

For this question there are 2 answers. The ball is 54 ft in the air when 2 and 3 seconds have gone by.

EXAMPLE 2: John and Alex are kicking a soccer ball back and forth to each other. The equation below represents the height of the ball at a specific point in time.
$s = -4t^2 + 24t$, where t = amount of time in seconds and s = height in feet

Question 1: How long does it take for the soccer ball to come back down to the ground?

Solution 1: Looking at this problem you can see that no value was given, but one was indirectly given. The question asks when will the ball come back down. This is just another way of asking, "When will the height of the ball be zero?" The value 0 will be used for s. Substitute 0 back in for s and solve.

$0 = -4t^2 + 24t$	Factor out the greatest common factor.
$0 = -4t\,(t-6)$	Set each factor to 0 and solve.
$t = 0 \quad t = 6$	It is clear that the ball is on the ground at 0 seconds, so the value of $t = 0$ is not the answer and the second value of t is used instead. It takes the ball 6 seconds to go up into the air and then come back down to the ground.

Question 2: What is the highest point the ball will go?

Solution 2: This is asking what is the vertex of the equation. You will need to use the vertex formula. As a reminder, the quadratic equation is defined as $y = ax^2 + bx + c$, where $a \neq 0$. The quadratic equation can also be written as a function of x by substituting $f(x)$ for y, such as $f(x) = ax^2 + bx + c$. To find the point of the vertex of the graph, you must use the formula below.

$$\text{vertex} = \left(\frac{-b}{2a}, f\left(\frac{-b}{2a}\right)\right)$$

where $f\left(\frac{-b}{2a}\right)$ is the quadratic equation evaluated at the value $-\frac{b}{2a}$. To do this, plug $-\frac{b}{2a}$ in for x.

To use the vertex equation, put the original equation in quadratic form and find a and b.

$s = -4t^2 + 24t = -4t^2 + 24t + 0.$

Since a is the coefficient of t^2, $a = -4$. b is the coefficient of t, so $b = 24$.

Find the solution to $f\left(-\frac{b}{2a}\right)$ by substituting the values of a and b from the equation into the expression.

$$-\frac{b}{2a} = -\left(\frac{24}{(2 \times -4)}\right) = -\left(\frac{24}{-8}\right) = -(-3) = 3$$

Find the solution to $f\left(\frac{-b}{2a}\right)$. We know that $-\frac{b}{2a} = 3$, so we need to find $f(3)$.

To do this, we must substitute 3 into the quadratic equation for x.

$f(t) = -4t^2 + 24t$

$f(3) = -4\,(3)^2 + 24\,(3) = -4\,(9) + 72 = -36 + 72 = 36$

The vertex equals (3, 36). This means at 3 seconds, the ball is 36 feet in the air. Therefore, the highest the soccer ball will go is 36 ft.

Solve the following quadratic problems.

1. Eric is at the top of a cliff that is 500 ft from the ocean's surface. He is waiting for his friend to climb up and meet him. As he is waiting he decides to start casually tossing pebbles off the side of the cliff. The equation that represents the height of his pebbles tosses is
$s = -t^2 + 5t + 500$, where s = distance in feet and t = time in seconds.

   A. How long does it take the pebble to hit the water?

   B. If fifteen seconds have gone by, what is the height of the pebble from the ocean?

   C. What is the highest point the pebble will go?

2. Devin is practicing golf at the driving range. The equation that represents the height of his ball is $s = -0.5t^2 + 12t$, where s = distance in feet and t = time in seconds.

   A. What is the highest her ball will ever go?

   B. If the ball is at 31.5 ft in the air, how many seconds has gone by?

   C. How long will it take for the ball to hit the ground?

3. Jack throws a ball up in the air to see how high he can get it to go. The equation that represents the height of the ball is $s = -2t^2 + 20t$, where s = distance in feet and t = time in seconds.

   A. How high will the ball be at 7 seconds?

   B. If the ball is 48 feet in the air, how many seconds have gone by?

   C. How long does it take for the ball to go up and come back down to the ground?

4. Kali is jumping on her super trampoline, getting as high as she possibly can. The equation to represent her height is $s = -5t^2 + 20t$, where s = distance in feet and t = time in seconds.

   A. What is the highest Kali can jump on her trampoline?

   B. How high will Kali be at 4 seconds?

   C. If Kali is 18.75 ft in the air, then how many seconds have gone by?

CHAPTER 7 REVIEW

Factor the following polynomials completely.

1. $8x - 18$	7. $2b^2 - 2b - 12$	13. $9w^2 - 54w - 63$
2. $16b^3 + 8b$	8. $3a^3 + 4a^2 + 9a + 12$	14. $12x^2 + 27x$
3. $15a^3 + 40$	9. $27y^2 + 42y - 5$	15. $2a^4 - 32$
4. $20y^6 - 12y^4$	10. $12b^2 + 25b - 7$	16. $21c^2 + 41c + 10$
5. $5a - 15a^2$	11. $6y^2 + 30y + 36$	17. $2b^3 - 24 + 16b - 3b^2$
6. $4y^2 - 36$	12. $2b^2 + 6b - 20$	18. $-2a - 25a^2 + 10a^3 + 5$

Factor and solve each of the following quadratic equations.

1. $16b^2 - 25 = 0$	6. $y^2 = 21 - 4y$	11. $4x^2 + 19x - 5 = 0$
2. $a^2 - a - 30 = 0$	7. $y^2 - 7y + 8 = 16$	12. $8x^2 = 6x + 2$
3. $x^2 - x = 6$	8. $6x^2 + x - 2 = 0$	13. $2y^2 - 6y - 20 = 0$
4. $100x^2 - 49 = 0$	9. $3y^2 + y - 2 = 0$	14. $-6x^2 + 7x - 2 = 0$
5. $81y^2 = 9$	10. $b^2 + 2b - 8 = 0$	15. $y^2 + 3y - 18 = 0$

Using the quadratic formula, find both solutions for the variable.

16. $x^2 + 10x - 11 = 0$	18. $b^2 + 9b + 18 = 0$	20. $a^2 - 8a - 48 = 0$
17. $y^2 - 14y + 40 = 0$	19. $y^2 - 12y - 13 = 0$	21. $x^2 + 2x - 63 = 0$

CHAPTER 8: ALGEBRA WORD PROBLEMS

Standards 2.12.6 & 2.12.7

An equation states that two mathematical expressions are equal. In working with word problems, the words that mean equal are **equals, is, was, is equal to, amounts to,** and other expressions with the same meaning. To translate a word problem into an algebraic equation, use a variable to represent the unknown or unknowns the problem is looking for.

In the following example, let *n* be the number you are looking for.

EXAMPLE: Four more than twice a number is two less than three times the number.

Step 1: Translation: $4 + 2n = 3n - 2$

Step 2: Now solve:

$$\begin{aligned} 4 + 2n &= 3n - 2 \\ -2n & \quad -2n \\ 4 &= n - 2 \\ +2 & \quad +2 \\ 6 &= n \end{aligned}$$

The number is 6.
Substitute the number back into the original equation to check.

Translate the following word problems into equations and solve.

1. Seven less than twice a number is eleven. Find the number.
2. Four more than three times a number is one less than four times the number. What is the number?
3. The sum of three times a number and the number was 24. What was the number?
4. Negative 16 is the sum of five and a number. Find the number.
5. Negative 20 is equal to ten minus the product of six and a number. What is the number?
6. Two less than twice a number equals the number plus 15. What is the number?
7. The difference between three times a number and 21 is three. What is the number?
8. Eighteen is fifteen less than the product of a number and three. What is the number?
9. Six more than twice a number is four times the difference between three and the number. What is the number?
10. Four less than twice a number is five times the sum of one and the number. What is the number?

GEOMETRY WORD PROBLEMS

The perimeter of a geometric figure is the distance around the outside of the figure.

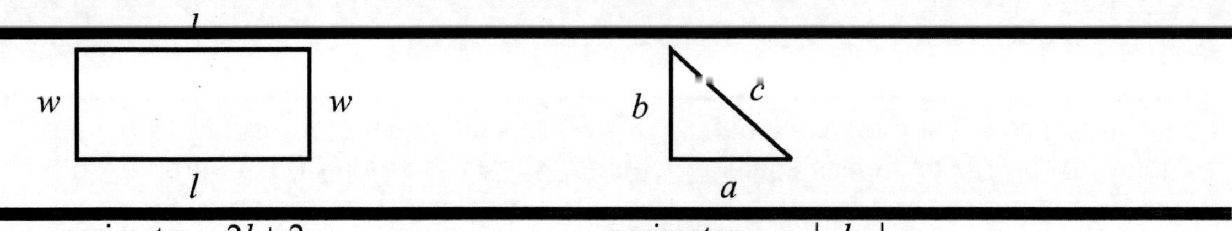

perimeter = $2l + 2w$ perimeter = $a + b + c$

EXAMPLE 1: The perimeter of a rectangle is 44 feet. The length of the rectangle is 6 feet more than the width. What is the measure of the width?

Step 1: Let the variable be the length of the unknown side.
width = w length = $6 + w$

Step 2: Use the equation for the perimeter of a rectangle as follows:
$2l + 2w$ = perimeter of a rectangle.
$2(w + 6) + 2w = 44$

Step 3: Solve for w.

Solution: width = 8 feet

EXAMPLE 2: The perimeter of a triangle is 26 feet. The second side is twice as long as the first. The third side is 1 foot longer than the second side. What is the length of the 3 sides?

Step 1: Let x = first side $2x$ = second side $2x + 1$ = third side

Step 2: Use the equation for perimeter of a triangle as follows:
sum of the length of the sides = perimeter of a triangle.
$x + 2x + 2x + 1 = 26$

Step 3: Solve for x. $5x + 1 = 26$ so $x = 5$

Solution: first side $x = 5$ second side $2x = 10$ third side $2x + 1 = 11$

Solve the following geometry word problems.

1. The length of a rectangle is 6 times longer than the width. The perimeter is 42. What is the width?
2. The length of a rectangle is 4 more than twice the width. The perimeter is 28. What is the length?
3. The perimeter of a triangle is 24 feet. The second side is two feet longer than the first. The third side is two feet longer than the second. What are the lengths of the sides?

4. In an isosceles triangle, two sides are equal. The third side is two less than twice the length of the two equal sides. The perimeter is 38. What are the lengths of the three sides?

5. The sum of the measures of the angles of a triangle is 180°. The second angle is twice the measure of the first angle. The third angle is three times the measure of the second angle. Find the measure of each angle.

6. The sum of the measures of the angles of a triangle is 180°. The second angle of a triangle is twice the measure of the first angle. The third angle is 4 more than 5 times the first. What are the measures of the three angles?

AGE PROBLEMS

EXAMPLE: Tara is twice as old as Gwen. Their sister, Amy, is 5 years older than Gwen. If the sum of their ages is 29 years, find each of their ages.

Step 1: We want to find each of their ages, so there are three unknowns. Tara is twice as old as Gwen, and Amy is older than Gwen, so Gwen is the youngest. Let x be Gwen's age. From the problem we can see that:

Gwen = x
Tara = $2x$
Amy = $x + 5$

} The sum of their ages is 29

Step 2: Set up the equation, and solve for x.

$$x + 2x + x + 5 = 29$$
$$4x + 5 = 29$$
$$4x = 29 - 5$$
$$x = \frac{24}{4}$$
$$x = 6$$

Solutions:

Gwen's age (x) = 6
Tara's age ($2x$) = 12
Amy's age ($x + 5$) = 11

Solve the following age problems.

1. Carol is 25 years older than her cousin Amanda. Cousin Bill is 3 times as old as Amanda. The sum of their ages is 90. Find each of their ages.

2. Derrick is 5 less than twice as old as Brandon. The sum of their ages is 43. How old are Derrick and Brandon?

3. Beth's mom is 6 times older than Beth. Beth's dad is 7 years older than Beth's mom. The sum of their ages is 72. How old are each of them?

4. Annie is 2 years more than three times as old as her son, Jared. If the difference between their ages is 26, how old are Annie and Jared?

5. Eileen is 6 years older than Karen. John is three times as old as Karen. The sum of their ages is 56. How old are Eileen, Karen and John?

6. Taylor is 18 years younger than Jim. Andrew is twice as old as Taylor. The sum of their ages is 26. How old are Taylor, Jim, and Andrew?

The following problems work in the same way as the age problems. There are two or three items of different weight, distance, number, or size. You are given the total and asked to find the amount of each item.

7. Three boxes have a total weight of 640 pounds. Box A weighs twice as much as Box B. Box C weighs 30 pounds more than Box A. How much do each of the boxes weigh?

8. There are 158 students registered for American History classes. There are twice as many students registered in second period as first period. There are 10 less than three times as many students in third period as in first period. How many students are in each period?

9. Cindy earns $2 less than three times as much as Olivia. Shane earns twice as much as Cindy. Together they earn $594 per week. How much does each person earn per week?

10. Ellie, the elephant, eats 4 times as much as Popcorn, the pony. Zac, the zebra, eats twice as much as Popcorn. Altogether, they eat 490 kilograms of feed per week. How much feed does each of them require each week?

11. The school cafeteria served three kinds of lunches today to 225 students. The students chose the cheeseburgers three times more often than the grilled cheese sandwiches. There were twice as many grilled cheese sandwiches sold as fish sandwiches. How many of each lunch were served?

12. Three friends drove west into Illinois. Kyle drove half as far as Jamaal. Conner drove 4 times as far as Kyle. Altogether, they drove 357 miles. How far did each friend drive?

13. Patricia is taking collections for this year's Feed the Hungry Project. So far she has collected $200 more from Company A than from Company B and $800 more from Company C than from Company A. Until now, she has collected $3,000. How much did Company C give?

14. For his birthday, Torin got $25.00 more from his grandmother than from his uncle. His uncle gave him $10.00 less than his cousin. Torin received $290.00 in total. How much did he receive from his cousin?

15. Cassidy loves black and yellow jelly beans. She noticed when she was counting them that number of yellow jelly beans was 8 less than 3 times the number of black jelly beans. How many black jelly beans did she have?

16. Karen Holloway planted a garden with red and white rose bushes. Because she was studying to be a botanist, she counted the number of blossoms on each bush. She counted 5 times as many red blossoms as white blossoms. In total, she counted 1,680 blossoms. How many red blossoms did she count?

INEQUALITY WORD PROBLEMS

Inequality word problems involve staying under a limit or having a minimum goal one must meet.

EXAMPLE: A contestant on a popular game show must earn a minimum of 800 points by answering a series of questions worth 40 points each per category in order to win the game. The contestant will answer questions from each of four categories. Her results for the first three categories are as follows: 160 points, 200 points, and 240 points. Write an inequality which describes how many points, (p), the contestant will need on the last category in order to win.

Step 1: Add to find out how many points she already has. $160 + 200 + 240 = 600$

Step 2: Subtract the points she already has from the minimum points she needs. $800 - 600 = 200$. She must get at least 200 points in the last category to win. If she gets more than 200 points, that is okay, too. To express the number of points she needs, use the following inequality statement:

$p \geq 200$ The points she needs must be greater than or equal to 200.

Solve each of the following problems using inequalities.

1. Stella wants to place her money in a high interest money market account. However, she needs at least \$1000 to open an account. Each month, she sets aside some of her earnings in a savings account. In January through July, she added the following amounts to her savings: \$121, \$206, \$138, \$212, \$109, and \$134. Write an inequality which describes the amount of money she can set aside in July to qualify for the money market account.

2. A high school band program will receive \$2000.00 for selling \$10,000.00 worth of coupon books. Six band classes participate in the sales drive. Classes 1-5 collect the following amounts of money: \$1,400, \$2,600, \$1,800, \$2,450, and \$1,550. Write an inequality which describes the amount of money the sixth class must collect so that the band will receive \$2,000.

3. A small elevator has a maximum capacity of 1,000 pounds before the cable holding it in place snaps. Six people get on the elevator. Five of their weights follow: 146, 180, 130, 262, and 135. Write an inequality which describes the amount the sixth person can weigh without snapping the cable.

4. A teacher told a small high school class of 9 students they would receive a pizza party if their class average was 92% or higher on the next exam. Students 1-8 scored the following on the exam: 86, 91, 98, 83, 97, 89, 99, and 96. Write an inequality which describes the score the ninth student must make for the class to qualify for the pizza party.

5. Raymond wants to spend his entire credit limit on his credit card. His credit limit is \$2000. He purchases items costing \$600, \$800, \$50, \$168, and \$3. Write an inequality which describes the amounts Raymond can put on his credit card for his next purchases.

CHAPTER 8 REVIEW

Solve each of the following problems.

1. Deanna is five more than six times older than Ted. The sum of their ages is 47. How old is Ted?

2. Ross is six years older than twice his sister Holly's age. The difference in their ages is 18 years. How old is Holly?

3. The perimeter of a triangle is 43 inches. The second side is three inches longer than the first side. The third side is one inch longer than the second. Find the length of each side.

4. The perimeter of a rectangle is 80 feet. The length of the rectangle is 2 feet less than 5 times the width. What is the length and width of the rectangle?

5. Joe, Craig, and Mike have a combined weight of 429 pounds. Craig weighs 34 pounds more than Joe. Mike weighs 13 pounds more than Craig. How many pounds does Craig weigh?

6. Tracie and Marcia drove to Florida to see Marcia's sister in Jacksonville. Tracie drove one hour more than three times as much as Marcia. The trip took a total of 17 driving hours. How many hours did Tracie drive?

7. Ralph and Larry entered a pie eating contest. Ralph ate 2 less than twice as many pies as Larry. They ate a total of 28 pies. How many pies did Larry eat?

8. Lena and Jodie are sisters and together they have 68 bottles of nail polish. Lena bought 5 more than half the bottles. How many did Jodie buy?

9. Janet and Artie wanted to play tug of war. Artie pulls with 150 pounds of force while Janet pulls with 40 pounds of force. In order to make this a fair contest, Janet enlists the help of her friends Trudi, Sherri, and Bridget who pull with 30, 25, and 40 pounds respectively. Write an inequality describing the minimum amount Janet's fourth friend, Tommy, must pull to beat Artie.

10. Jim takes great pride in decorating his float for the homecoming parade for his high school. With the $5,000 he has to spend, Jim bought 5,000 carnations at $0.25 each, 4,000 tulips at $0.50 each, and 300 irises at $0.90 each. Write an inequality which describes how many roses, r, Jim can buy if roses cost $0.80 each.

11. Pete Williams wants to sell some or all of his shares of stock in a company. He purchased the 80 shares for $0.50 last month, and the shares are now worth $4.50 each. Write an inequality which describes how much profit, p, Pete can make by selling his shares.

12. Connie drove for 2 hours at a constant speed of 55 miles per hour. How many total miles did she travel?

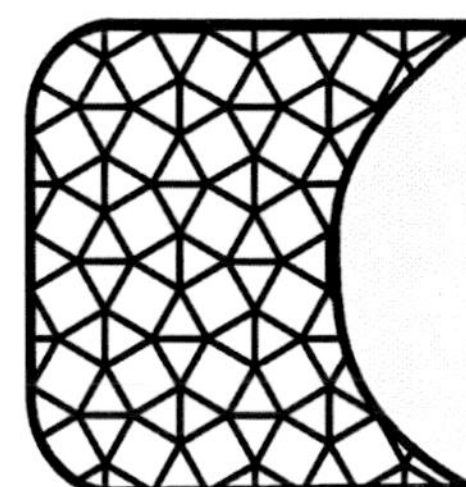

CHAPTER 9: SYSTEMS OF EQUATIONS

Standards 2.12.5 and 4.12.5

SYSTEMS OF EQUATIONS

Two linear equations considered at the same time are called a **system** of linear equations. The graph of a linear equation is a straight line. The graphs of two linear equations can show that the lines are **parallel**, **intersecting**, or **collinear**. Two lines that are **parallel** will never intersect and have no ordered pairs in common. If two lines are **intersecting**, they have one point in common, and in this chapter, you will learn to find the ordered pair for that one point. If the graph of two linear equations is the same line, the lines are said to be **collinear**.

If you are given a system of two linear equations, and you put both equations in slope-intercept form, you can immediately tell if the graph of the lines will be **parallel**, **intersecting**, or **collinear**.

If two linear equations have the same slope and the same y-intercept, then they are both equations for the same line. They are called **collinear** or **coinciding** lines. A line is made up of an infinite number of points extending infinitely far in two directions. Therefore, collinear lines have an infinite number of points in common.

EXAMPLE: $2x + 3y = -3$ Or in slope-intercept form the equations become : $y = -\frac{2}{3}x - 1$

$4x + 6y = -6$ $y = -\frac{2}{3}x - 1$

In slope-intercept form, we notice that both slopes equal $-\frac{2}{3}$ and both y-intercepts equal -1. They are collinear lines.
If two linear equations have the same slope but different y-intercepts, they are **parallel** lines. Parallel lines never touch each other, so they have no points in common.

If two linear equations have different slopes, then they are intersecting lines and share exactly one point in common.

The chart below summarizes what we know about the graphs of two equations in slope-intercept form.

y-Intercepts	Slopes	Graphs	Number of Solutions
same	same	collinear	infinite
different	same	distinct parallel lines	none (they never touch)
same or different	different	intersecting lines	exactly one

For the pairs of equations below, put each equation in slope-intercept form, and tell whether the graphs of the lines will be collinear, parallel, or intersecting.

1. $x - y = -1$ __________
   $-x + y = 1$

2. $x - 2y = 4$ __________
   $-x + 2y = 6$

3. $y - 2 = x$ __________
   $x + 2 = y$

4. $x = y - 1$ __________
   $-x = y - 1$

5. $2x + 5y = 10$ __________
   $4x + 10y = 20$

6. $x + y = 3$ __________
   $x - y = 1$

7. $2y = 4x - 6$ __________
   $-6x + y = 3$

8. $x + y = 5$ __________
   $2x + 2y = 10$

9. $2x = 3y - 6$ __________
   $4x = 6y - 6$

10. $2x - 2y = 2$ __________
    $3y = -x + 5$

11. $x = -y$ __________
    $x = 4 - y$

12. $2x = y$ __________
    $x + y = 3$

13. $x = y + 1$ __________
    $y = x + 1$

14. $x - 2y = 4$ __________
    $-2x + 4y = -8$

15. $2x + 3y = 4$ __________
    $-2x + 3y = 4$

16. $2x - 4y = 1$ __________
    $-6x + 12y = 3$

17. $-3x + 4y = 1$ __________
    $6x + 8y = 2$

18. $x + y = 2$ __________
    $5x + 5y = 10$

19. $x + y = 4$ __________
    $x - y = 4$

20. $y = -x + 3$ __________
    $x - y = 1$

FINDING COMMON SOLUTIONS FOR INTERSECTING LINES

When two lines intersect, they share exactly one point in common.

EXAMPLE: $3x + 4y = 20$ and $4x + 2y = 12$

Put each equation in slope-intercept form.

$$\begin{array}{ll} 3x + 4y = 20 & 2y - 4x = 12 \\ 4y = -3x + 20 & 2y = 4x + 12 \\ y = -\frac{3}{4}x + 5 & y = 2x + 6 \end{array}$$

slope-intercept form

Straight lines with different slopes are **intersecting lines**. Look at the graph of the lines on the same Cartesian plane.

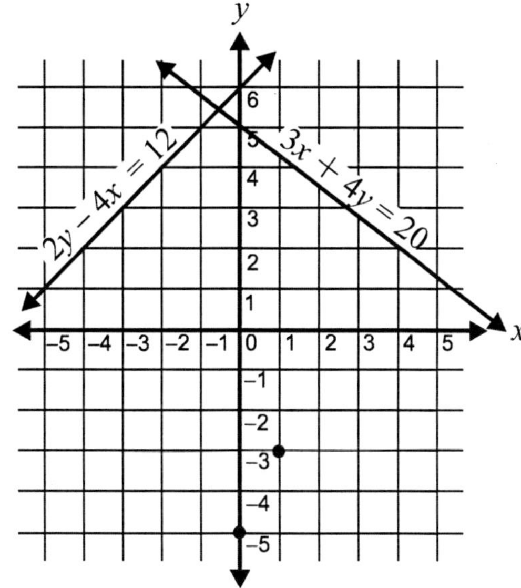

You can see from looking at the graph that the intersecting lines share one point in common. However, it is hard to tell from looking at the graph what the coordinates are for the point of intersection. To find the exact point of intersection, you can use the **substitution method** to solve the system of equations algebraically.

SOLVING SYSTEMS OF EQUATIONS BY SUBSTITUTION

You can solve systems of equations algebraically by using the substitution method.

EXAMPLE: Find the point of intersection of the following two equations:

Equation 1: $x - y = 3$
Equation 2: $2x + y = 9$

Step 1: Solve one of the equations for x or y. Let's choose to solve equation 1 for x.
Equation 1: $x - y = 3$
$x = y + 3$

Step 2: Substitute the value of x from equation 1 in place of x in equation 2.
Equation 2: $2x + y = 9$
$2(y + 3) + y = 9$
$2y + 6 + y = 9$
$3y + 6 = 9$
$3y = 3$
$y = 1$

Step 3: Substitute the solution for y back in equation 1 and solve for x.
Equation 1: $x - y = 3$
$x - 1 = 3$
$x = 4$

Step 4: The solution set is (4, 1). Substitute in one or both of the equations to check.

Equation 1: $x - y = 3$
$4 - 1 = 3$
$3 = 3$

Equation 2: $2x + y = 9$
$2(4) + 1 = 9$
$8 + 1 = 9$
$9 = 9$

The point (4,1) is common to both equations. This is the **point of intersection**.

For each of the following pairs of equations, find the point of intersection (the common solution) using the substitution method.

1. $x + 2y = 8$
   $2x - 3y = 2$

2. $x - y = -5$
   $x + y = 1$

3. $x - y = 4$
   $x + y = 2$

4. $x - y = -1$
   $x + y = 9$

5. $-x + y = 2$
   $x + y = 8$

6. $x + 4y = 10$
   $x + 5y = 12$

7. $2x + 3y = 2$
   $4x - 9y = -1$

8. $x + 3y = 5$
   $x - y = 1$

9. $-x = y - 1$
   $x = y - 1$

10. $x - 2y = 2$
    $2y + x = -2$

11. $5x + 2y = 1$
    $2x + 4y = 10$

12. $3x - y = 2$
    $5x + y = 6$

13. $2x + 3y = 3$
    $4x + 5y = 5$

14. $x - y = 1$
    $-x - y = 1$

15. $x = y + 3$
    $y = 3 - x$

SOLVING SYSTEMS OF EQUATIONS BY ADDING OR SUBTRACTING

You can solve systems of equations algebraically by adding or subtracting an equation from another equation or system of equations.

EXAMPLE 1: Find the point of intersection of the following two equations.

Equation 1: $x + y = 10$

Equation 2: $-x + 4y = 5$

Step 1: Eliminate one of the variables by adding the two equations together. Since the x has the same coefficient in each equation, but opposite signs, it will cancel nicely by adding.

$$x + y = 10$$
$$+ (-x + 4y = 5)$$

Add each like term together.

$$0 + 5y = 15$$

Simplify.

$$5y = 15$$

Divide both sides by 5.

$$y = 3$$

Step 2: Substitute the solution for y back into an equation, and solve for x.

Equation 1: $x + y = 10$ Substitute 3 for y.

$x + 3 = 10$ Subtract 3 from both sides.

$x = 7$

Step 3: The solution set is (7, 3). Substitute in both of the equations to check.

Equation 1:

$$x + y = 10$$
$$7 + 3 = 10$$
$$10 = 10$$

Equation 2:

$$-x + 4y = 5$$
$$-(7) + 4(3) = 5$$
$$-7 + 12 = 5$$
$$5 = 5$$

The point (7, 3) is the point of intersection.

EXAMPLE 2: Find the point of intersection of the following two equations:

Equation 1: $3x - 2y = -1$

Equation 2: $-4y = -x - 7$

Step 1: Put the variables on the same side of each equation. Take equation 2 out of y-intercept form.

$-4y = -x - 7$ Add x to both sides.

$x - 4y = -x + x - 7$ Simplify.

$x - 4y = -7$

Step 2: Add the two equations together to cancel one variable. Since each variable has the same sign and different coefficients, we have to multiply one equation by a negative number so one of the variables will cancel. Equation 1's y variable has a coefficient of 2, and if multiplied by –2, the y will have the same variable as the y in equation 2, but a different sign. This will cancel nicely when added.

$-2(3x - 2y = -1)$ Multiply by –2.

$-6x + 4y = 2$

Step 3: Add the two equations.

$$\begin{aligned} -6x + 4y &= 2 \\ + \quad (x - 4y &= -7) \\ \hline -5x + 0 &= -5 \\ -5x &= -5 \\ x &= 1 \end{aligned}$$

Add equation 2 to equation 1.
Simplify.
Divide both sides by –5.

Step 4: Substitute the solution for x back into an equation and solve for y.

Equation 1:

$3x - 2y = -1$	Substitute 1 for x.
$3(1) - 2y = -1$	Simplify.
$3 - 2y = -1$	Subtract 3 from both sides.
$3 - 3 - 2y = -1 - 3$	Simplify.
$-2y = -4$	Divide both sides by –2.
$y = 2$	

Step 5: The solution set is (1, 2). Substitute in both equations to check.

Equation 1:	Equation 2:
$3x - 2y = -1$	$-4y = -x - 7$
$3(1) - 2(2) = -1$	$-4(2) = -1 - 7$
$3 - 4 = -1$	$-8 = -8$
$-1 = -1$	

The point (1, 2) is the point of intersection.

For each of the following pairs of equations, find the point of intersection by adding the 2 equations together. Remember you might need to change the coefficients and/or signs of the variables before adding.

1. $x + 2y = 8$
   $-x - 3y = 2$

2. $x - y = 5$
   $2x + y = 1$

3. $x - y = -1$
   $x + y = 9$

4. $3x - y = -1$
   $x + y = 13$

5. $-x + 4y = 2$
   $x + y = 8$

6. $x + 4y = 10$
   $x + 7y = 16$

7. $2x - y = 2$
   $4x - 9y = -3$

8. $x + 3y = 13$
   $5x - y = 1$

9. $-x = y - 1$
   $x = y - 1$

10. $x - y = 2$
    $2y + x = 5$

11. $5x + 2y = 1$
    $4x + 8y = 20$

12. $3x - 2y = 14$
    $x - y = 6$

13. $2x + 3y = 3$
    $3x + 5y = 5$

14. $x - 4y = 6$
    $-x - y = -1$

15. $x = 2y + 3$
    $y = 3 - x$

CHAPTER 9 REVIEW

For each pair of equations below, tell whether the graphs of the lines will be collinear, parallel, or intersecting.

1. $y = 4x + 1$
   $y = 4x - 3$

2. $y - 4 = x$
   $2x + 8 = 2y$

3. $x + y = 5$
   $x - y = -1$

4. $2y - 3x = 6$
   $4y = 6x + 8$

5. $5y = 3x - 7$
   $4x \quad 3y = \quad 7$

6. $2x - 2y = 2$
   $y - x = -1$

Find the common solution for each of the following pairs of equations, using the substitution method.

7. $x - y = 2$
   $x + 4y = -3$

8. $x + y = 1$
   $x + 3y = 1$

9. $4y = -2x + 4$
   $-x = -2y - 4$

10. $2x + 8y = 20$
    $5y = 12 - x$

11. $x = y - 3$
    $-x = y + 3$

12. $-2x + y = -3$
    $x - y = 9$

Find the point of intersection for each pair of equations by adding and/or subtracting the two equations.

13. $2x + y = 4$
    $3x - y = 6$

14. $x + 2y = 3$
    $x + 5y = 0$

15. $x + y = 1$
    $y = x + 7$

16. $2x + 4y = 5$
    $3x + 8y = 9$

17. $2x - 2y = 7$
    $3x - 5y = \frac{5}{2}$

18. $x - 3y = -2$
    $y = -\frac{1}{3}x + 4$

CHAPTER 10: RELATIONS AND FUNCTIONS

Standard 2.12.4

RELATIONS

A **relation** is a set of ordered pairs. The set of the first members of each ordered pair is called the **domain** of the relation. The set of the second members of each ordered pair is called the **range**.

EXAMPLE: State the domain and range of the following relation.

{(2, 4), (3, 7), (4, 9), (6, 11)}

Solution: **Domain:** {2, 3, 4, 6} the first member of each ordered pair
Range: {4, 7, 9, 11} the second member of each ordered pair

State the domain and range for each relation.

1. {(2, 5), (9, 12), (3, 8), (6, 7)}
2. {(12, 4), (3, 4), (7, 12), (26, 19)}
3. {(4, 3), (7, 14), (16, 34), (5, 11)}
4. {(2, 45), (33, 43), (98, 9), (43, 61), (67, 54)}
5. {(78, 14), (29, 67), (84, 49), (16, 18), (98, 46)}
6. {(–8, 16), (23, –7), (–4, –9), (16, –8), (–3, 6)}
7. {(–7, –4), (–3, 16), (–4, 17), (–6, –8), (–8, 12)}
8. {(–1, –2), (3, 6), (–7, 14), (–2, 8), (–6, 2)}
9. {(0, 9), (–8, 5), (3, 12), (–8, –3), (7, 18)}
10. {(58, 14), (44, 97), (74, 32), (6, 18), (63, 44)}
11. {(–7, 0), (–8, 10), (–3, 11), (–7, –32), (–2, 57)}
12. {(18, 34), (22, 64), (94, 36), (11, 18), (91, 45)}

When given an equation in two variables, the **domain** is the set of ***x*** values that satisfies the equation. The **range** is the set of ***y*** values that satisfies the equation.

EXAMPLE: Find the range of the relation $3x = y + 2$ for the domain $\{-1, 0, 1, 2, 3\}$.

Solution: Solve the equation for each value of x given. The result, the y values, will be the range.

Given:

x	y
−1	
0	
1	
2	
3	

Solution:

x	y
−1	−5
0	−2
1	1
2	4
3	7

The range is $\{-5, -2, 1, 4, 7\}$.

Find the range of each relation for the given domain.

Relation	**Domain**	**Range**
1. $y = 5x$	$\{1, 2, 3, 4\}$	
2. $y = \lvert x \rvert$	$\{-3, -2, -1, 0, 1\}$	
3. $y = 3x + 2$	$\{0, 1, 3, 4\}$	
4. $y = -\lvert x \rvert$	$\{-2, -1, 0, 1, 2\}$	
5. $y = -2x + 1$	$\{0, 1, 3, 4\}$	
6. $y = 10x - 2$	$\{-2, -1, 0, 1, 2,\}$	
7. $y = 3\lvert x \rvert + 1$	$\{-2, -1, 0, 1, 2,\}$	
8. $y - x = 0$	$\{1, 2, 3, 4\}$	
9. $y - 2x = 0$	$\{1, 2, 3, 4\}$	
10. $y = 3x - 1$	$\{0, 1, 3, 4\}$	
11. $y = 4x + 2$	$\{0, 1, 3, 4\}$	
12. $y = 2\lvert x \rvert - 1$	$\{-2, -1, 0, 1, 2,\}$	

DETERMINING DOMAIN AND RANGE FROM GRAPHS

The domain is all of the x values that lie on the function in the graph from the lowest x value to the highest x value. The range is all of the y values that lie on the function in the graph from the lowest y to the highest y.

EXAMPLE: Find the domain and range of the graph.

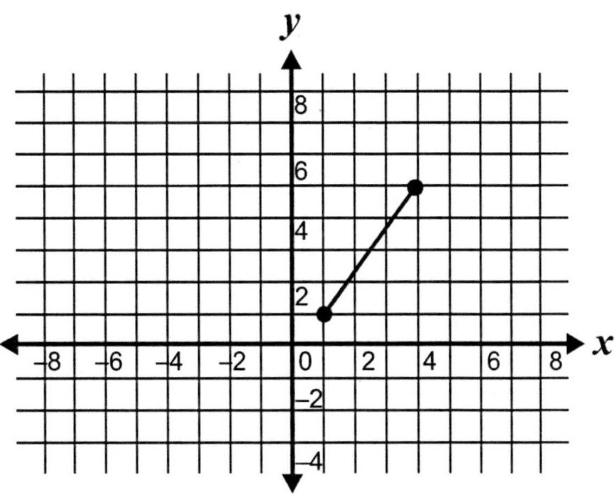

Step 1: First find the lowest x value depicted on the graph. In this case it is 1. Then find the highest x value depicted on the graph. The highest value of x on the graph is 4. The domain must contain all of the values between the lowest x value and the highest x value. The easiest way to write this is $1 \leq \text{Domain} \leq 4$ or $1 \leq x \leq 4$.

Step 2: Do the same process for the range, but this time look at the lowest and highest y values. The answer is $1 \leq \text{Range} \leq 5$ or $1 \leq y \leq 5$.

Find the domain and range of each graph below. Write your answers in the line provided.

1.

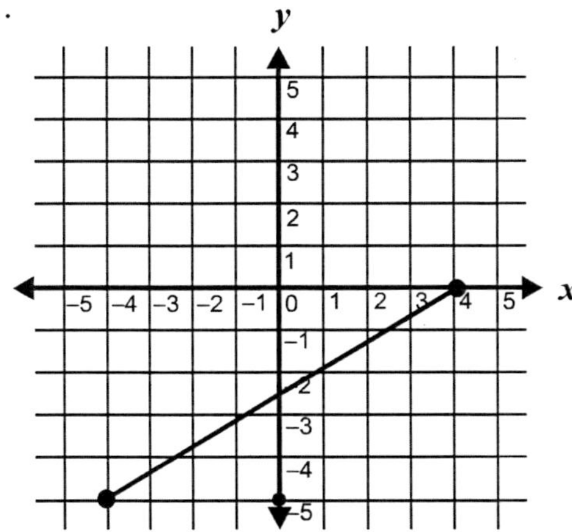

____

2.

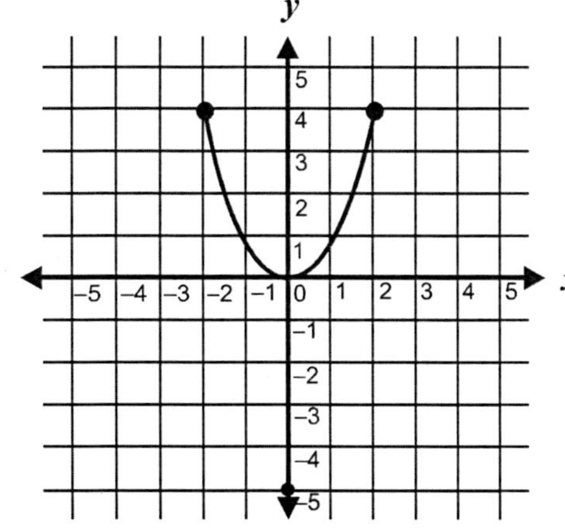

____

3.

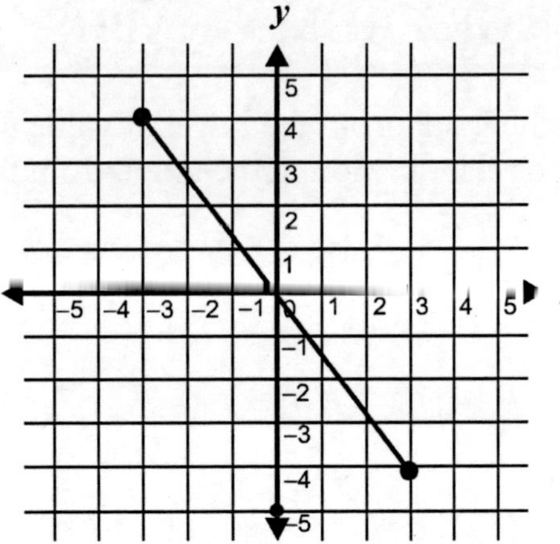
y
x
5
4
3
2
1
–5
–4
–3
–2
–1
0
1
2
3
4
5
–1
–2
–3
–4
–5

6.

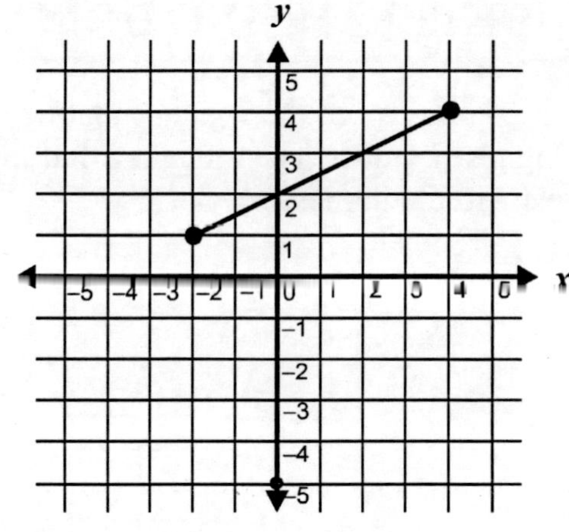
y
x
5
4
3
2
1
–5
–4
–3
–2
–1
0
–1
–2
–3
–4
–5

4.

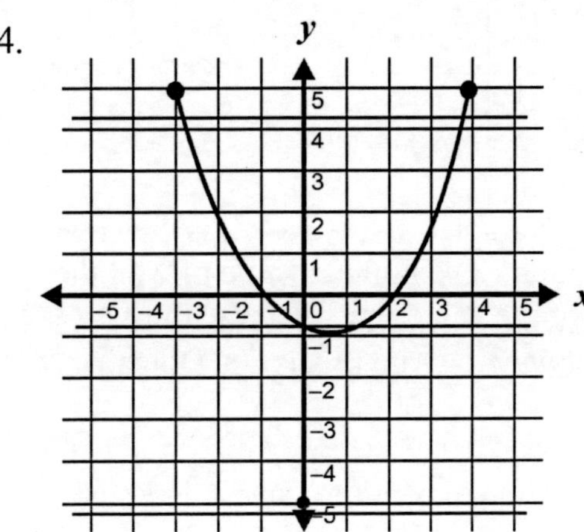
y
x
5
4
3
2
1
–5
–4
–3
–2
–1
0
1
2
3
4
5
–1
–2
–3
–4
–5

7.

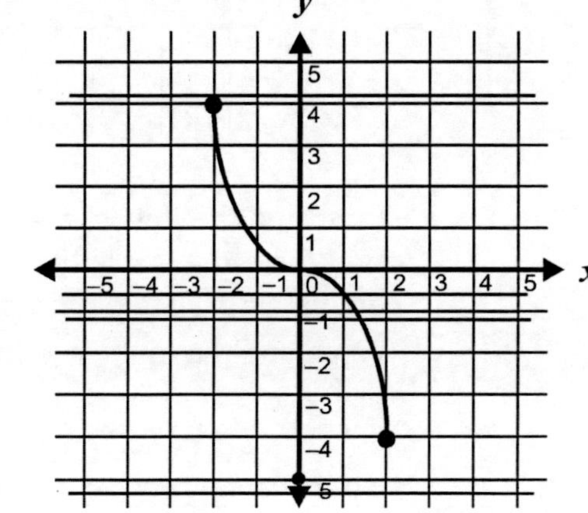
y
x
5
4
3
2
1
–5
–4
–3
–2
–1
0
1
2
3
4
5
–1
–2
–3
–4
–5

5.

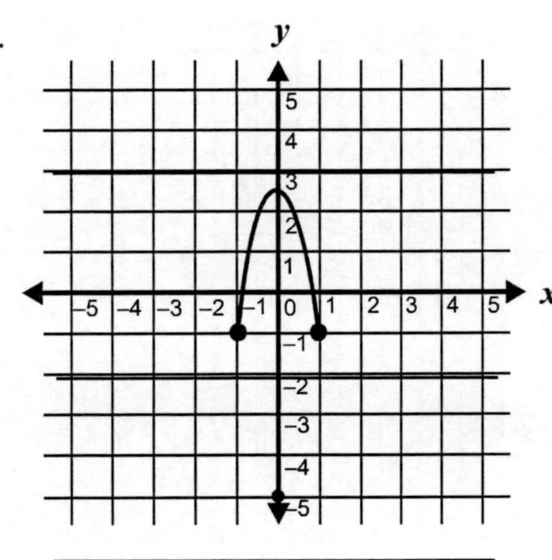
y
x
5
4
3
2
1
–5
–4
–3
–2
–1
0
1
2
3
4
5
–1
–2
–3
–4
–5

8.

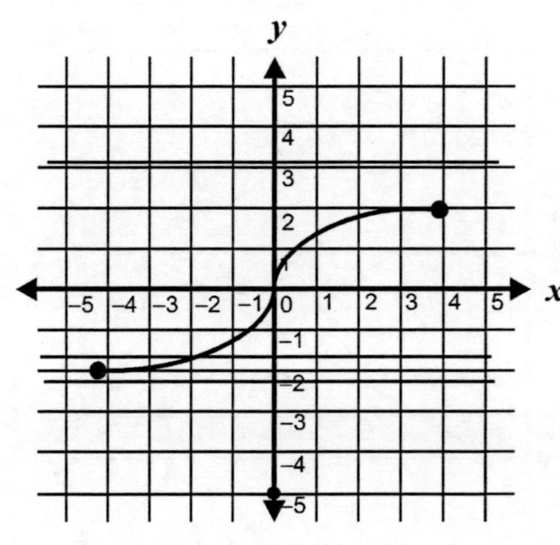
y
x
5
4
3
2
1
–5
–4
–3
–2
–1
0
1
2
3
4
5
–1
–2
–3
–4
–5

DOMAIN AND RANGE OF QUADRATIC EQUATIONS

The **domain** of a quadratic equation is the set of independent variables, or x values, over which the equation is defined. The **range** is the set of y values for which an equation given in two variables, x and y, is satisfied. A quadratic equation in the form of $y = x^2$ is represented by the following graph:

$y = x^2$

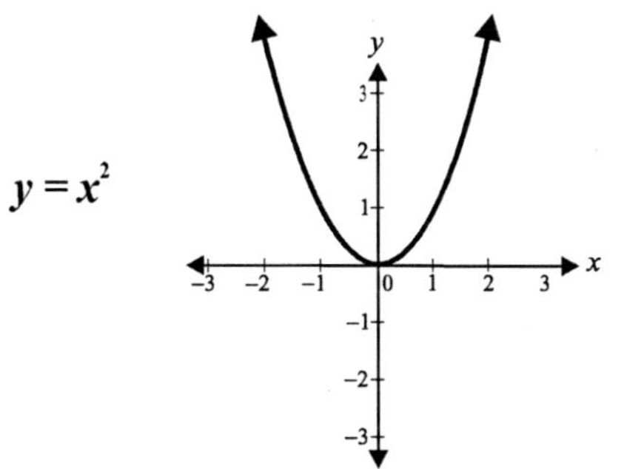

x	$y = x^2$
2	4
1	1
0	0
1	1
2	4
3	9

In this example the domain, or set of independent variables over which the equation is defined, will be all real numbers (positive and negative.) The range, however, will only include all **positive** real numbers. How would the graph be affected by multiplying x^2 by a constant, that is $y = ax^2$? If 'a' is a positive number greater than 1, the graph will be the same shape but will be taller and thinner. For example, let $a = 2$:

$y = 2x^2$

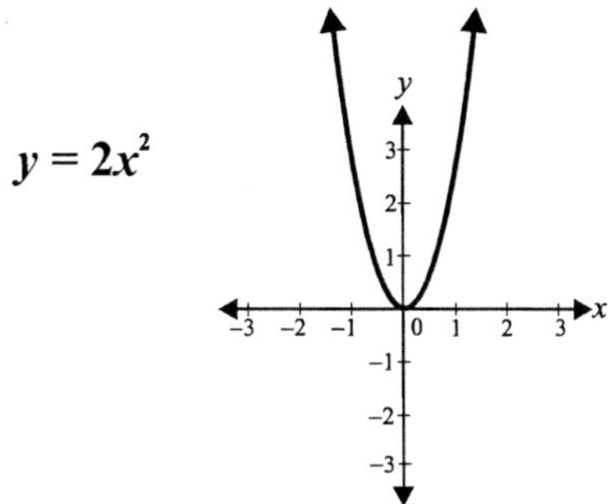

x	y
−2	8
−1	2
0	0
1	2
2	8
3	18

If 'a' is a negative number smaller than −1, the graph is the same shape (tall and thin), but is inverted. For example, let $a = -2$.

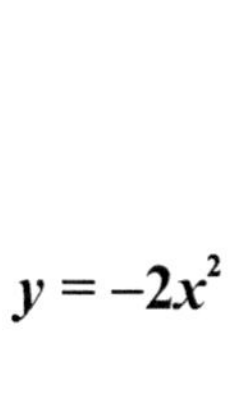

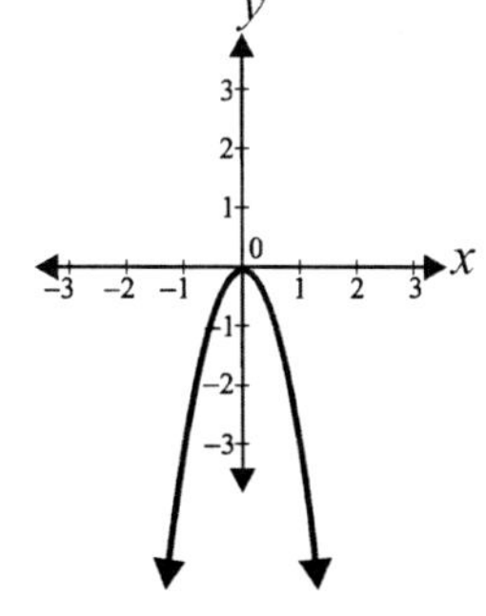

x	y
−2	−8
−1	−2
0	0
1	−2
2	−8
3	− 18

Using the same logic, you can see that when $0 < a < 1$ or $-1 < a < 0$, the graph will widen and flatten as shown in the figures below:

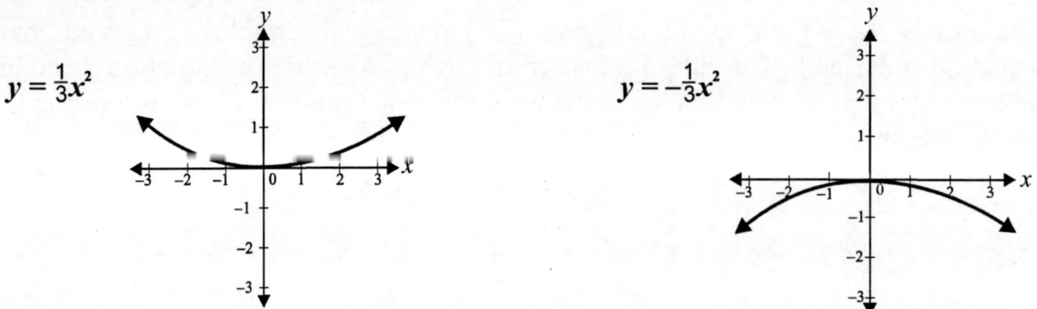

When a constant '*c*' is added to the equation, then the graph is shifted up (or down) the *y*-axis by the constant amount '*c*':

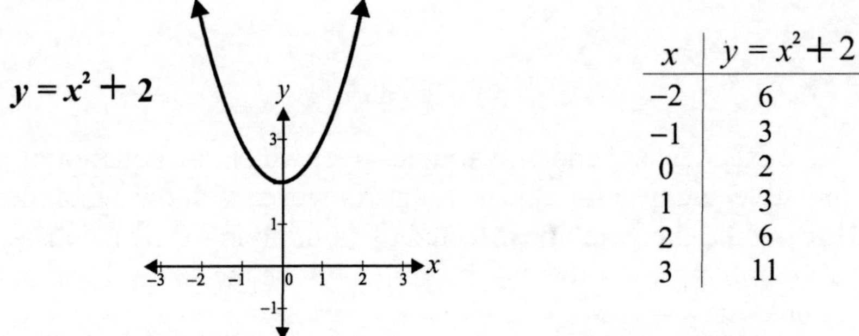

x	$y = x^2 + 2$
−2	6
−1	3
0	2
1	3
2	6
3	11

The magnitude of the constant '*a*' determines the width of the curve, the sign of '*a*' determines the orientation of the curve, and the constant '*c*' determines the *y*-intercept.

Roots of the Quadratic Equation

When factoring quadratic equations, the answer is often left in the form of two factors in parentheses multiplied together. These factors are called the **roots** of the quadratic equation. For example, the quadratic equation $2x^2 - 11x + 12 = 0$ can be factored as $(2x - 3)(x - 4) = 0$. In this example, $(2x - 3)$ and $(x - 3)$ are the **roots** of the equation. To find the **solution** or **solution set** to the equation, each of these roots must be set equal to zero, and then solved for x. In this case, the solutions are:

$$\begin{array}{r} 2x - 3 = 0 \\ +3\ +3 \\ \hline \frac{2x}{2} = \frac{3}{2} \end{array} \qquad \begin{array}{r} x - 4 = 0 \\ +4\ +4 \\ \hline x = 4 \end{array}$$

$x = \frac{3}{2}$ and $x = 4$ are the solutions of the equation.

The solution set $\{\frac{3}{2}, 4\}$ of the equation is derived from the roots of the equation. The solution(s) will satisfy the original equation when substituted and simplified.

Answer the following questions about the quadratic equation graphs.

1. Which of the following graphs has the largest value of a in $y = ax^2$?

A.

B.

C.

D.

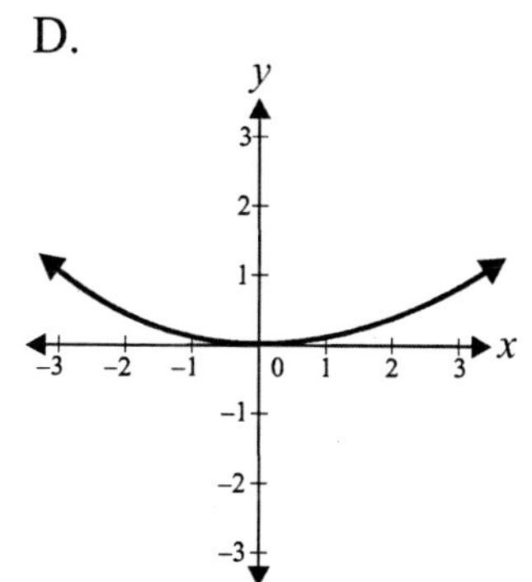

2. Fill in the tables and draw the graphs for the following equations:

A.

x	$y = -2x^2 + 2$
−3	
−2	
−1	
0	
1	
2	
3	

B.

x	$y = -\frac{1}{3}x^2 - 2$
−3	
−2	
−1	
0	
1	
2	
3	

What is the domain and range of each of these graphs? What is the y-intercept in each of the equations?

3. Identify the domain, range, and y-intercept for the following equations. Constructing a table and graph on another sheet of paper would be helpful.

A. $y = x^2 - 4$

B. $y = 4x^2 - 3$

C. $y = -\frac{1}{5}x^2 - 5$

D. $y = -2x^2 - 2$

E. $y = \frac{1}{2}x^2 + 3$

F. $y = \frac{1}{x^2}$

G. $\frac{y}{2} - 2 = 2x^2$

H. $7 + y = x^2$

I. $4 = x^2 + y$

4. Factor the quadratic equation $6x^2 + x - 2 = 0$. Show the roots of the equation and find the the solution set. Substitute one of the roots into the original equation, and verify that it does solve the equation.

DOMAIN AND RANGE OF ABSOLUTE VALUE EQUATIONS

The **domain** of an absolute value equation is the set of independent variables, or x values, over which the equation is defined. The **range** is the set of y values for which an equation given in two variables, x and y, is satisfied. An absolute value equation in the form of $y = |x|$ is represented by the following graph:

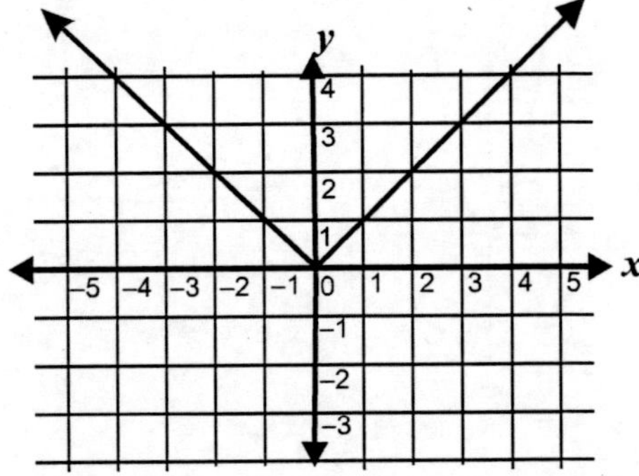

x	y
−2	2
−1	1
0	0
1	1
2	2
3	3

In this example the domain, or set of independent variables over which the equation is defined, will be all real numbers (positive and negative.) The range, however, will only include all **positive** real numbers. How would the graph be affected by multiplying x by coefficient such as 2 as in $y = |2x|$?

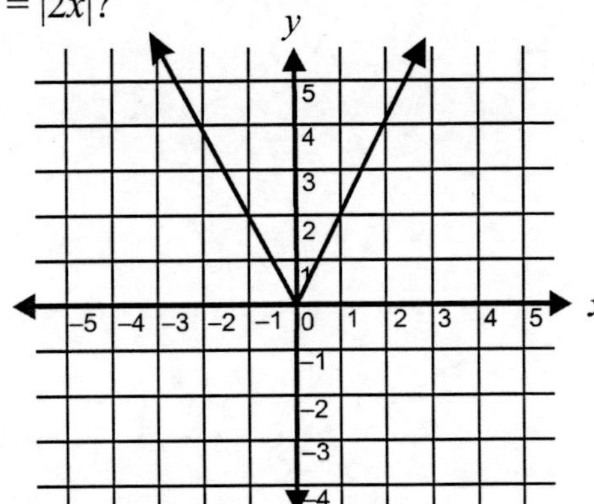

x	y
−2	4
−1	2
0	0
1	2
2	4

The domain remains the same, all positive and negative real numbers. The range remains the same, all positive real numbers.

What would happen if a constant such as 3 was added to the absolute value of x as in $y = |x + 3|$? How would the graph be affected?

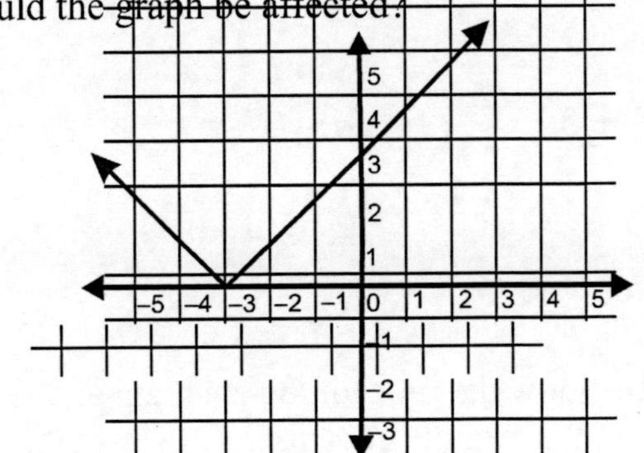

x	y
0	3
−1	2
−2	1
−3	0
−4	1
1	4
2	5

The domain remains the same, all positive and negative real numbers. The range remains the same, all positive real numbers.

For each of the following absolute value equations, graph the equation and state the domain and range.

1. $y = |x + 2|$

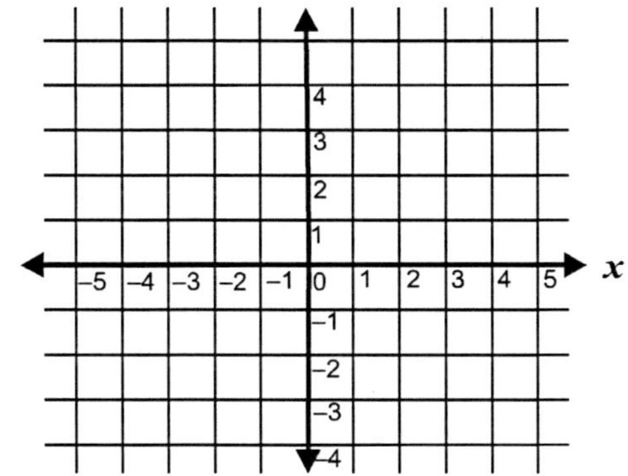

2. $y = |x - 1|$

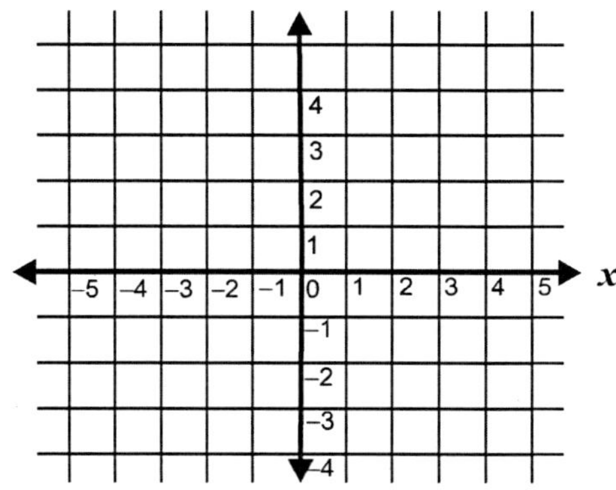

3. $y = |3x|$

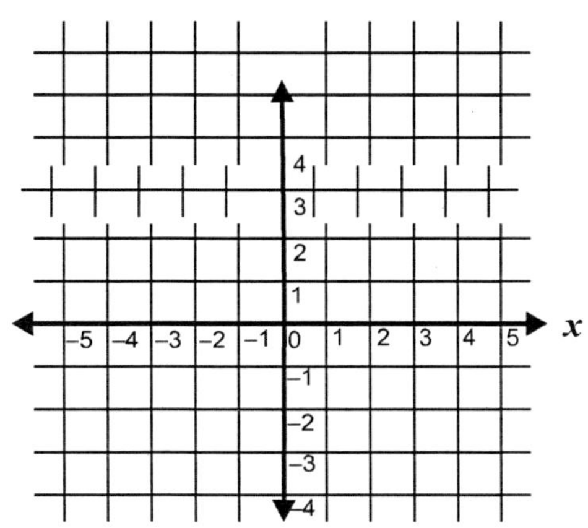

4. $y = |2x + 2|$

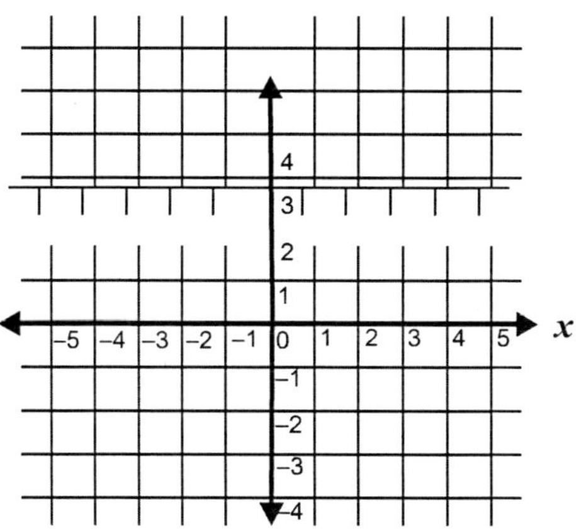

5. $y = -|x + 2|$

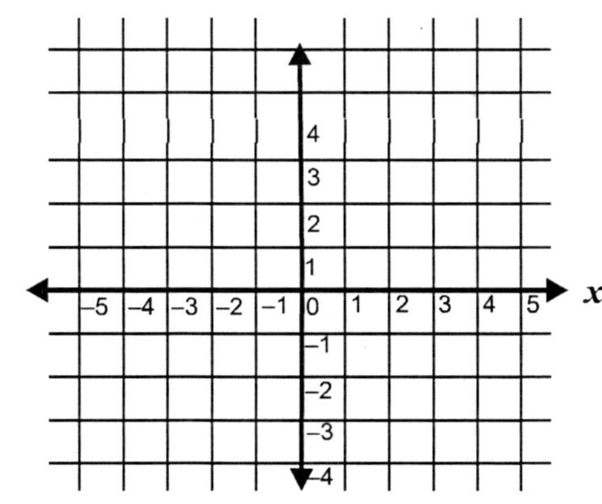

6. $y = -|2x|$

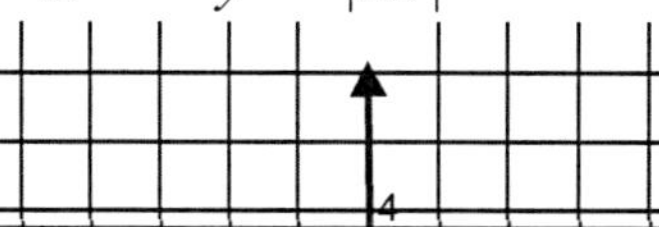
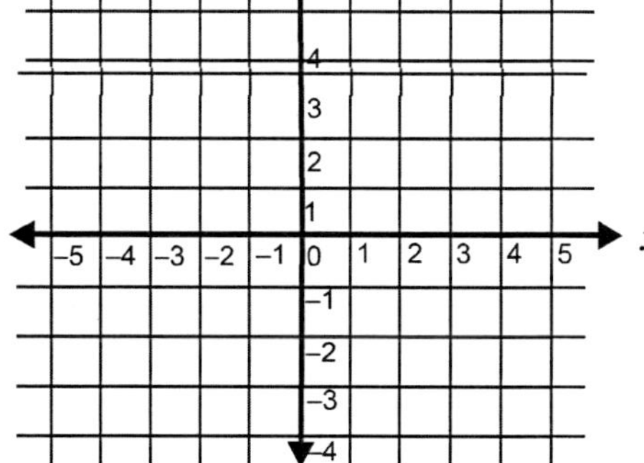

FUNCTIONS

Some relations are also **functions**. A relation is a function if **for every element in the domain, there is exactly one element in the range**. In other words, for each value for x there is only one unique value for y.

EXAMPLE 1: {(2, 4), (2, 5), (3, 4)} is **NOT** a function because in the first pair, 2 is paired with 4, and in the second pair, 2 is paired with 5. The 2 can be paired with only one number to be a function. In this example, the x value of 2 has more than one value for y: 4 and 5.

EXAMPLE 2: {(1, 2), (3, 2), (5, 6)} **IS** a function. Each first number is paired with only one second number. The 2 is repeated as a second number, but the relation remains a function.

Determine whether the ordered pairs of numbers below represent a function. Write "F" if it is a function. Write "NF" if it is not a function.

1. {(−1, 1), (−3, 3), (0, 0), (2, 2)} _______
2. {(−4, −3), (−2, −3), (−1,−3), (2, −3)} _______
3. {(5, −1), (2, 0), (2, 2), (5, 3)} _______
4. {(−3, 3), (0, 2), (1, 1), (2, 0)} _______
5. {(−2, −5), (−2, −1), (−2, 1), (−2, 3)} _______
6. {(0, 2), (1, 1), (2, 2), (4, 3)} _______
7. {(4, 2), (3, 3), (2, 2), (0, 3)} _______
8. {(−1, −1), (−2, −2), (3, −1), (3, 2)} _______
9. {(2, −2), (0, −2), (−2, 0), (1, −3)} _______
10. {(2, 1), (3, 2), (4, 3), (5, −1)} _______
11. {(−1, 0), (2, 1), (2, 4), (−2, 2)} _______
12. {(1, 4), (2, 3), (0, 2), (0, 4)} _______
13. {(0, 0), (1, 0), (2, 0), (3, 0)} _______
14. {(−5, −1), (−3, −2), (−4, −9), (−7, −3)} _______
15. {(8, −3), (−4, 4), (8, 0), (6, 2)} _______
16. {(7, −1), (4, 3), (8, 2), (2, 8)} _______
17. {(4, −3), (2, 0), (5, 3), (4, 1)} _______
18. {(2, −6), (7, 3), (−3, 4), (2, −3)} _______
19. {(1, 1), (3, −2), (4, 16), (1, −5)} _______
20. {(5, 7), (3, 8), (5, 3), (6, 9)} _______

FUNCTION NOTATION

Function notation is used to represent relations which are functions. Some commonly used letters to represent functions include *f*, *g*, *h*, *F*, *G*, and *H*.

EXAMPLE 1: $f(x) = 2x - 1$; find $f(-3)$

Find $f(-3)$ means replace x with -3 in the relation $2x - 1$.

$f(-3) = 2(-3) - 1$

$f(-3) = -6 - 1 = -7$

Solution: $f(-3) = -7$

EXAMPLE 2: $g(x) = 4 - 2x^2$: find $g(2)$

$g(2) = 4 - 2(2)^2 = 4 - 2(4) = 4 - 8 = -4$

Solution: $g(2) = -4$

Find solutions for each of the following:

1. $F(x) = 2 + 3x^2$; find $F(3)$
2. $f(x) = 4x + 6$; find $f(-4)$
3. $H(x) = 6 - 2x^2$; find $H(-1)$
4. $g(x) = -3x + 7$; find $g(-3)$
5. $f(x) = -5 + 4x$; find $f(7)$
6. $G(x) = 4x^2 + 4$; find $G(0)$
7. $f(x) = 7 - 6x$; find $f(-4)$
8. $h(x) = 2x^2 + 10$; find $h(5)$
9. $F(x) = 7 - 5x$; find $F(2)$
10. $f(x) = -4x^2 + 5$; find $f(-2)$

RECOGNIZING FUNCTIONS

Recall that a relation is a function with only one y value for every x value. Functions can be depicted in many ways including through graphs.

EXAMPLE 1:

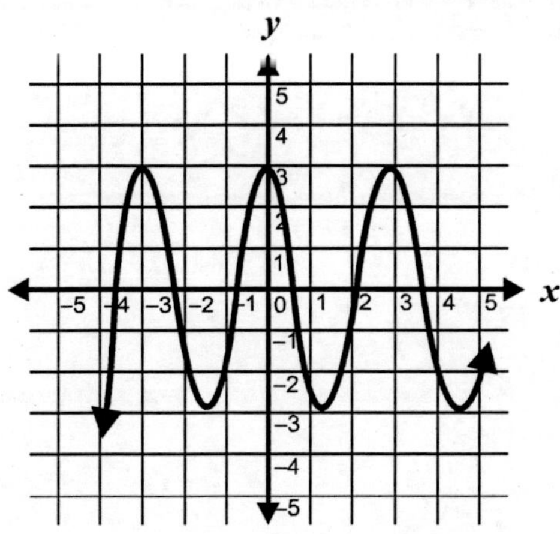

This graph **IS** a function because it has only one y value for each value of x.

EXAMPLE 2:

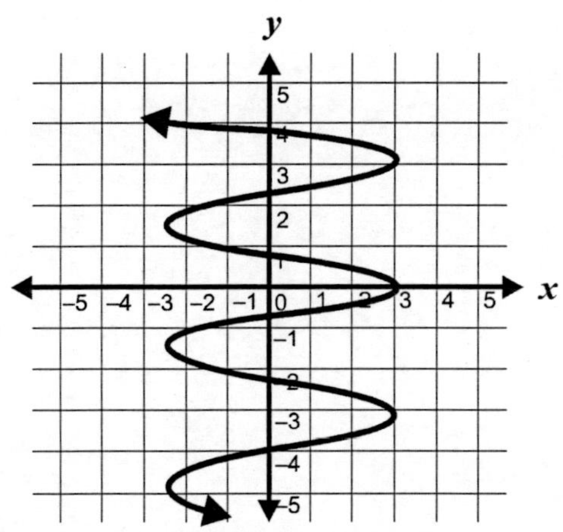

This graph is **NOT** a function because there are more than one y value for each value of x.

HINT: An easy way to determine a function from a graph is to do a vertical line test. First, draw a vertical line that crosses over the whole graph. If the line crosses the graph more than one time, then it is not a function. If it only crosses it once, it is a function. Take the example above.

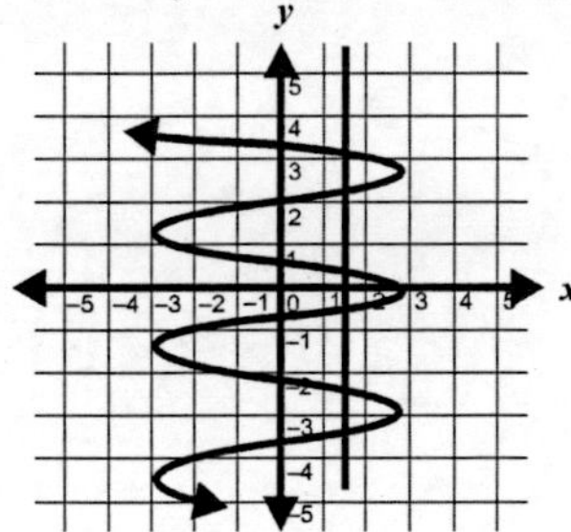

Since the vertical line passes over the graph six times, it is not a function.

Determine whether or not each of the following graphs is a function. If it is, write function on the line provided. If it is not a function, write NOT a function on the line provided.

1. 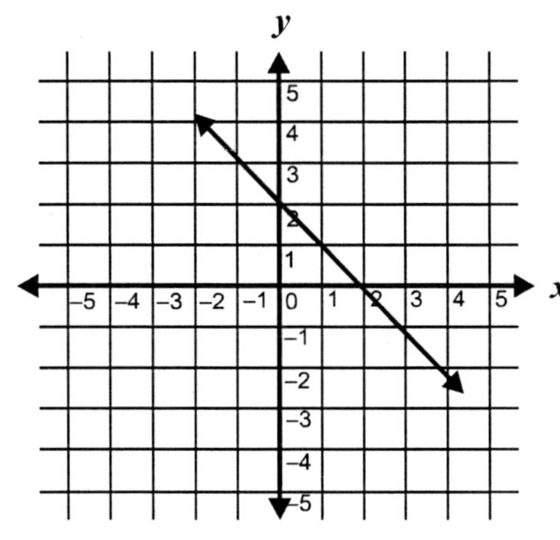

____

4. 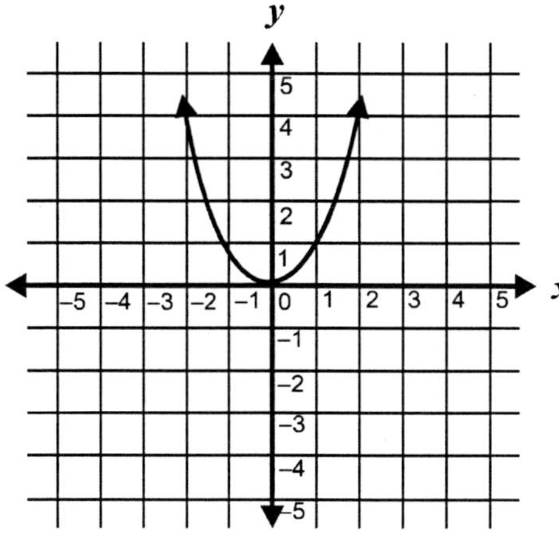

____

2. 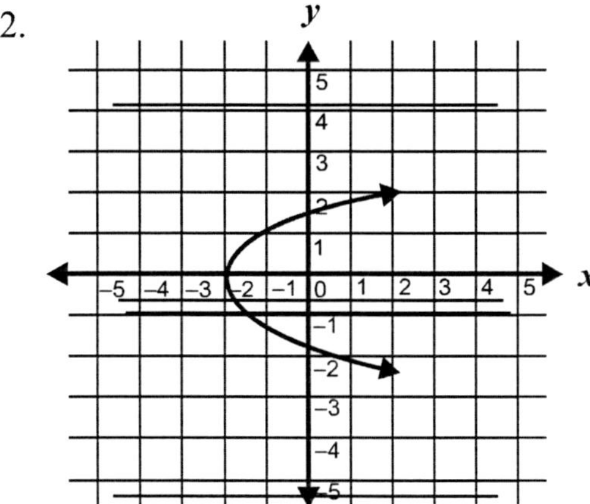

____

5. 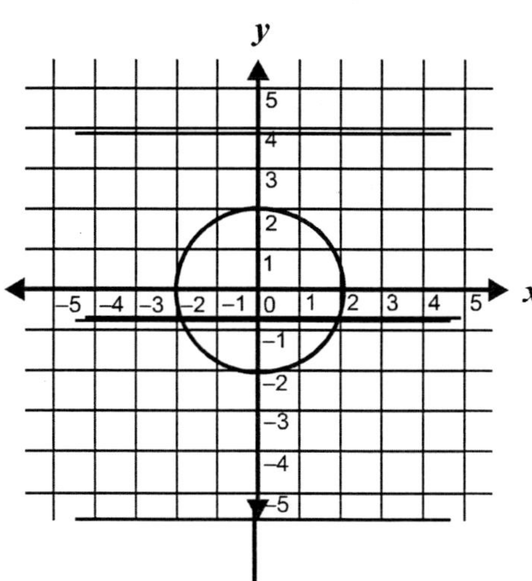

____

3. 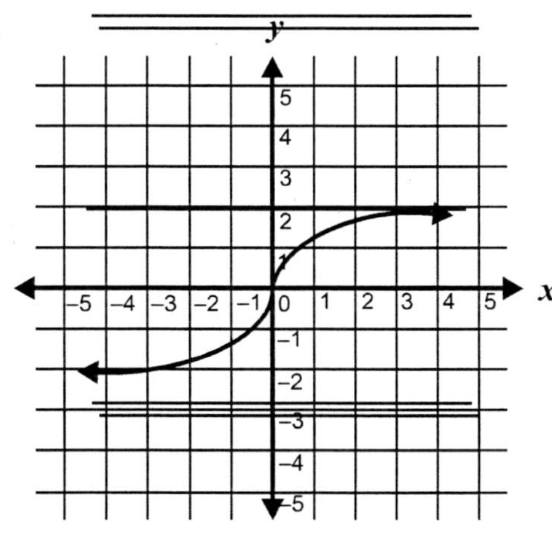

____

6. 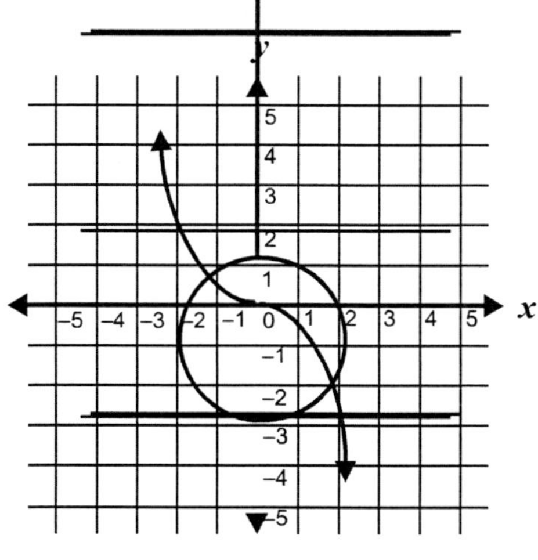

____

CHAPTER 10 REVIEW

1. What is the domain of the following relation?
   {(–1, 2), (2, 5), (4, 9), (6, 11)}

2. What is the range of the following relation?
   {(0, –2), (–1, –4), (–2, 6), (–3, –8)}

3. Find the range of the relation $y = 5x$ for the domain {0, 1, 2, 3, 4}.

4. Find the values of $M(y)$ of the relation $M(y) = 2(1.1)^y$ for the domain {2, 3, 4, 5, 6}.

5. Find the range of the following relation for the domain {0, 2, 6, 8, 10}.
   $B(t) = 600(.75)^t$

6. Find the range of the following relation for the domain {–8, –3, 7, 12, 17}.

   $y = \frac{3(x-2)}{5}$

   _____

7. Find the range of the following relation for the domain {–8, –4, 0, 4, 8}.

   $y = 10 - 2x$

   _____

8. Find the range of the following relation for the domain {–7, –1, 2, 5, 8}.

   $y = \frac{4+x}{3}$

For each of the following relations given in questions 9–14, write F if it is a function and NF if it is not a function.

9. {(0, –1), (–1, –2),(2, 4)} _____

10. {(1, 2), (2, 2), (3, 2)} _____

11. {(–1, 0), (0, 1), (1, 2), (2, 3)} _____

12. {(2, 1), (2, 2), (2, 3)} _____

13. {(1, 7), (2, 5), (3, 6), (2, 4)} _____

14. {(0, –1), (–1, –2), (–2,–3), (–3, –4)} _____

For questions 15–20, find the range of the following functions for the given value of the domain.

15. For $g(x) = 2x^2 - 4x$; find $g(-1)$

16. For $h(x) = 3x(x - 4)$; find $h(3)$

17. For $f(n) = \frac{1}{n+3}$; find $f(4)$

18. For $G(n) = \frac{2-n}{2}$; find $G(8)$

19. For $H(x) = 2x(x - 1)$; find $H(4)$

    _____

20. For $f(x) = 7x^2 + 3x - 2$, find $f(2)$

    _____

    _____

    _____

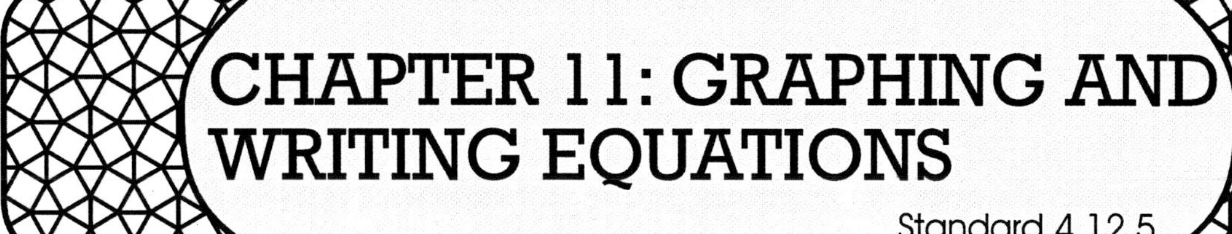

CHAPTER 11: GRAPHING AND WRITING EQUATIONS

Standard 4.12.5

GRAPHING LINEAR EQUATIONS

In addition to graphing ordered pairs, the Cartesian plane can be used to graph the solution set for an equation. Any equation with two variables that are both to the first power is called a **linear equation**. The graph of a linear equation will always be a straight line.

EXAMPLE 1: Graph the solution set for $x + y = 7$.

Step 1: Make a list of some pairs of numbers that will work in the equation.

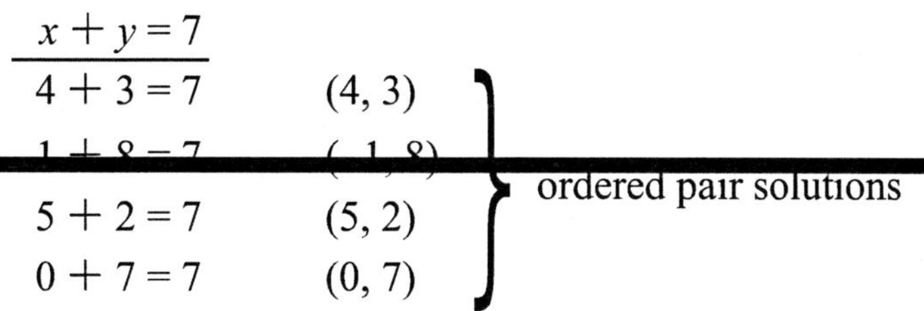

$x + y = 7$		
$4 + 3 = 7$	(4, 3)	ordered pair solutions
$-1 + 8 = 7$	(−1, 8)	
$5 + 2 = 7$	(5, 2)	
$0 + 7 = 7$	(0, 7)	

Step 2: Plot these points on a Cartesian plane.

Step 3: By passing a line through these points, we graph the solution set for $x + y = 7$.

This means that every point on this line is a solution to the equation $x + y = 7$. For example, (1, 6) is a solution; therefore, the line passes through the point (1, 6).

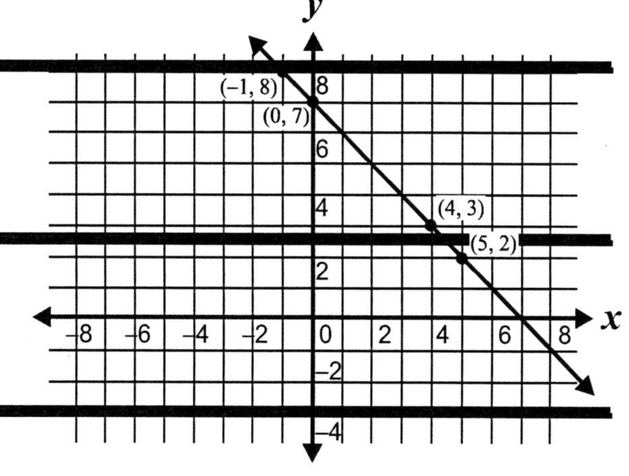

Make a table of solutions for each linear equation below. Then, plot the ordered pair solutions on graph paper. Draw a line through the points. (If one of the points does not line up, you have made a mistake.)

1. $x + y = 6$
2. $y = x + 1$
3. $y = x - 2$
4. $x + 2 = y$
5. $x - 5 = y$
6. $x - y = 0$

EXAMPLE 2: Graph the equation $y = 2x - 5$.

Step 1: This equation has 2 variables, both to the first power, so we know the graph will be a straight line.

Substitute some numbers for x or y to find pairs of numbers that satisfy the equation. For the above equation, it will be easier to substitute values of x in order to find the corresponding value for y. Record the values for x and y in a table.

	x	y
If x is 0, y would be –5	0	–5
If x is 1, y would be –3	1	–3
If x is 2, y would be –1	2	–1
If x is 3, y would be 1	3	1

Step 2: Graph the ordered pairs, and draw a line through the points.

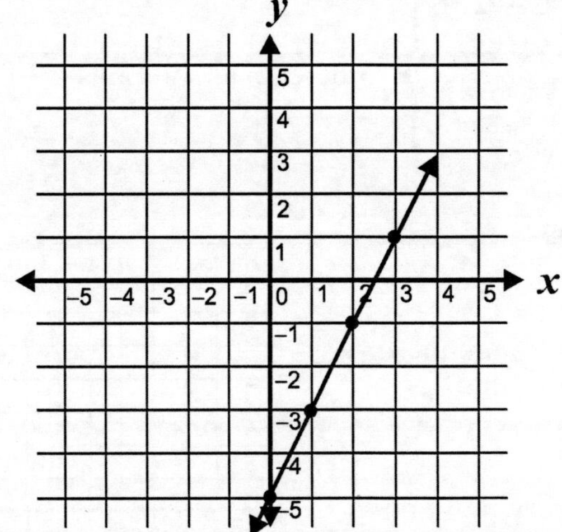

Find pairs of numbers that satisfy the equations below, and graph the line on graph paper.

1. $y = -2x + 2$
2. $2x - 2 = y$
3. $-x + 3 = y$
4. $y = x + 1$
5. $4x - 2 = y$
6. $y = 3x - 3$
7. $x = 4y - 3$
8. $2x = 3y + 1$
9. $x + 2y = 4$

GRAPHING HORIZONTAL AND VERTICAL LINES

The graph of some equations is a horizontal or a vertical line.

EXAMPLE 1: $y = 3$

Step 1: Make a list of ordered pairs that satisfy the equation $y = 3$.

x	y
0	3
1	3
2	3
3	3

No matter what value of x you choose, y is always 3.

Step 2: Plot these points on a Cartesian plane, and draw a line through the points.

The graph is a horizontal line.

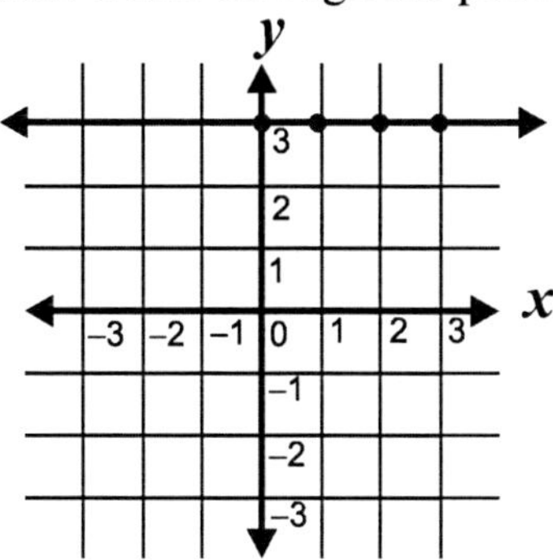

EXAMPLE 2: $2x + 3 = 0$

Step 1: For these equations with only one variable, find what x equals first.

$2x + 3 = 0$
$2x = -3$
$x = \frac{-3}{2}$

Step 2: Just like Example 1, find ordered pairs that satisfy the equation, plot the points, and graph the line.

x	y
$\frac{-3}{2}$	0
$\frac{-3}{2}$	1
$\frac{-3}{2}$	2
$\frac{-3}{2}$	3

No matter which value of y you choose, the value of x does not change.

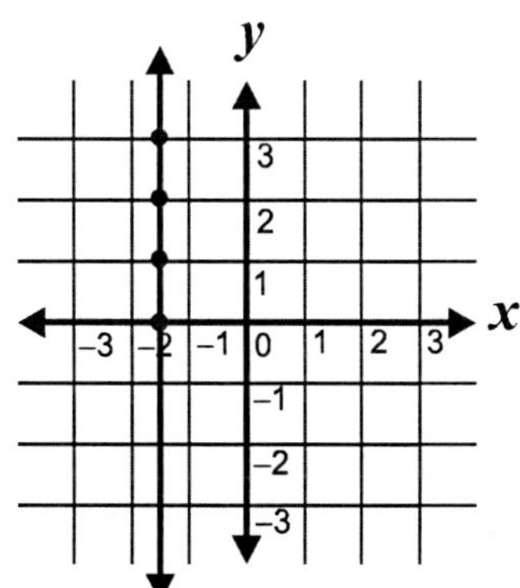

The graph is a vertical line.

Find pairs of numbers that satisfy the equations below, and graph the lines on graph paper.

1. $2y + 2 = 0$
2. $x = -4$
3. $3x = 3$
4. $y = 5$
5. $4x - 2 = 0$
6. $2x - 6 = 0$
7. $4y = 1$
8. $5x + 10 = 0$
9. $3y + 12 = 0$
10. $x + 1 = 0$
11. $2y - 8 = 0$
12. $3x = -9$
13. $x = -2$
14. $6y - 2 = 0$
15. $5x - 5 = 0$
16. $2y - 4 = 0$
17. $2y - 2 = 0$
18. $3x + 1 = 0$
19. $4y = -2$
20. $-2y = 6$
21. $-4x = -8$
22. $3y = -6$
23. $x = 2$
24. $4y = 8$

UNDERSTANDING SLOPE

The **slope** of a line refers to how steep a line is. Slope is also defined as the rate of change. When we graph a line using ordered pairs, we can easily determine the slope. Slope is often represented by the letter ***m***.

> The formula for slope of a line is: $m = \dfrac{y_2 - y_1}{x_2 - x_1}$ or $\dfrac{\text{rise}}{\text{run}}$

EXAMPLE 1: What is the slope of the following line that passes through the ordered pairs (–4, –3) and (1, 3)?

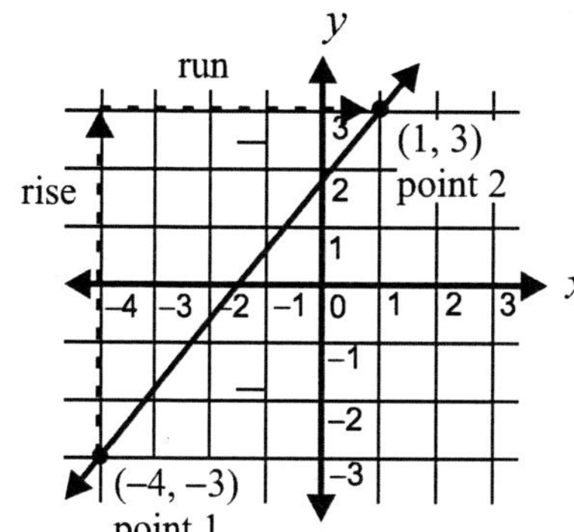

y_2 is 3, the y-coordinate of point 2.

y_1 is –3, the y-coordinate of point 1.

x_2 is 1, the x-coordinate of point 2.

x_1 is –4, the x-coordinate of point 1.

Use the formula for slope given above. $m = \dfrac{3-(-3)}{1-(-4)} = \dfrac{6}{5}$

The slope is $\frac{6}{5}$. This shows us that we can go up 6 (rise) and over 5 to the right (run) to find another point on the line.

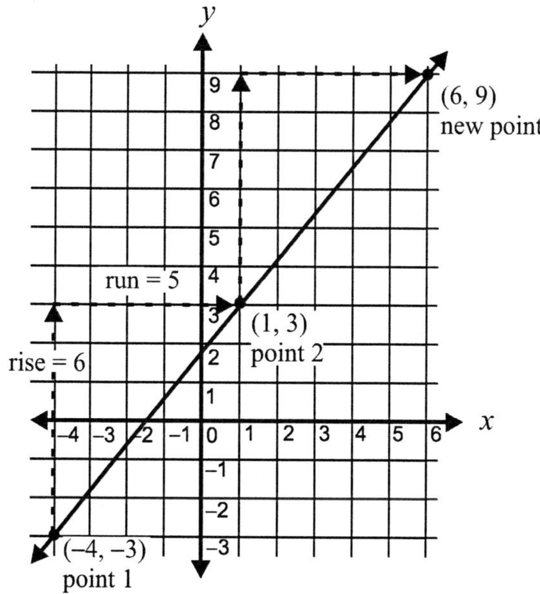

EXAMPLE 2: Find the slope of a line through the points (–2, 3) and (1, –2). It doesn't matter which pair we choose for point 1 and point 2. The answer is the same.

let point 1 be (–2, 3)
let point 2 be (1, –2)

$$\text{slope} = \frac{(y_2 - y_1)}{(x_2 - x_1)} = \frac{-2 - 3}{1 - (-2)} = \frac{-5}{3}$$

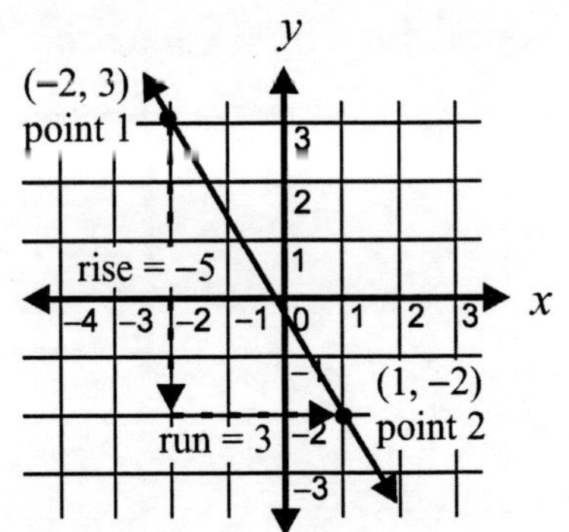

When the slope is negative, the line will slant left. For this example, the line will go **down** 5 units and then over 3 to the **right**.

EXAMPLE 3: What is the slope of a line that passes through (1, 1) and (3, 1)?

$$\text{slope} = \frac{1 - 1}{3 - 1} = \frac{0}{2} = 0$$

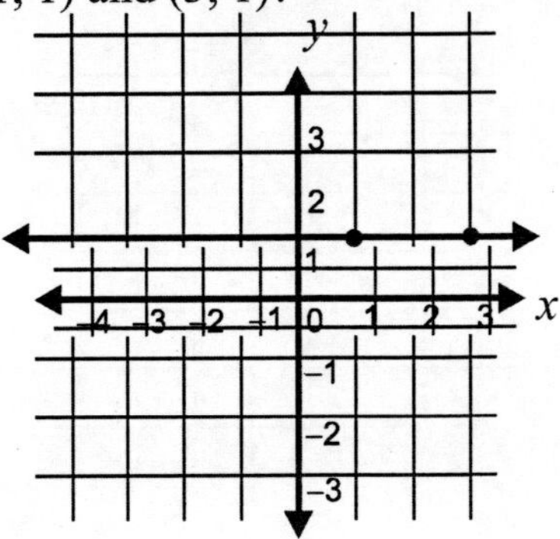

When $y_2 - y_1 = 0$, the slope will equal 0, and the line will be horizontal.

EXAMPLE 4: What is the slope of a line that passes through (2, 1) and (2, –3)?

$$\text{slope} = \frac{-3 - 1}{2 - 2} = \frac{4}{0} = \text{undefined}$$

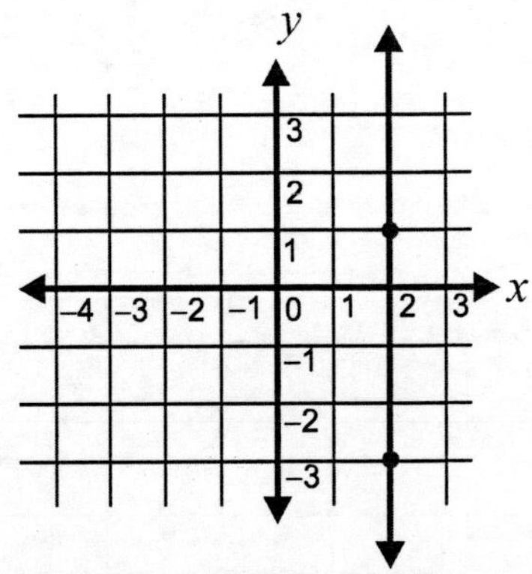

When $x_2 - x_1 = 0$, the slope is undefined, and the line will be vertical.

The following lines summarize what we know about slope.

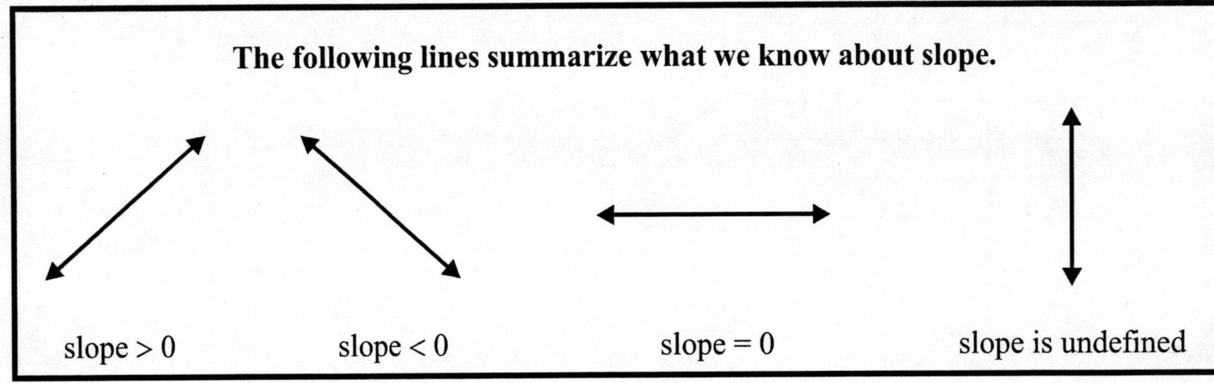

slope > 0 slope < 0 slope = 0 slope is undefined

Find the slope of the line that goes through the following pairs of points. Use the formula slope = $\frac{y_2 - y_1}{x_2 - x_1}$. Then, using graph paper, graph the line through the two points, and label the rise and the run. (See Examples 1 and 2.)

1. (2, 3) (4, 5)
2. (1, 3) (2, 5)
3. (−1, 2) (4, 1)
4. (1, −2) (4, −2)
5. (3, 0) (3, 4)
6. (3, 2) (−1, 8)
7. (4, 3) (2, 4)
8. (2, 2) (1, 5)
9. (3, 4) (1, 2)
10. (3, 2) (3, 6)
11. (6, −2) (3, −2)
12. (1, 2) (3, 4)
13. (−2, 1) (−4, 3)
14. (5, 2) (4, −1)
15. (1, −3) (−2, 4)
16. (2, −1) (3, 5)
17. (2, 4) (5, 3)
18. (5, 2) (2, 5)
19. (4, 5) (6, 6)
20. (2, 1) (−1, −3)

SLOPES OF PERPENDICULAR LINES

Once you have found the slope of a line, determining the slope of a line perpendicular to it is done in two simple steps:

1. multiply the slope by –1
2. invert the slope

EXAMPLE: The solid line on the graph at right has a slope of $\frac{3}{4}$.

Step 1: $(-1)\,\frac{3}{4} = -\frac{3}{4}$

Step 2: Taking the inverse of the negative Slope, $-\frac{3}{4}$, gives $-\frac{4}{3}$.

The dotted line on the graph at right has a slope of $-\frac{4}{3}$ and is perpendicular to the solid line with a slope of $\frac{3}{4}$.

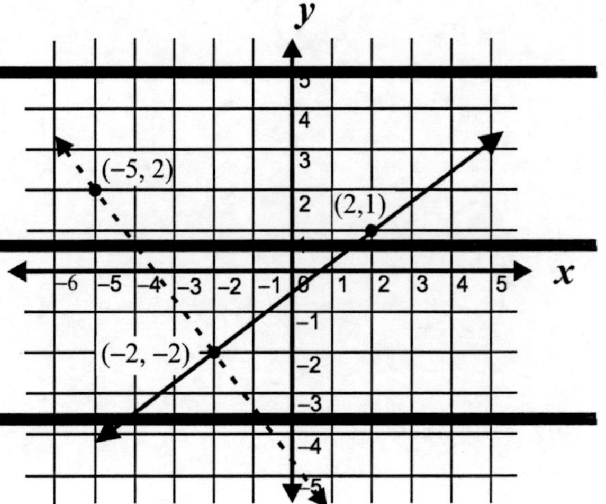

1. Find the slope of the line perpendicular to the following line, and draw it in as a dotted line:

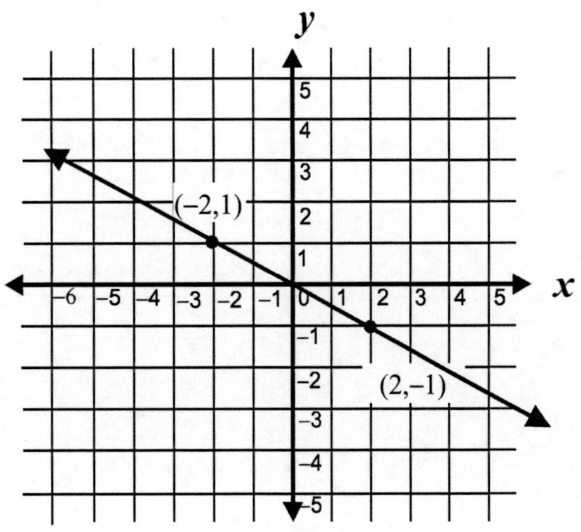

2. Find the slope of a line perpendicular to lines with the following slopes.

A. $-\frac{4}{5}$

B. $\frac{3}{2}$

C. $\frac{5}{7}$

D. $-\frac{6}{5}$

E. $-\frac{7}{9}$

F. $\frac{1}{4}$

G. -4

SLOPE OF PARALLEL LINES

Parallel lines have the same slope. Parallel lines have a different y-intercept. Knowing this, the line $y = \frac{-2}{3}x + 3$ would be parallel to the line above.

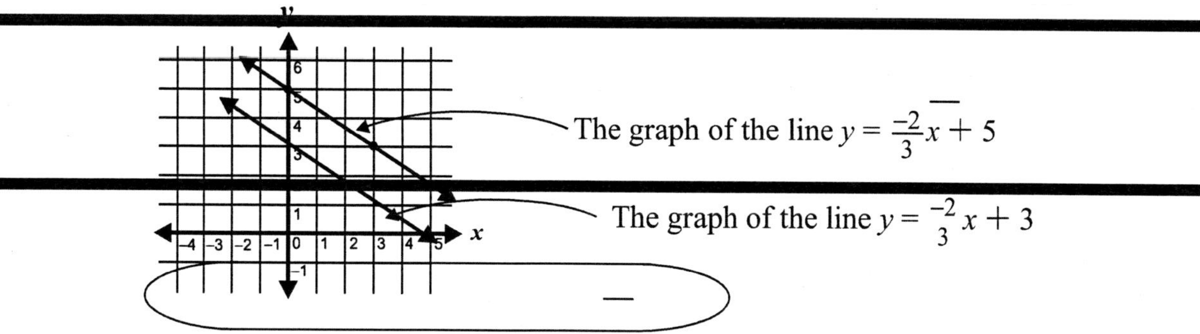

SLOPE-INTERCEPT FORM OF A LINE

An equation that contains two variables, each to the first degree, is a **linear equation**. The graph for a linear equation is a straight line. To put a linear equation in slope-intercept form, solve the equation for y. This form of the equation shows the slope and the y-intercept. Slope-intercept form follows the pattern of $y = mx + b$. The "m" represents slope, and the "b" represents the y-intercept. The y-intercept is the point at which the line crosses the y-axis.

When the slope of a line is not 0, the graph of the equation shows a **direct variation** between y and x. When y increases, x increases in a certain proportion. The proportion stays constant. The constant is called the **slope** of the line.

EXAMPLE: Put the equation $2x + 3y = 15$ in slope-intercept form. What is the slope of the line? What is the y-intercept? Graph the line.

Step 1: Solve for y:

$$2x + 3y = 15$$
$$-2x \qquad -2x$$
$$\frac{3y}{3} = \frac{-2x}{3} + \frac{15}{3}$$

slope-intercept form: $y = \frac{-2}{3}x + 5$

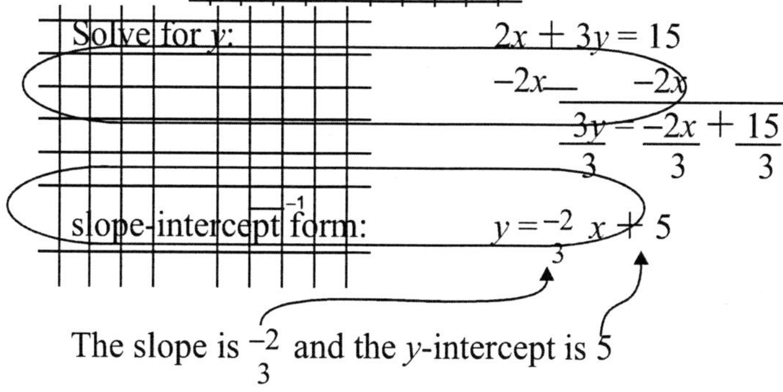

The slope is $\frac{-2}{3}$ and the y-intercept is 5

Step 2: Knowing the slope and the y-intercept, we can graph the line.

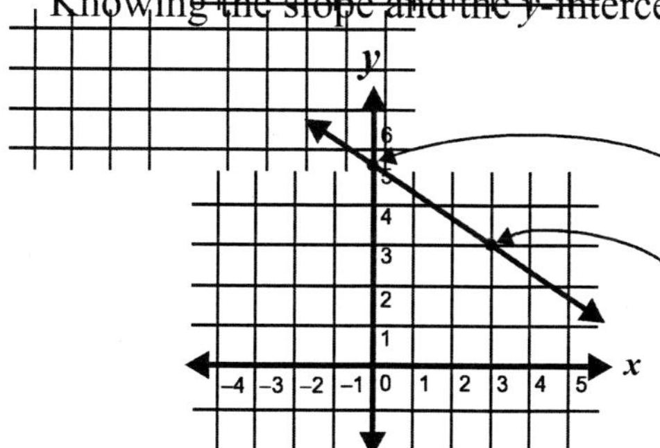

The y-intercept is 5, so the line passes through the point (0, 5) on the y-axis.

The slope is $\frac{-2}{3}$, so go down 2 and over 3 to the right to get a second point which is also a solution to the equation.

Put each of the following equations in slope-intercept form by solving for *y*. On your graph paper, graph the line using the slope and *y*-intercept.

1. $4x - y = 5$
2. $2x + 4y = 16$
3. $3x - 2y = 10$
4. $x + 3y = -12$
5. $6x + 2y = 0$
6. $8x - 5y = 10$
7. $-2x + y = 4$
8. $-4x + 3y = 12$
9. $-6x + 2y = 12$
10. $x - 5y = 5$
11. $3x - 2y = -10$
12. $3x + 4y = 2$
13. $-x = 2 + 4y$
14. $2x = 4y - 2$
15. $6x - 3y = 9$
16. $4x + 2y = 8$
17. $6x - y = 4$
18. $-2x - 4y = 8$
19. $5x + 4y = 16$
20. $6 = 2y - 3x$

VERIFY THAT A POINT LIES ON A LINE

To know whether or not a point lies on a line, substitute the coordinates of the point into the formula for the line. If the point lies on the line, the equation will be true. If the point does not lie on the line, the equation will be false.

EXAMPLE 1: Does the point (5, 2) lie on the line given by the equation $x + y = 7$?

Solution: Substitute 5 for x and 2 for y in the equation. $5 + 2 = 7$. Since this is a true statement, the point (5, 2) does lie on the line $x + y = 7$.

EXAMPLE 2: Does the point (0, 1) lie on the line given by the equation $5x + 4y = 16$?

Solution: Substitute 0 for x and 1 for y in the equation. $5x + 4y = 16$. Does $5(0) + 4(1) = 16$? No, it equals 4, not 16. Therefore, the point (0, 1) is not on the line given by the equation $5x + 4y = 16$.

For each point below, state whether or not it lies on the line given by the equation that follows the point coordinates.

1. (2, 4) $6x - y = 8$
2. (1, 1) $6x - y = 5$
3. (3, 8) $-2x + y = 2$
4. (9, 6) $-2x + y = 0$
5. (3, 7) $x - 5y = -32$
6. (0, 5) $-6x - 5y = 3$
7. (2, 4) $4x + 2y = 16$
8. (9, 1) $3x - 2y = 29$
9. (6, 8) $6x - y = 28$
10. (−2, 3) $x + 2y = 4$
11. (4, −1) $-x - 3y = -1$
12. (−1, −3) $2x + y = 1$

GRAPHING A LINE KNOWING A POINT AND SLOPE

If you are given a point of a line and the slope of a line, the line can be graphed.

EXAMPLE 1: Given that line l has a slope of $\frac{4}{3}$ and contains the point (2, –1), graph the line.

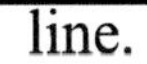

Step 1: Plot and label the point (2, –1) on a Cartesian plane.

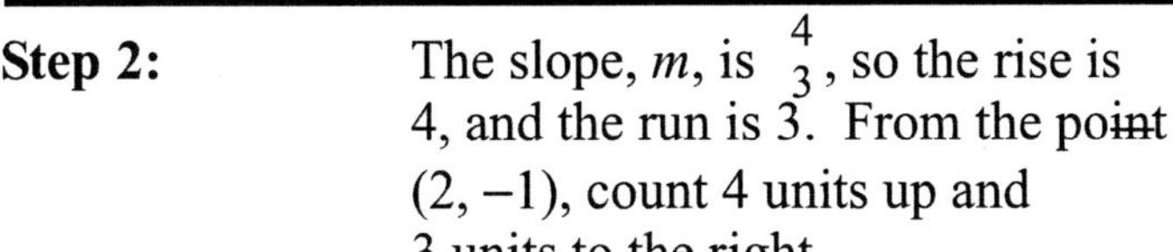

Step 2: The slope, m, is $\frac{4}{3}$, so the rise is 4, and the run is 3. From the point (2, –1), count 4 units up and 3 units to the right.

Step 3: Draw the line through the two points.

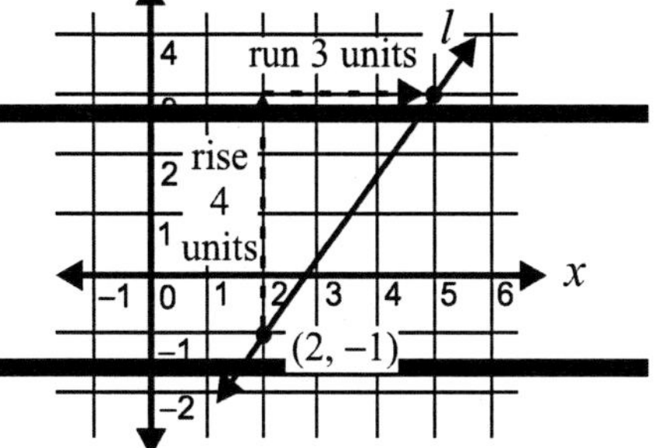

EXAMPLE 2: Given a line that has a slope of $\frac{-1}{4}$ and passes through the point (–3, 2), graph the line.

Step 1: Plot the point (–3, 2).

Step 2: Since the slope is negative, go **down** 1 unit and over 4 to get a second point.

Step 3: Graph the line through the two points.

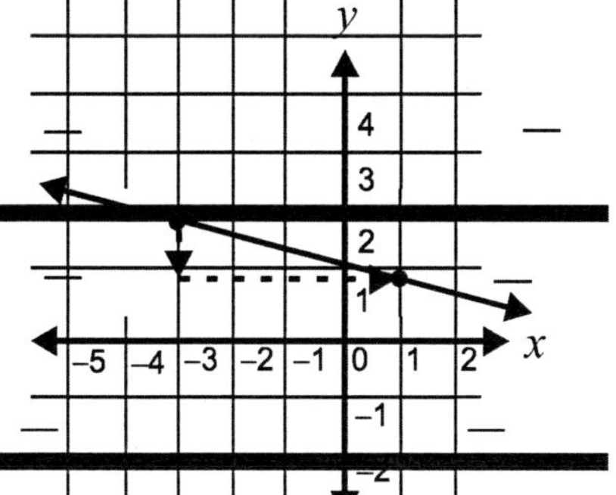

Graph a line on your own graph paper for each of the following problems. First, plot the point. Then, use the slope to find a second point. Draw the line formed from the point and the slope.

1. (2, –2), $m = \frac{3}{4}$
2. (3, –4), $m = \frac{1}{2}$
3. (1, 3), $m = \frac{-1}{3}$
4. (2, –4), $m = 1$
5. (3, 0), $m = \frac{-1}{2}$
6. (–2, 1), $m = \frac{4}{3}$
7. (–4, –2), $m = \frac{1}{2}$
8. (1, –4), $m = \frac{3}{4}$
9. (2, –1), $m = \frac{-1}{2}$
10. (5, –2), $m = \frac{1}{4}$
11. (–2, –3), $m = \frac{2}{3}$
12. (4, –1), $m = \frac{-1}{3}$
13. (–1, 5), $m = \frac{2}{5}$
14. (–2, 3), $m = \frac{3}{4}$
15. (4, 4), $m = \frac{-1}{2}$
16. (3, –3), $m = \frac{-3}{4}$
17. (–2, 5), $m = \frac{1}{3}$
18. (–2, –3), $m = \frac{-3}{4}$
19. (4, –3), $m = \frac{2}{3}$
20. (1, 4), $m = \frac{-1}{2}$

FINDING THE EQUATION OF A LINE USING TWO POINTS OR A POINT AND SLOPE

If you can find the slope of a line and know the coordinates of one point, you can write the equation for the line. You know the formula for the slope of a line is:

$$m = \frac{y_2 - y_1}{x_2 - x_1} \text{ or } \frac{y_2 - y_1}{x_2 - x_1} = m$$

Using algebra, you can see that if you multiply both sides of the equation by $x_2 - x_1$, you get:

$$y - y_1 = m(x - x_1) \longleftarrow \text{point-slope form of an equation}$$

EXAMPLE: Write the equation of the line passing through the points (–2, 3) and (1, 5).

Step 1: First, find the slope of the line using the two points given.

$$m = \frac{y_2 - y_1}{x_2 - x_1} = \frac{5 - 3}{1 - (-2)} = \frac{2}{3}$$

Step 2: Pick one of the points to use in the point-slope equation. For point (–2, 3), we know $x_1 = -2$ and $y_1 = 3$, and we know $m = \frac{2}{3}$. Substitute these values into the point-slope form of the equation.

$$y - y_1 = m(x - x_1)$$

$$y - 3 = \tfrac{2}{3}[x - (-2)]$$

$$y - 3 = \tfrac{2}{3}x + \tfrac{4}{3}$$

$$y = \tfrac{2}{3}x + \tfrac{13}{3}$$

Use the point-slope formula to write an equation for each of the following lines.

1. $(1, -2),\ m = 2$
2. $(-3, 3),\ m = \frac{1}{3}$
3. $(4, 2),\ m = \frac{1}{4}$
4. $(5, 0),\ m = 1$
5. $(3, -4),\ m = \frac{1}{2}$
6. $(-1, 4)\ (2, -1)$
7. $(2, 1)\ (-1, -3)$
8. $(-2, 5)\ (-4, 3)$
9. $(-4, 3)\ (2, -1)$
10. $(3, 1)\ (5, 5)$
11. $(-3, 1),\ m = 2$
12. $(-1, -2),\ m = \frac{4}{3}$
13. $(2, -5),\ m = -2$
14. $(-1, 3),\ m = \frac{1}{3}$
15. $(0, -2),\ m = -\frac{3}{2}$

IDENTIFYING GRAPHS OF LINEAR EQUATIONS

Match each equation below with the graph of the equation.

A:	$x = y$	D:	$y = 4$	G:	$x = -2y$
B:	$x = -4$	E:	$x = \frac{1}{2}y$	H:	$4x = y$
C:	$y = -x$	F:	$y = -4$	I:	$-2x = y$

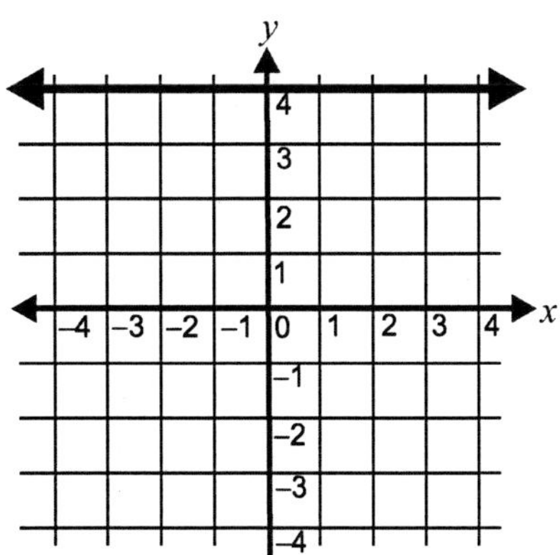

1. ________

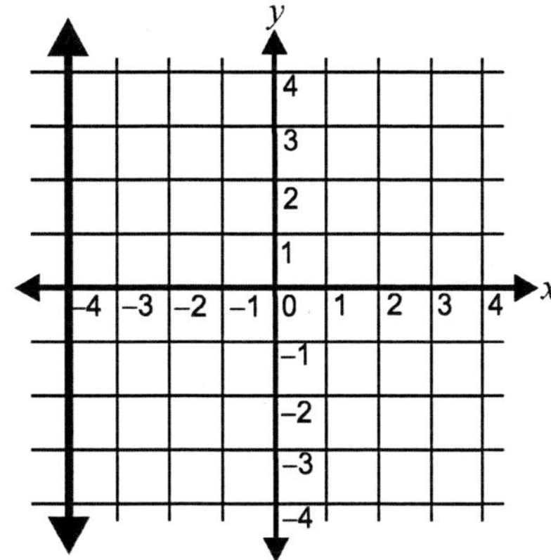

2. ________

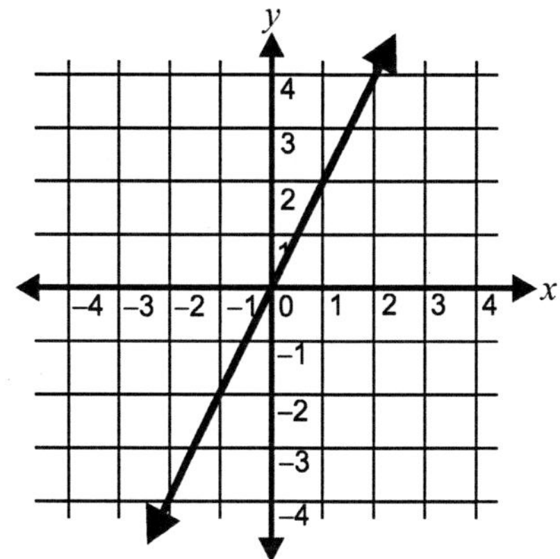

3. ________

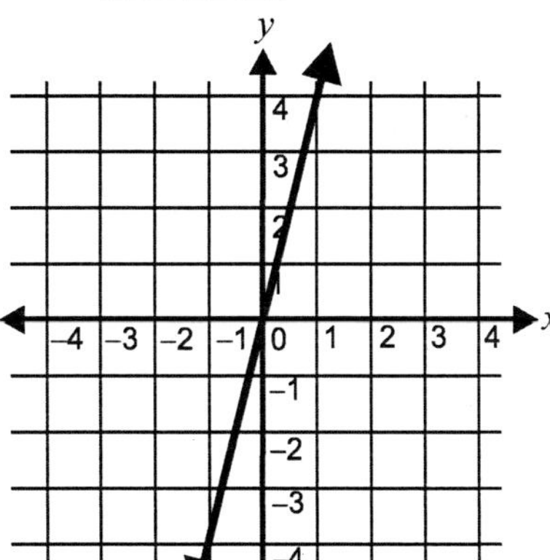

4. ________

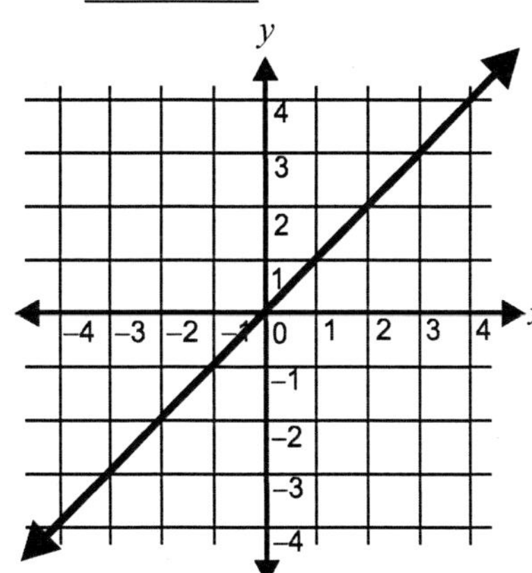

5. ________

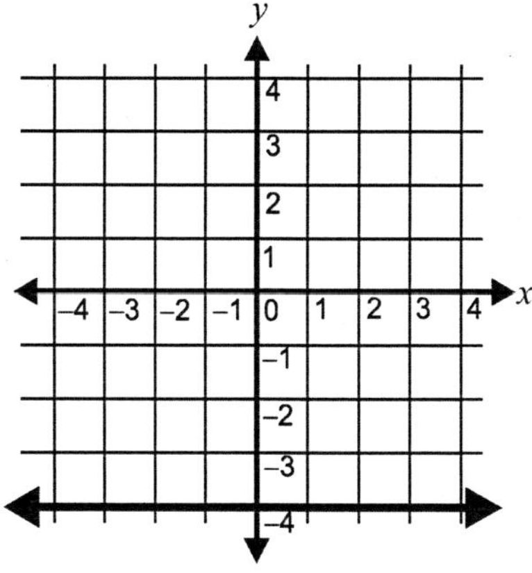

6. ________

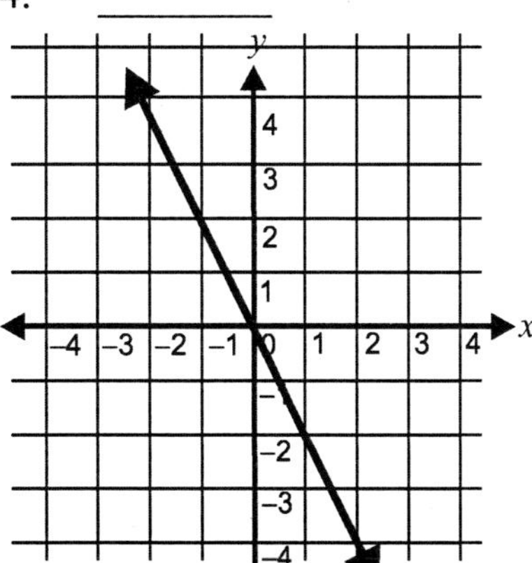

7. ________

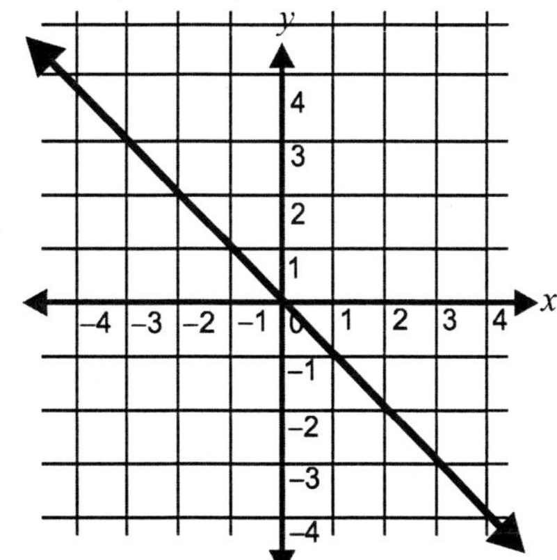

8. ________

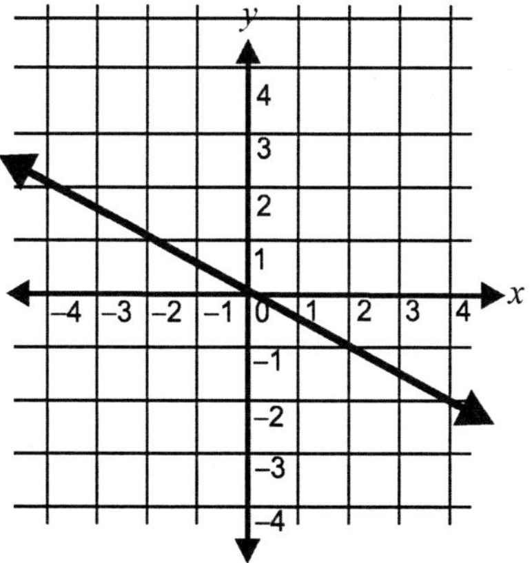

9. ________

Match each equation below with the graph of the equation.

A:	$y = 4x$	D:	$x - 2y = 6$	G:	$y = 3x + 2$
B:	$y = -4x$	E:	$y = 3x - 1$	H:	$x + 2y = 6$
C:	$4x + y = 4$	F:	$2x + 3y = 6$	I:	$y = x - 3$

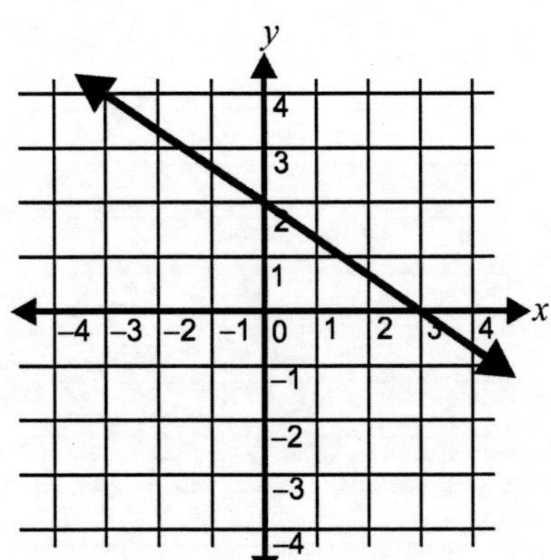

1. ________

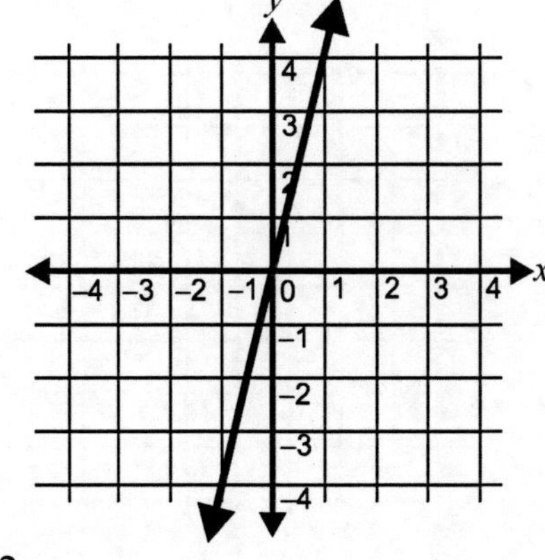

2.  ________

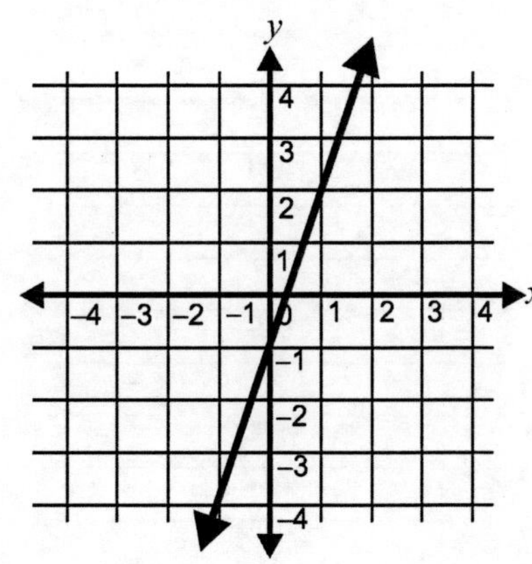

3.  ________

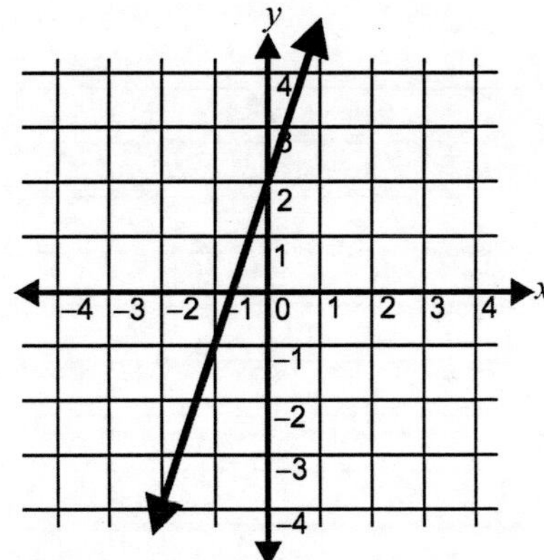

4.  ________

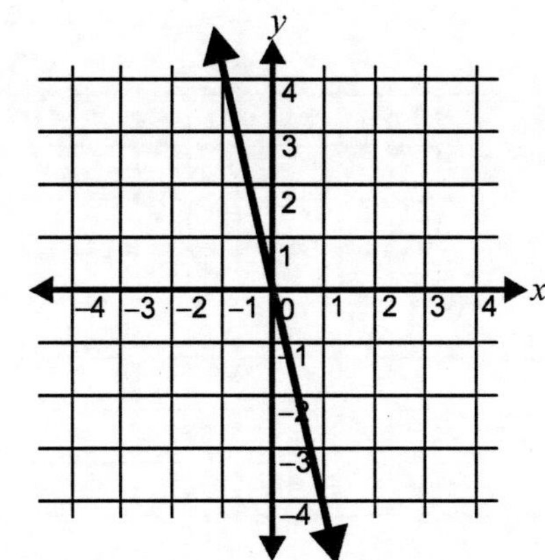

5.  ________

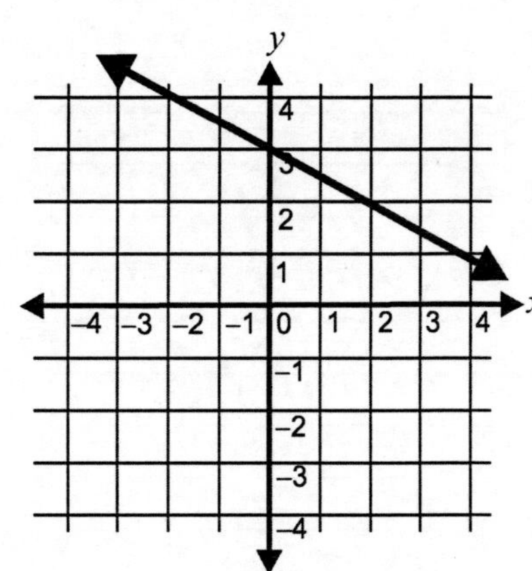

6. ________

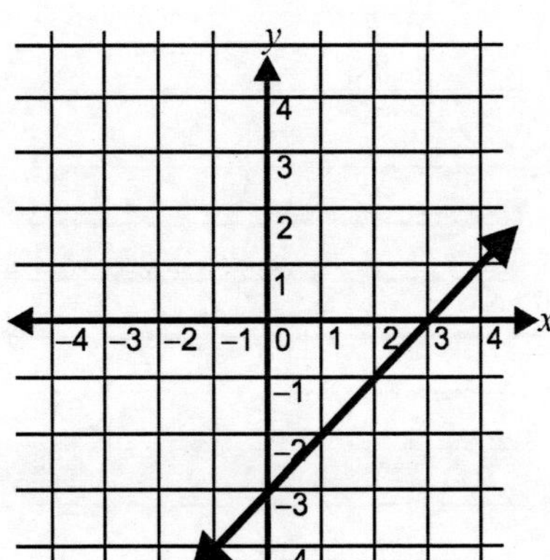

7. ________

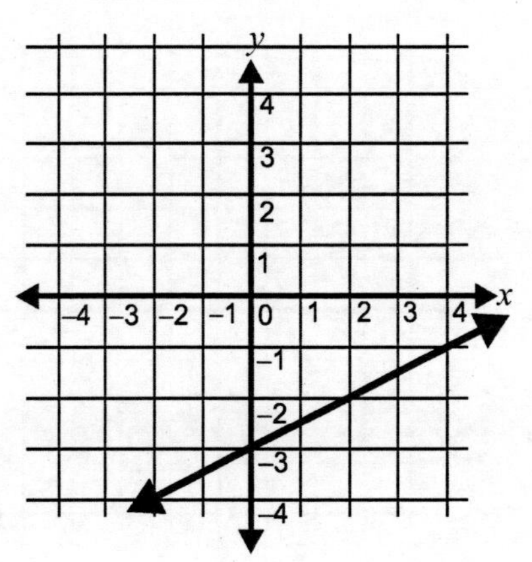

8. ________

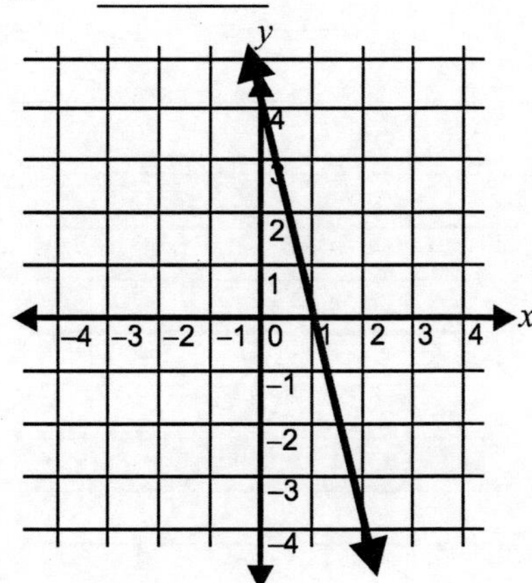

9. ________

CHAPTER 11 REVIEW

1. Graph the solution set for the linear equation: $x - 3 = y$.

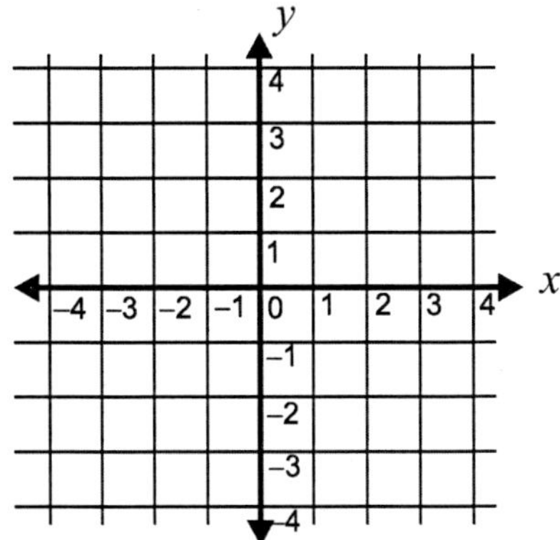

2. Which of the following is not a solution of $3x = 5y - 1$?

   A. (3, 2)
   B. (7, 4)
   C. $(-\frac{1}{3}, 0)$
   D. (–2, –1)

3. (–2, 1) is a solution for which of the following equations?

   A. $y + 2x = 4$
   B. $-2x - y = 5$
   C. $x + 2y = -4$
   D. $2x - y = -5$

4. Graph the equation $2x - 4 = 0$.

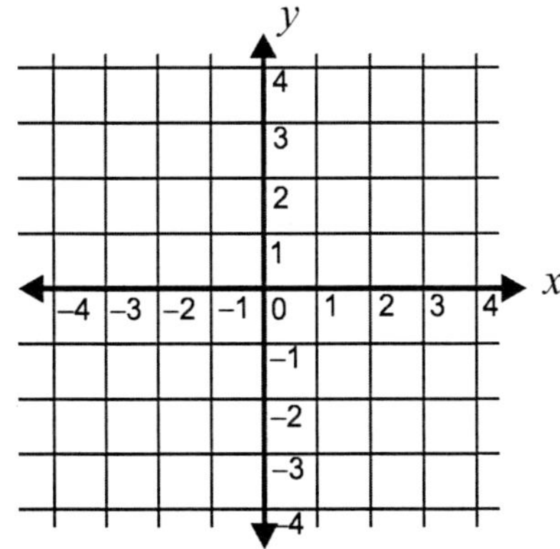

5. What is the slope of the line that passes through the points (5, 3) and (6, 1)?

6. What is the slope of the line that passes through the points (–1, 4) and (–6, –2)?

7. What is the x-intercept for the following equation?

$$6x - y = 30$$

8. What is the y-intercept for the following equation?

$$4x + 2y = 28$$

9. Graph the equation $3y = 9$.

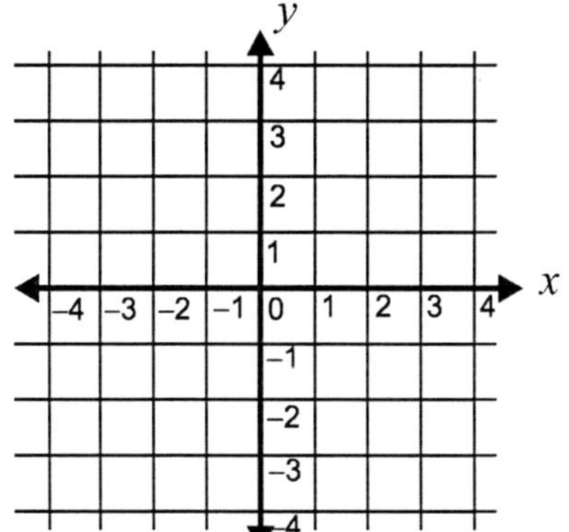

10. Write the following equation in slope-intercept form:

$$3x = -2y + 4$$

11. What is the slope of the line $y = -\frac{1}{2}x + 3$?

12. What is the x-intercept of the line $y = 5x + 6$?

13. What is the y-intercept of the line $y - \frac{2}{3}x + 2 = 0$?

14. Graph the line which has a slope of 2 and a y-intercept of -3.

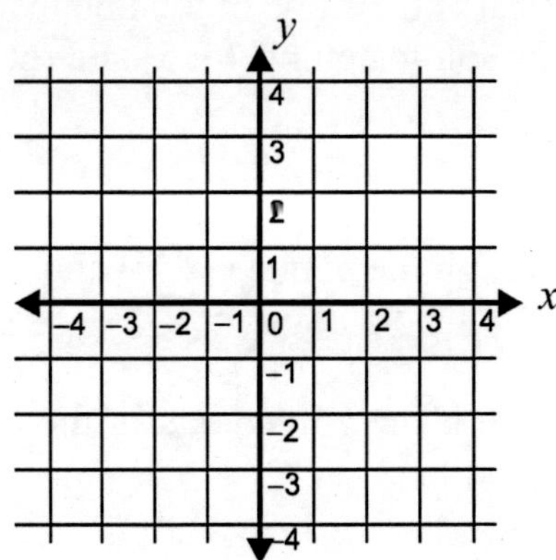

15. Graph the line which has a slope of -2 and a y-intercept of -3.

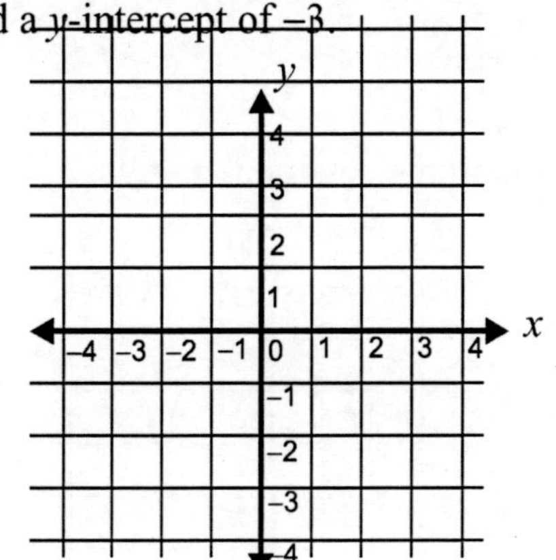

16. Which of the following points does ***not*** lie on the line $y = 3x - 2$?

A. $(0, -2)$
B. $(1, 1)$
C. $(-1, 5)$
D. $(2, 4)$

17. Which of the following points lies on the line $2y = -x + 1$?

A. $(\frac{1}{2}, 0)$
B. $(2, -\frac{1}{2})$
C. $(0, 1)$
D. $(-1, -1)$

18. Graph the equation $-x = 6 + 2y$.

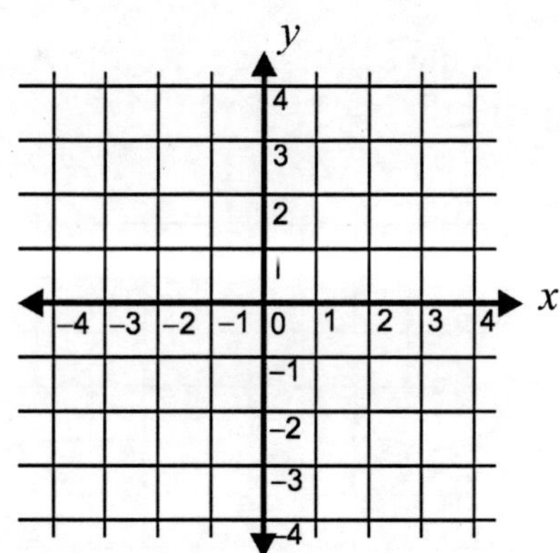

19. Find the equation of the line which contains the point $(0, 2)$ and has a slope of $\frac{3}{4}$.

20. Which is the graph of $x - 3y = 6$?

A.

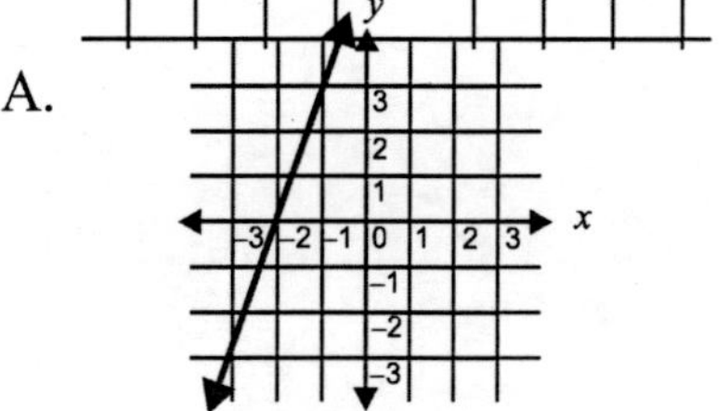

B.

C.

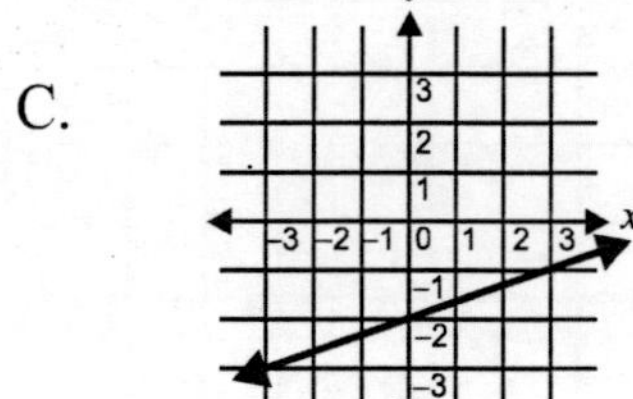

D.

21. Which of the following is the graph of the line which has a slope of −2 and a y-intercept of (0, 3)?

A. 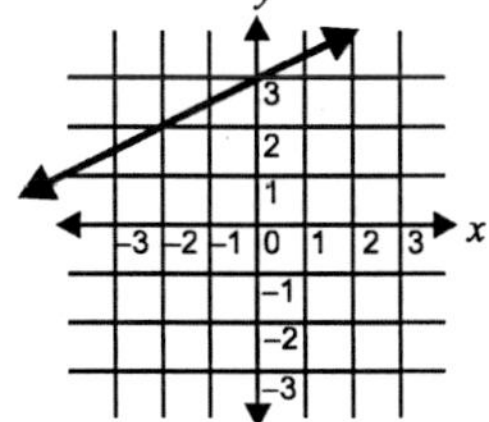

B. 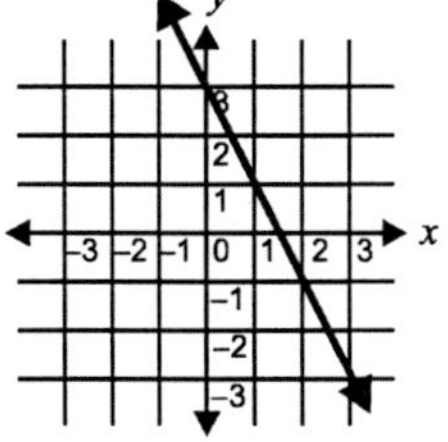

C. 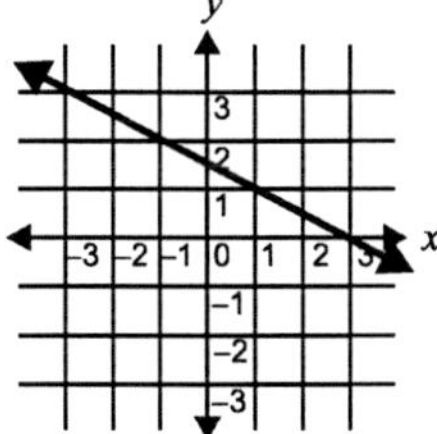

D. 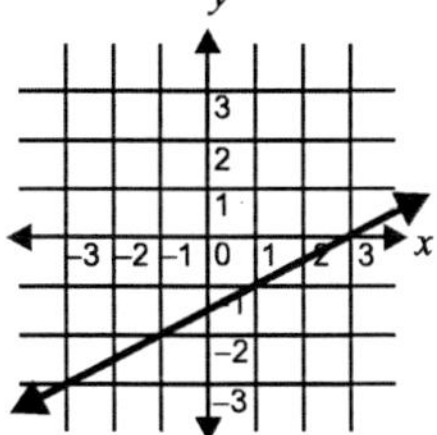

22. Given that a line contains the point (2, 3) and has a slope of $-\frac{1}{2}$, graph the line.

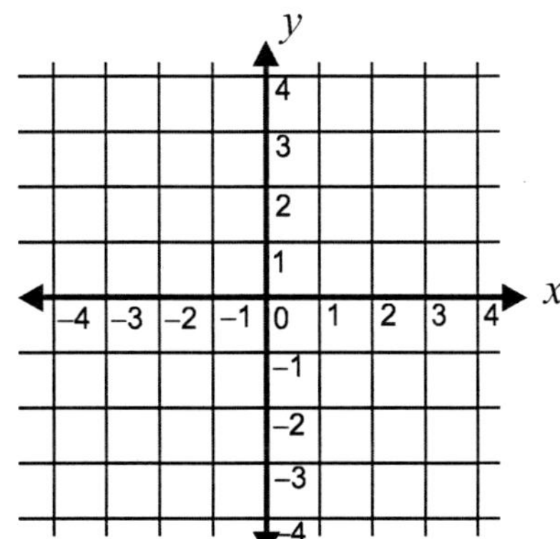

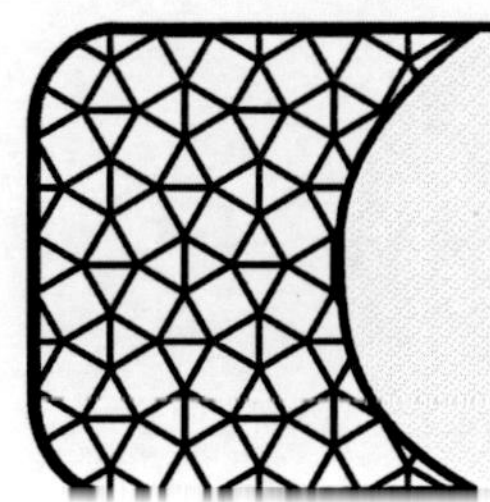

CHAPTER 12: MEASUREMENT

Standards 3.12.2 and 3.12.3

PRECISION, ERROR, AND TOLERANCE

Precision refers to how small a scale is being used to make a measurement. The smaller the scale, the more precise the measurement. For example, a measurement made to the nearest eighth of an inch is more precise than a measurement that is rounded to the nearest inch.

PRECISION

The precision of a measurement is indicated by how many decimal points are used or how small a fraction is used. For example, a ruler marked in sixteenths of an inch will measure more precisely than a ruler marked in quarter inches. Or, a container marked in ounces will measure more precisely than a container marked in cups.

EXAMPLE 1: Look at the ruler below. How precise could you measure a line segment with this ruler?

1 inch 2 3 4 5 6

Answer: The ruler is marked in sixteenths of an inch so the precision of a measurement made with this ruler would be to the nearest sixteenth of an inch.

EXAMPLE 2: Look at the 2 thermometers below. Which one would give you a more precise temperature reading?

1

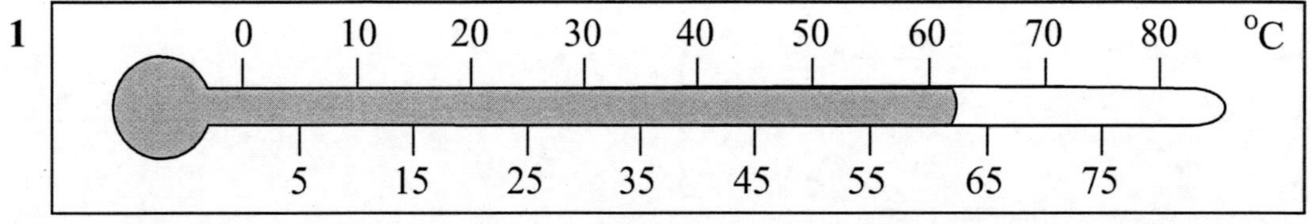

2

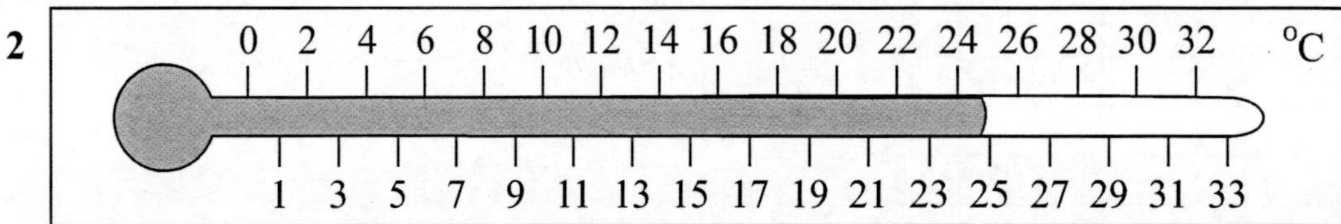

Answer: The second thermometer would give a more precise temperature reading because it measures to the nearest 1 degree. The first thermometer only measures to the nearest 5 degrees.

EXAMPLE 3: Which of the following measurements is most precise?

A. $3\frac{1}{4}$ in. B. $3\frac{1}{16}$ in. C. 3 in. D. $3\frac{1}{32}$ in.

Answer: Choice D is the most precise measurement because it is given in the smallest scale. This choice indicates that the ruler used to measure was marked in thirty-seconds of an inch. Remember, the smallest fraction indicates the greatest precision. Choice A indicates that the instrument used measured to the nearest one-fourth of an inch, choice B to the nearest sixteenth, and choice C to the nearest inch.

Special Note: What if choice A was written as $3\frac{8}{32}$? Then the precision of choice A, and the precision of choice D would be the same. They would both show that the ruler measured to the nearest thirty-second of an inch. **When showing precision in a measurement, you do not simplify the fraction.**

EXAMPLE 4: Which of the following measurements is most precise?

A. 2.8623 cm B. 2.8 cm C. 2.87 cm D. 2 cm

Answer: Choice A is the most precise because it is given to the nearest ten-thousandth. This choice indicates that the instrument used to take this measurement (probably not a ruler) could read to the nearest ten-thousandth of an inch.

Special Note: What if choice D were written as 2.0000 cm? Then the precision of choice D would be the same as the precision of choice A.

Circle the most precise measurement.

	A	B	C	D	E
1.	A. $3\frac{1}{4}$ in.	B. $3\frac{1}{8}$ in.	C. $9\frac{1}{2}$ in.	D. $7\frac{7}{16}$ in.	E. $5\frac{3}{8}$ in.
2.	A. $6\frac{2}{8}$ cups	B. $4\frac{1}{4}$ cups	C. $6\frac{4}{16}$ cups	D. $7\frac{8}{32}$ cups	E. $4\frac{5}{16}$ cups
3.	A. 2.90 L	B. 7 L	C. 24 L	D. 24.3 L	E. 71.9 L
4.	A. 1 oz.	B. 0.0150 oz.	C. 0.015 oz.	D. 1.015 oz.	E. 1.5 oz.
5.	A. 42.6 m	B. 8.4 m	C. 2 cm	D. 21 cm	E. 9.8 cm
6.	A. 91 kg	B. 3 g	C. 37.2 g	D. 21.98 kg	E. 8.4 kg

7. Which would give you a more precise measurement: a graduated cylinder marked in mL or a graduated cylinder marked in dL?

8. Which would give you a more precise measurement: a pressure gauge that measured in pounds or a pressure gauge that measured in tenths of pounds?

9. Which would give you a more precise measurement: a ruler marked in centimeters or a ruler marked in millimeters?

10. Which would give you a more precise measurement: a machine that measures to the nearest tenth of an inch or one that measures to the nearest one-hundredth of an inch?

ERROR

Since no instrument can measure with perfect precision, all measurements are subject to a certain amount of **error**. This error does not refer to human error but to error in rounding. The more precise the scale, the smaller the possible error.

For example, measure the line segment below.

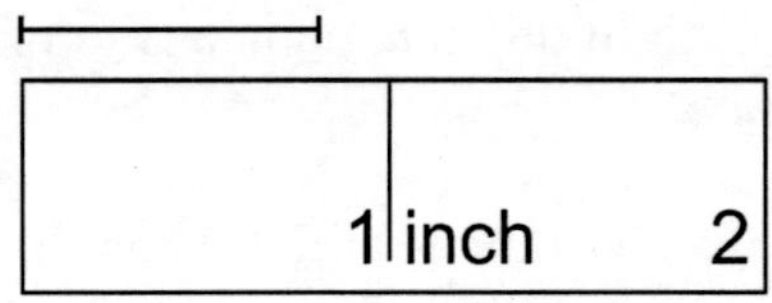

According to this ruler, the line segment is close to 1 inch. The ruler shown measures to the nearest inch, so the **precision** of the measurement can only be to the inch. In other words, you have to **round** to the nearest inch. The result of rounding the measurement is called **error of measurement** because the true measurement is really a little less than the stated measurement of 1 inch.

So, what is the maximum error in a measurement? Look at the line segment above. Since the line segment is more than half way between 0 and 1, you round to 1. Therefore, the error in measuring line segments with this particular ruler will be half of the distance between 0 and 1. The maximum error is half an inch.

A line segment falling within the gray area shown below would be rounded to 1 inch. Any line segment shorter than half an inch would be rounded to zero, and any line segment longer than one and a half inches would be rounded to two inches.

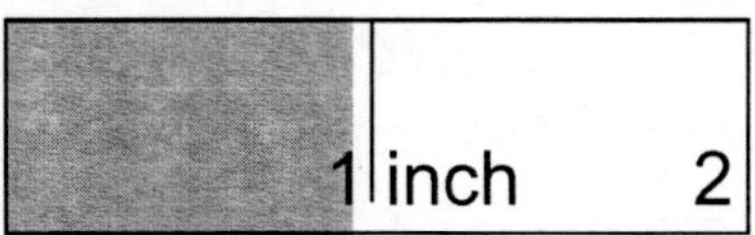

EXAMPLE 1: What is the measurement of the line segment below? What is the precision of the ruler? What is the maximum error of a line segment measured with this ruler?

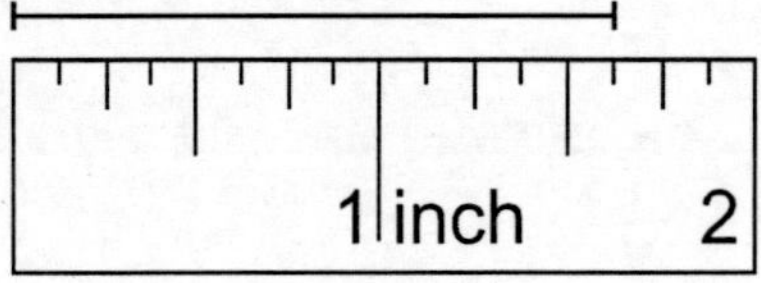

Answer: The line segment is $1\frac{5}{8}$ inches. The precision of the ruler is to the nearest one-eighth of an inch. The maximum error of this ruler is one-half of one-eighth of an inch which is $\frac{1}{16}$ of an inch.

The maximum error is one-half the unit of measure.

EXAMPLE 2: A map gives a scale in increments of ten miles. What is the maximum error when measuring between two cities on the map?

Answer: Since the scale is ten miles, the maximum error would be 5 miles.

Find the maximum error for the following problems.

1. a ruler marked in sixteenth of an inch
2. a ruler marked in millimeters
3. a graduated cylinder marked in milliliters
4. a map scale marked in 50 mile increments
5. a thermometer marked in 5 degree increments
6. a scale marked in tenths of a pound
7. a micrometer marked in one-hundredths of an inch
8. exit sign numbers designated by the nearest mile

TOLERANCE

So, where does tolerance fit into the picture? Tolerance refers to the probability of the largest error in measurement. When two people measure the same distance, most likely there will be some minor difference in their readings. The same is true if you use two different instruments to measure a length. If you use a plastic ruler with painted lines and numbers and then a metal ruler with the increments etched with a laser to measure a line segment, you will probably get two different readings. You are uncertain which reading is correct. When scientists measure things, they advertise their uncertainty by stating the tolerance with a $\pm$ sign. Tolerance is $\frac{1}{2}$ distance between the largest measurement and the smallest measurement.

EXAMPLE 1: A desk was measured with two different yardsticks.
Yardstick A measured the desk at 29.8 inches wide.
Yardstick B measured the desk at 30 inches wide,
What is the tolerance of the measurement?

Tolerance = $\frac{1}{2}$ (largest measurement – smallest measurement)

$\frac{1}{2}$ (30 – 29.8)
$\frac{1}{2}$ (0.2)

Tolerance = $\pm$ 0.1

EXAMPLE 2: The weight of a medium apple is 5.34 ounces. A different medium apple weighs 5.5 ounces. Based on these two apples, what is the tolerance for the weight of a medium apple?

1. Find the difference in the two measurements.

   $5.5 - 5.34 = 0.16$

2. Find the tolerance.

   $\frac{1}{2}(0.16) = 0.08$ The tolerance is $\pm$ 0.08 ounce

Find the tolerance in the measurements of the following questions.

1. Vince measured the length of board with two different tape measures. One measured $3\frac{1}{2}$ ft. The other measured the same board at 3 ft. $5\frac{1}{2}$ inches. What is the tolerance in the length of the board?

2. The Bouncy Ball Company makes rubber balls for children. On Monday, the manager Measured the circumference of a ball to be 3.54 cm. On Tuesday he measured a ball to be 3.60 cm. What is the tolerance for the measurement of a ball?

3. Frank's meat packing company packages hamburger in 1 lb packages. The owner of the company weighed package A and found it to be 16.2 ounces. He weighed package B and found it to be 15.93 ounces. What is the tolerance for the weight of a package of hamburger?

CHAPTER 12 REVIEW

Complete the following table.

	Measurement	Unit	Precision	Maximum Error
1	310 miles	10 miles	10 miles	5 miles
2	200 meters	10 meters		
3	10.0 miles	.1 miles		
4	25 degrees	5 degrees		
5	$\frac{3}{8}$ inches	$\frac{1}{8}$ inch		
6	5 feet	1 foot		
7	600 grams	100 grams		
8	2.0 millimeters	0.1 mm		
9	64 ounces	1 ounce		
10	8500 pounds	100 lb		

For questions 11-18, circle the most precise measurement.

11. A. 3 cm B. 300 cm C. 30 cm D. 0.0003 cm E. 0.030 cm
12. A. 4 feet B. 44 inches C. 1/2 inch D. 6 inches E. 60 inches
13. A. 5 tsp. B. 4 Tbsp. C. 1 cup D. 1 pint E. 1 gallon
14. A. 468 lbs. B. 16 oz. C. 5 oz. D. 200 lbs. E. 3.67 tons
15. A. 28 lbs. B. 2.6 oz. C. 53 lbs. D. 43 tons E. 19.3 oz.
16. A. 39 g B. 76 kg C. 210 mg D. 2.1 g E. 78 kg

17. A. Lake Michigan measured in liters
    B. a swimming pool measured in liters
    C. the Atlantic Ocean measured in liters
    D. a glass of water measured in liters
    E. Lake Michigan measured in kiloliters

18. A. a marathon measured in hours
    B. a marathon measured in hours and minutes
    C. a marathon measured in hours, minutes, and seconds
    D. a marathon measured in hours, minutes, seconds, and hundredths of seconds
    E. the 100-meter dash measured in seconds and hundredths of seconds

Fnd the tolerance of the following measurements.

	Measurement A	Measurement B
19.	23 mm	21 mm
20.	15 in.	14.8 in.
21.	191 g	201 g

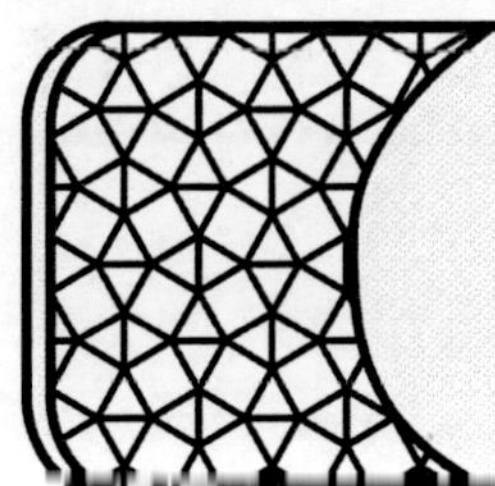

CHAPTER 13: ANGLES

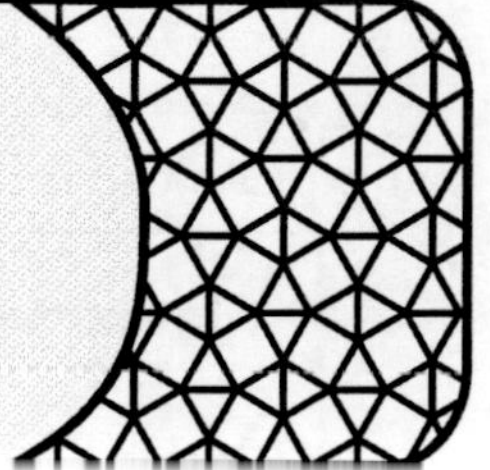

Standards 3.12.5 and 4.12.1 and 4.12.6

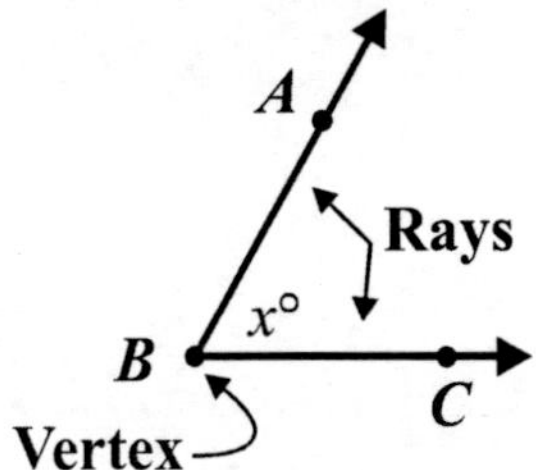

Angles are made up of two rays with a common endpoint. Rays are named by the endpoint *B* and another point on the ray. Ray $\overrightarrow{BA}$ and ray $\overrightarrow{BC}$ share a common endpoint.

Angles are usually named by three capital letters. The middle letter names the vertex. The angle to the left can be named $\angle ABC$ or $\angle CBA$. An angle can also be named by a lower case letter between the sides, $\angle x$, or by the vertex alone, $\angle B$.

A protractor, , is used to measure angles. The protractor is divided evenly into a half circle of 180 degrees (180°). When the middle of the bottom of the protractor is placed on the vertex, and one of the rays of the angle is lined up with 0°, the other ray of the angle crosses the protractor at the measure of the angle. The angle below has the ray pointing left lined up with 0° (the outside numbers), and the other ray of the angle crosses the protractor at 55°. The angle measures 55°.

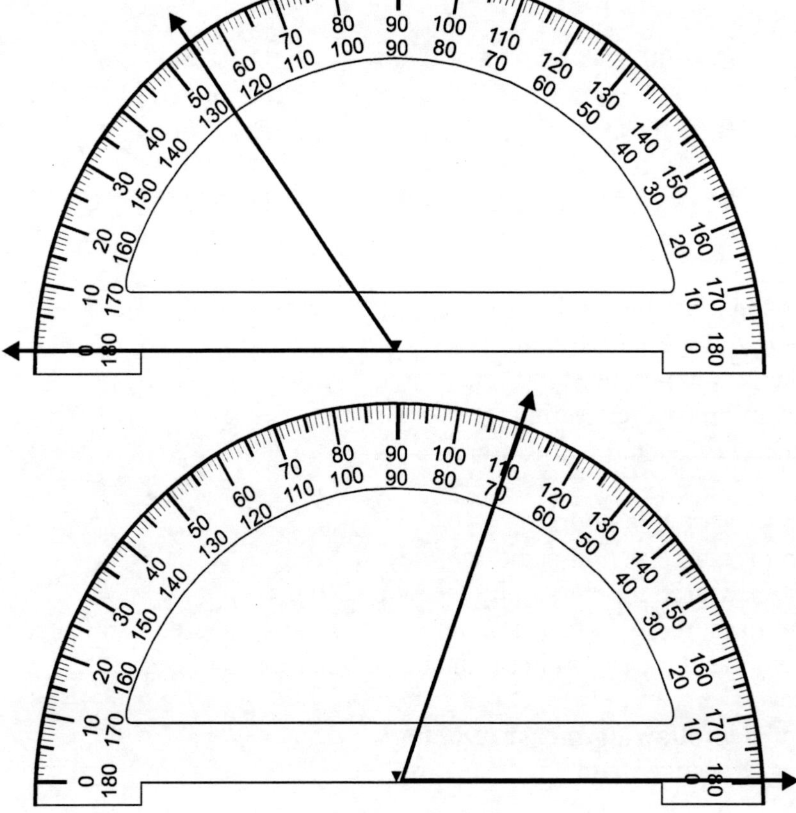

The angle above has the ray pointing right lined up with 0° using the inside numbers. The other ray crosses the protractor and measures the angle at 70°.

TYPES OF ANGLES

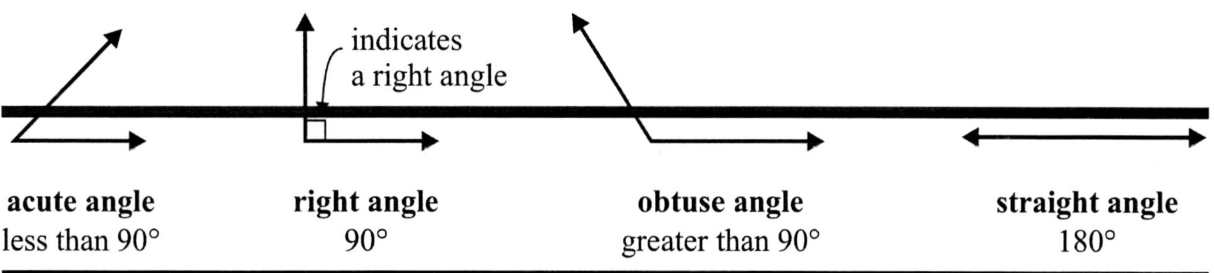

acute angle	**right angle**	**obtuse angle**	**straight angle**
less than 90°	90°	greater than 90°	180°

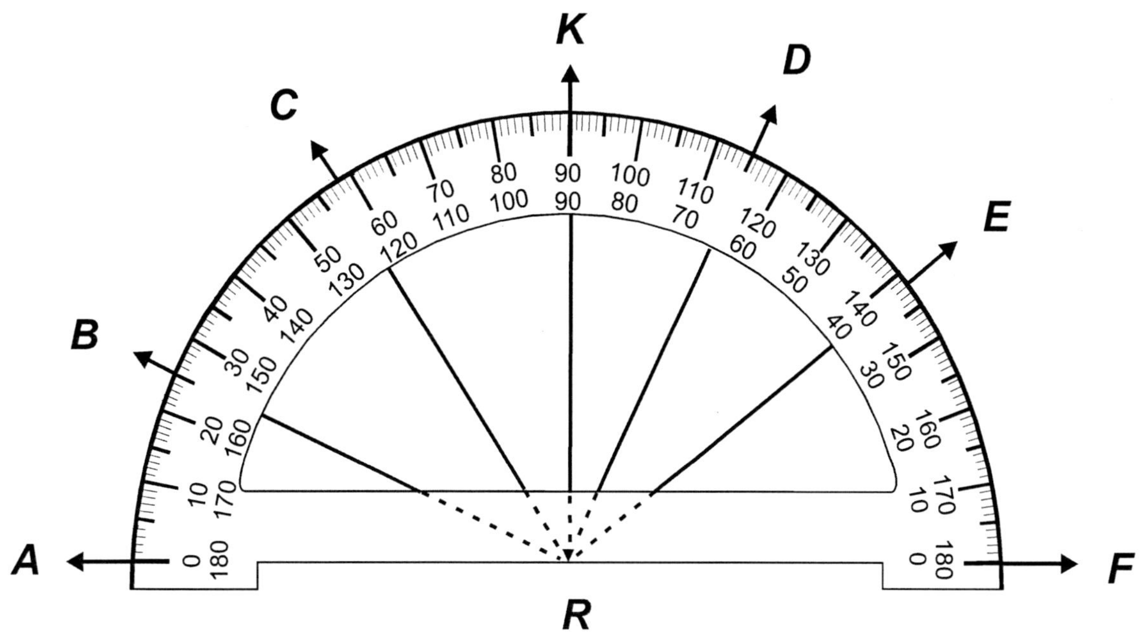

Using the protractor above, find the measure of the following angles. Then, tell what type of angle it is: acute, right, obtuse, or straight.

	Measure	Type of Angle
1. What is the measure of angle *ARF*?	__________	__________
2. What is the measure of angle *CRF*?	__________	__________
3. What is the measure of angle *BRF*?	__________	__________
4. What is the measure of angle *ERF*?	__________	__________
5. What is the measure of angle *ARB*?	__________	__________
6. What is the measure of angle *KRA*?	__________	__________
7. What is the measure of angle *CRA*?	__________	__________
8. What is the measure of angle *DRF*?	__________	__________
9. What is the measure of angle *ARD*?	__________	__________
10. What is the measure of angle *FRK*?	__________	__________

MEASURING ANGLES

Estimate the measure of the following angles. Then, use your protractor to record the actual measure.

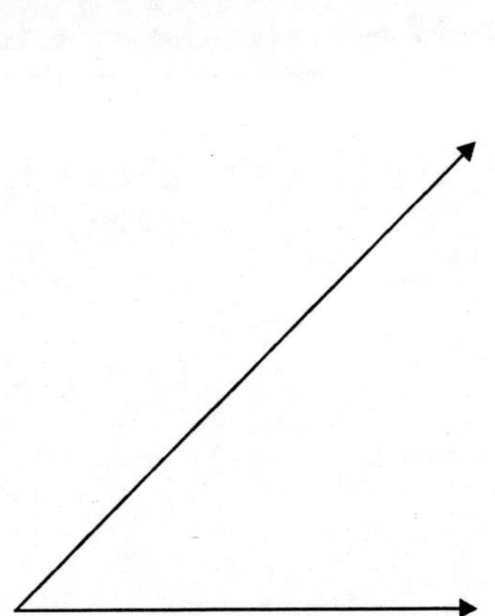

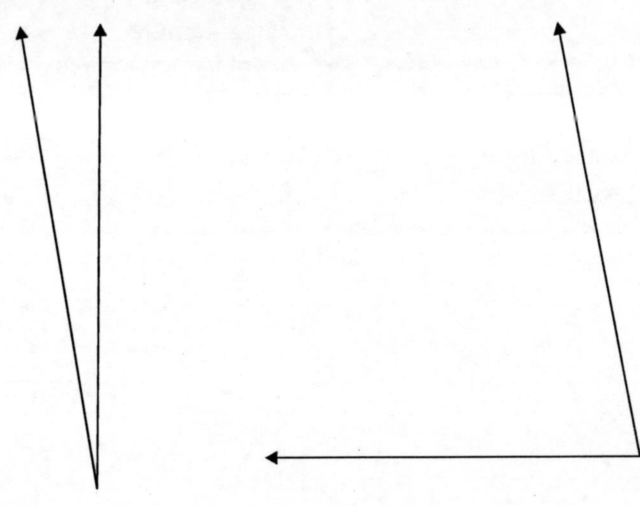

1. Estimate = ________ °
   Measure = ________ °

4. Estimate = ________ °
   Measure = ________ °

7. Estimate = ________ °
   Measure = ________ °

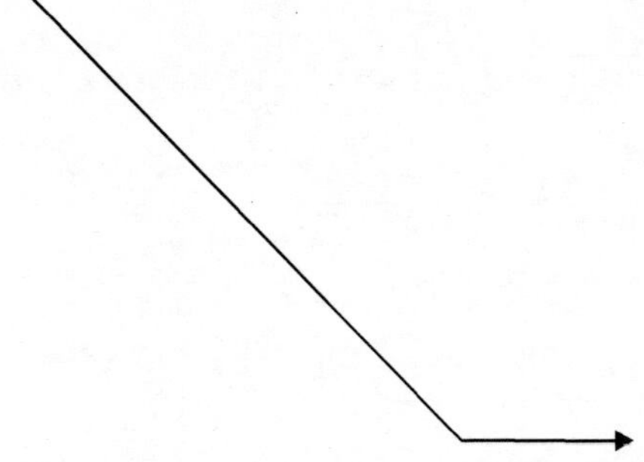

2. Estimate = ________ °
   Measure = ________ °

5. Estimate = ________ °
   Measure = ________ °

8. Estimate = ________ °
   Measure = ________ °

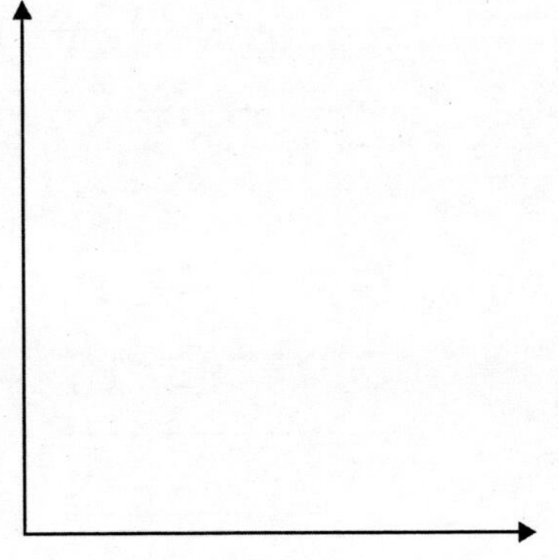

3. Estimate = ________ °
   Measure = ________ °

6. Estimate = ________ °
   Measure = ________ °

9. Estimate = ________ °
   Measure = ________ °

CENTRAL ANGLES

In this section, you will learn about central angles and why they are important when making circle graphs.

A central angle is the angle formed by each "piece of the pie." The vertex of a central angle is in the center of the circle. Look at the diagram on the right.

Central Angle

In circle graphs, the percentages have to add up to 100%, and the angles for each "piece of pie" must add up to 360°. Notice the pie graph to the right. Each percent of the pie is marked with a corresponding angle measure.

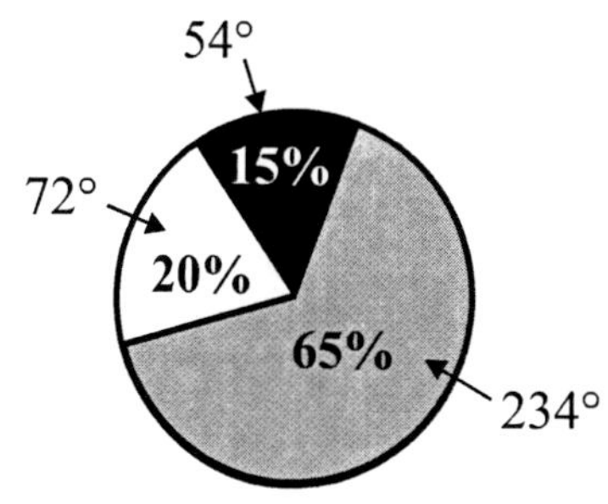

EXAMPLE: Consider the following data:

Survey of fiction reading preferences among 300 high school students	
Romance novels	70
Science fiction	60
Fantasy	90
Historical novels	40
Poetry	10
Mysteries	30

Step 1: First, you need to find what percent of the total each type of fiction represents. Divide each item in the survey by the total number of students. For example, according to the chart, 70 out of 300 students prefer romance novels, so to find the percent, divide 70 by 300. Rounding to the nearest percent, you get 23%. Repeat for each type of fiction.

Step 2: Multiply the percent of each type of fiction by 360° to figure out how many degrees each "piece of pie" should be. For romance novels, 23% of 360° is 83° (rounded to the nearest degree).

	Number of Students	% of total	Central angle of the circle
Romance novels	70	23%	83°
Science fiction	60	20%	72°
Fantasy	90	30%	108°
Historical novels	40	13%	47°
Poetry	12	4%	14°
Mysteries	30	10%	36°
	300	100%	360°

Complete the following exercises on central angles.

A dartboard has been divided into wedges according to the following color percentages.

blue	30%
red	25%
yellow	10%
green	35%

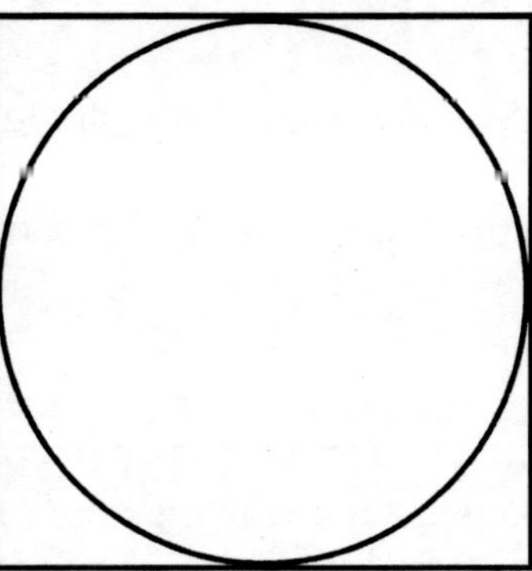

1. Find the central angle measurements of each color.

2. Complete the dartboard by drawing each color wedge to scale. Label each color on the dartboard and indicate the measure of each central angle.

The students at Maverick High School voted on Teacher-of-the-Year. The Student Council tallied the 720 total votes and created a pie chart.

Mr. Perry	252
Mrs. Nance	180
Miss Murphy	144
Mr. Bard	87
Mr. Olson	36
All Others	21

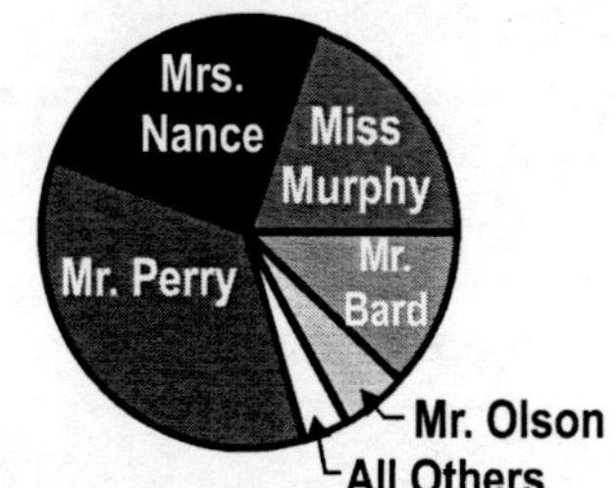

3. Calculate the percent of the votes that each teacher received.

4. Calculate the measures of each central angle on the pie chart.

Other central angle type questions involve a clock. These are very similar to the pie chart.

EXAMPLE: The second hand on a clock goes from the 11 to the 2. How many degrees has the second hand traveled?

Step 1: You know that there are 60 seconds in a minute. The second hand traveled 15 seconds out of 60 as it went from the 11 to the 2. $15 \div 60 = 0.25$ or 25%.

Step 2: Multiply the percent by 360° just like you did for the pie chart. $0.25 \times 360 = 90°$. So, the second hand traveled 90°.

Calculate how many degrees the second hand travels for the following:

5.	12 to 3	_____	8.	6 to 3	_____
6.	9 to 11	_____	9.	12 to 1	_____
7.	4 to 10	_____	10.	3 to 5	_____

ADJACENT ANGLES

Adjacent angles are two angles that have the same vertex and share one ray. They do not share space inside the angles.

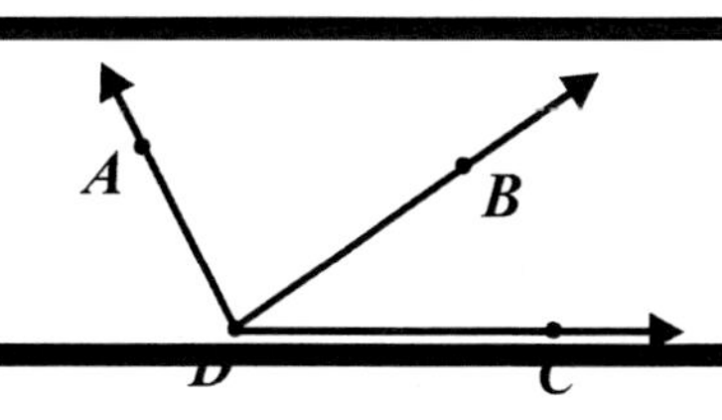

In this diagram, $\angle ADB$ is **adjacent** to $\angle BDC$.

However, $\angle ADB$ is **not adjacent** to $\angle ADC$ because adjacent angles do not share any space inside the angle.

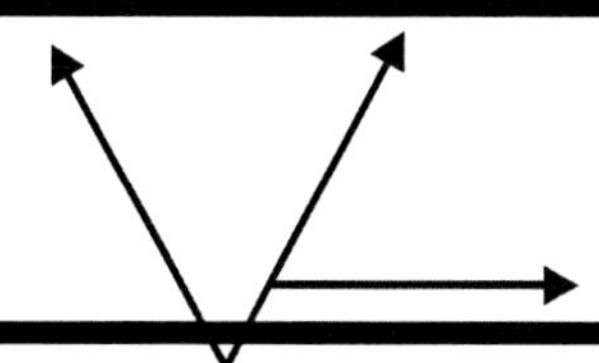

These two angles are **not adjacent.** They share a common ray but do not share the same vertex.

For each diagram below, name the angle that is adjacent to it.

1. 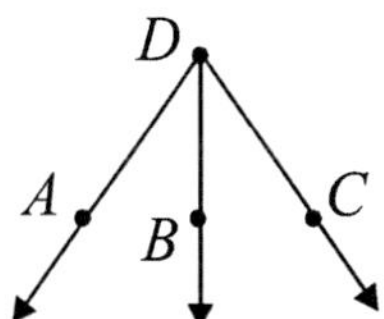

$\angle CDB$ is adjacent to $\angle$ ______

2. 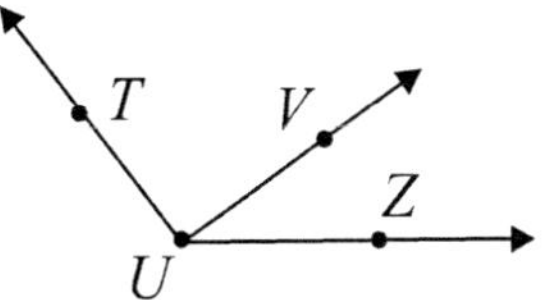

$\angle TUV$ is adjacent to $\angle$ ______

3. 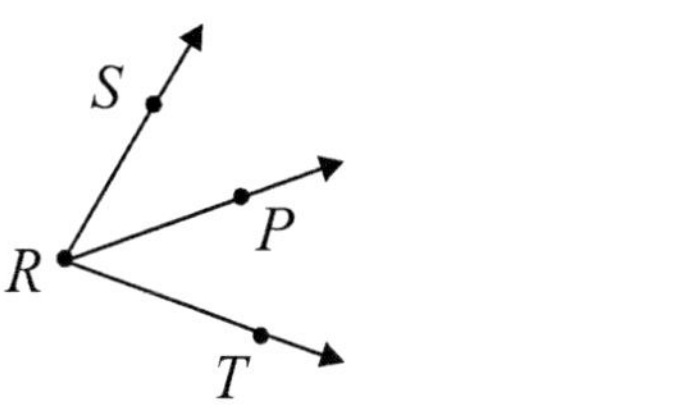

$\angle SRP$ is adjacent to $\angle$ ______

4. 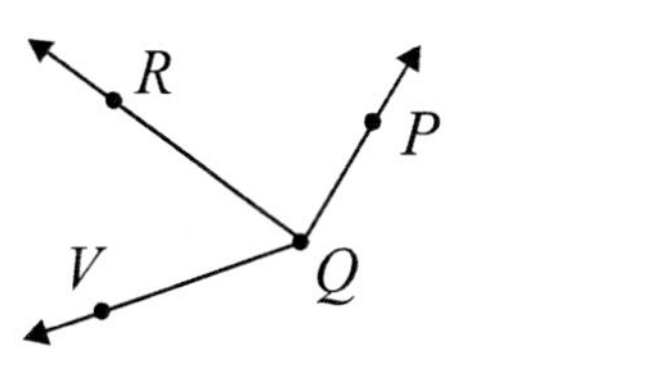

$\angle PQR$ is adjacent to $\angle$ ______

5. 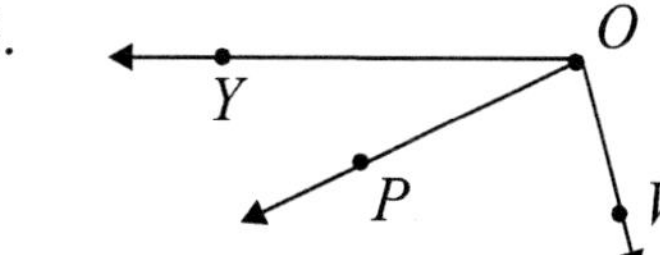

$\angle YOP$ is adjacent to $\angle$ ______

6. 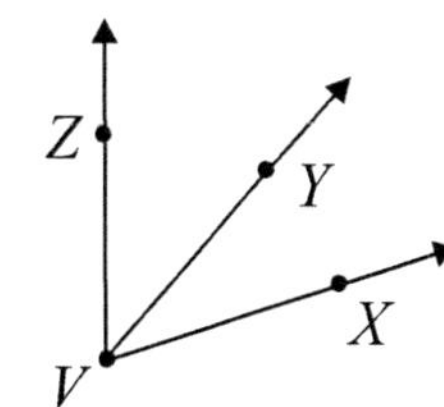

$\angle XVY$ is adjacent to $\angle$ ______

7. 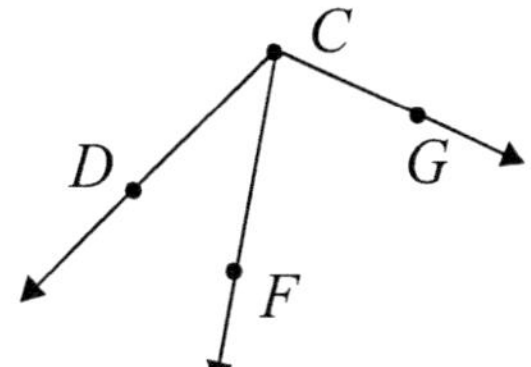

$\angle DCF$ is adjacent to $\angle$ ______

8. 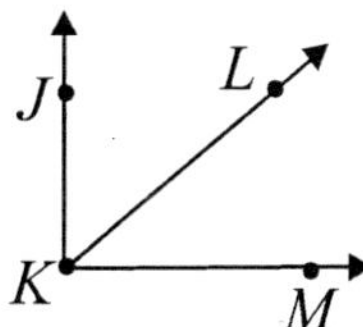

$\angle JKL$ is adjacent to $\angle$ ______

VERTICAL ANGLES

When two lines intersect, two pairs of vertical angles are formed. Vertical angles are not adjacent. Vertical angles have the same measure.

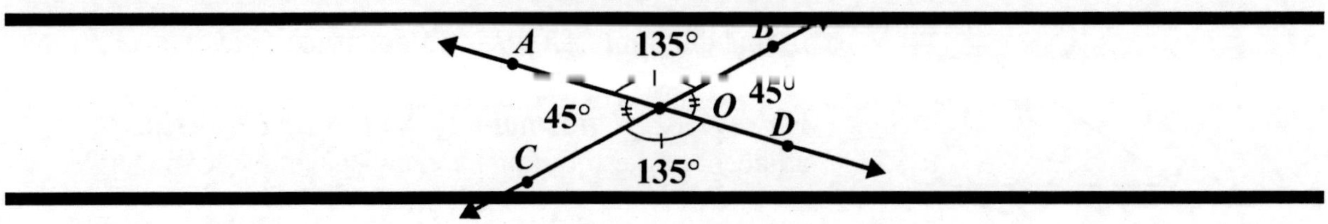

∠*AOB* and ∠*COD* are vertical angles. ∠*AOC* and ∠*BOD* are vertical angles. **Vertical angles** are **congruent**. Congruent means they have the same measure.

In the diagrams below, name the second angle in each pair of vertical angles.

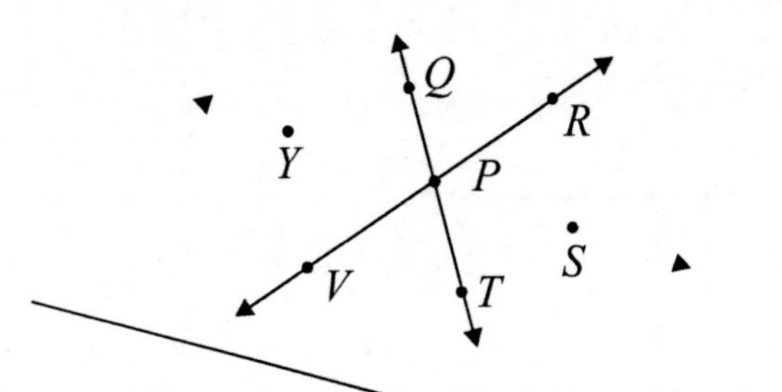

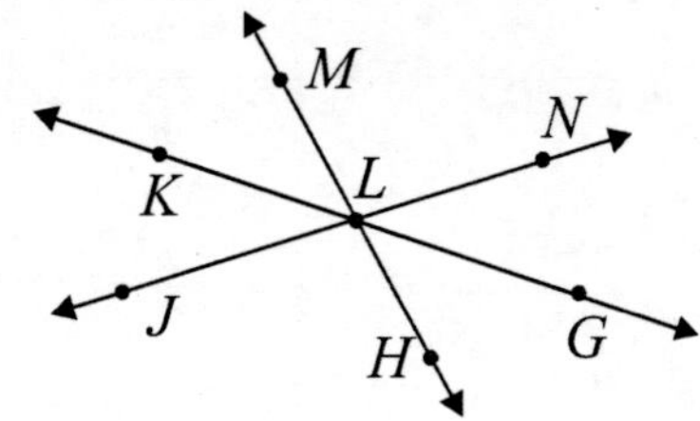

1. ∠*YPV* ______
2. ∠*QPR* ______
3. ∠*SPT* ______
4. ∠*VPT* ______
5. ∠*RPT* ______
6. ∠*VPS* ______
7. ∠*MLN* ______
8. ∠*KLH* ______
9. ∠*GLN* ______
10. ∠*GLM* ______
11. ∠*KLM* ______
12. ∠*HLG* ______

Use the information given to find the measure of each unknown vertical angle.

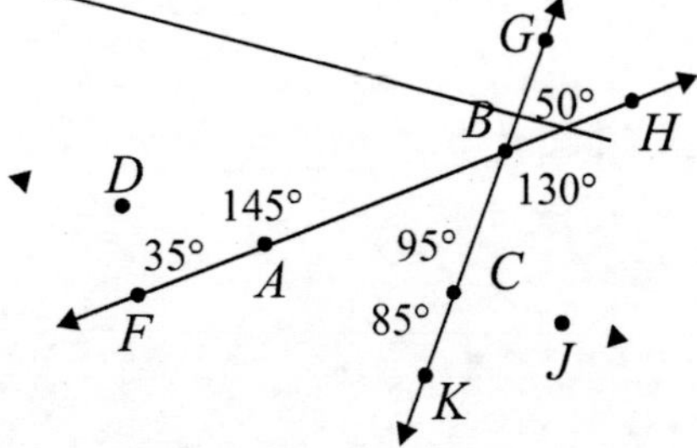

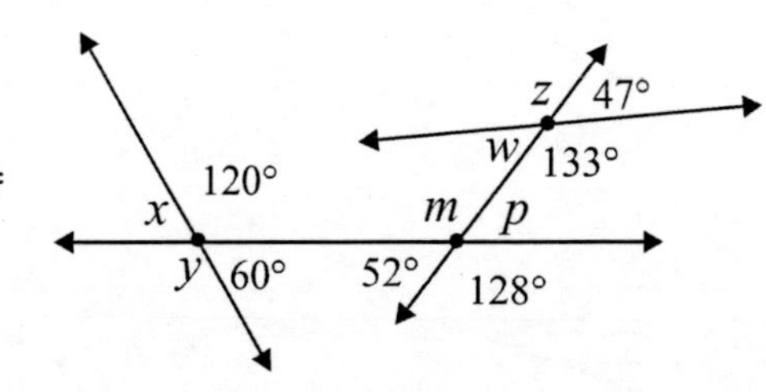

13. ∠*CAF* = ______
14. ∠*ABC* = ______
15. ∠*KCJ* = ______
16. ∠*ABG* = ______
17. ∠*BCJ* = ______
18. ∠*CAB* = ______
19. ∠*x* = ______
20. ∠*y* = ______
21. ∠*z* = ______
22. ∠*w* = ______
23. ∠*m* = ______
24. ∠*p* = ______

COMPLEMENTARY AND SUPPLEMENTARY ANGLES

Two angles are **complementary** if the sum of the measures of the angles is 90°.
Two angles are **supplementary** if the sum of the measures of the angles is 180°.

The angles may be adjacent but do not need to be.

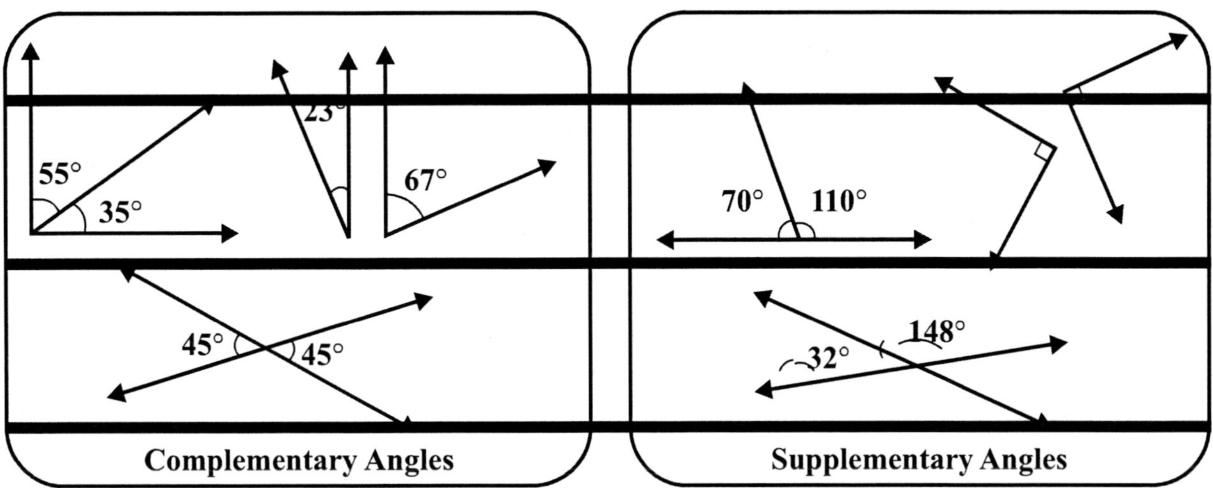

Calculate the measure of each unknown angle.

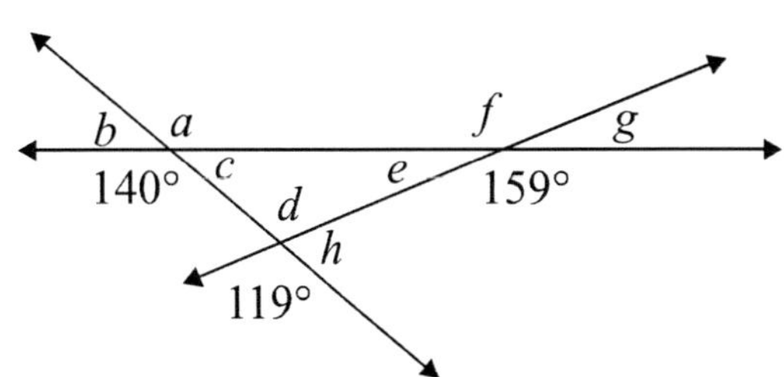

1. $\angle a$ = ______
2. $\angle b$ = ______
3. $\angle c$ = ______
4. $\angle d$ = ______
5. $\angle e$ = ______
6. $\angle f$ = ______
7. $\angle g$ = ______
8. $\angle h$ = ______

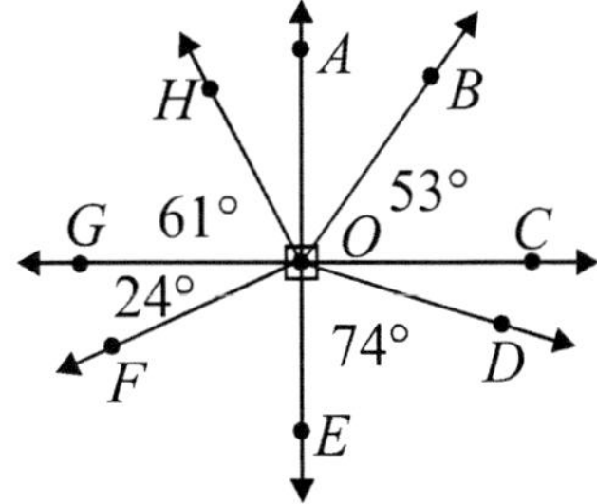

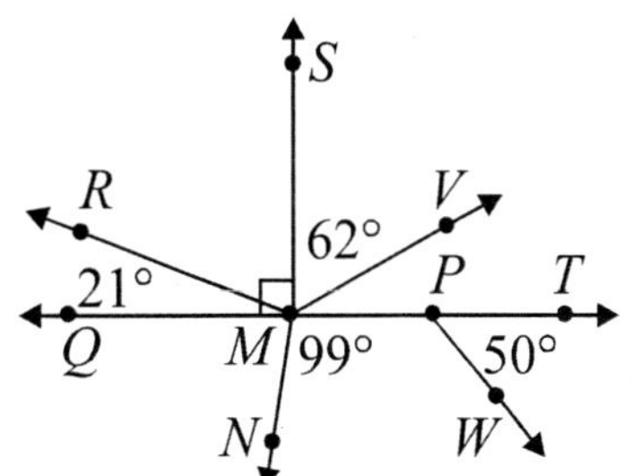

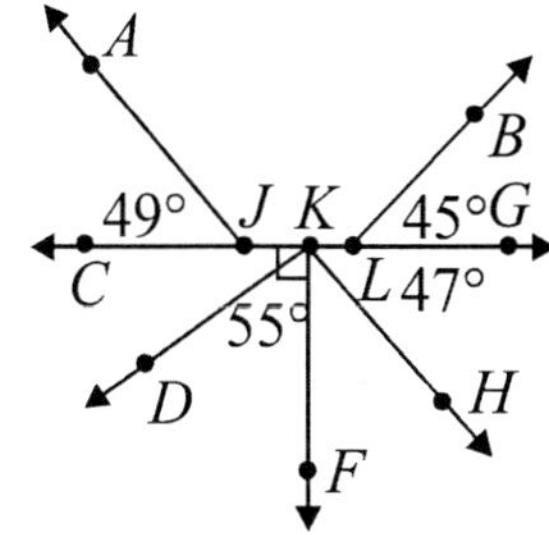

9. $\angle AOB$ = ______
10. $\angle COD$ = ______
11. $\angle EOF$ = ______
12. $\angle AOH$ = ______
13. $\angle RMS$ = ______
14. $\angle VMT$ = ______
15. $\angle QMN$ = ______
16. $\angle WPQ$ = ______
17. $\angle AJK$ = ______
18. $\angle CKD$ = ______
19. $\angle FKH$ = ______
20. $\angle BLC$ = ______

CORRESPONDING, ALTERNATE INTERIOR, AND ALTERNATE EXTERIOR ANGLES

If two parallel lines are intersected by a **transversal**, a line passing through both parallel lines, the **corresponding angles** are congruent.

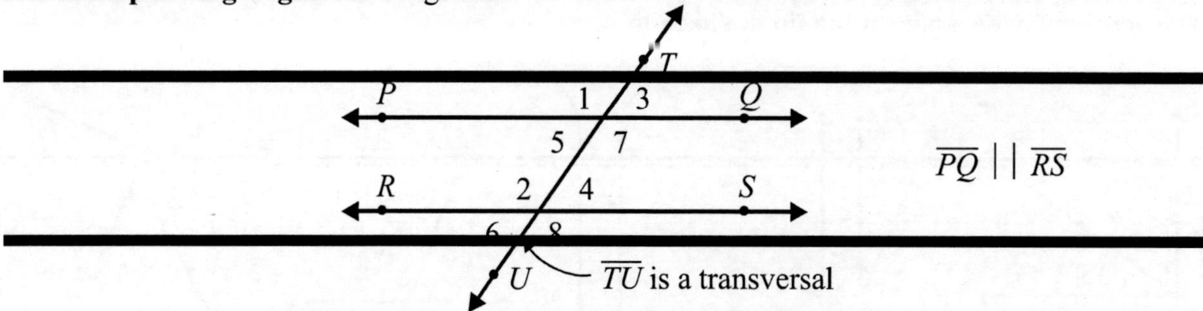

$\angle 1$ and $\angle 2$ are congruent. They are corresponding angles.
$\angle 3$ and $\angle 4$ are congruent. They are corresponding angles.
$\angle 5$ and $\angle 6$ are congruent. They are corresponding angles.
$\angle 7$ and $\angle 8$ are congruent. They are corresponding angles.

Alternate interior angles are also congruent. They are on opposite sides of the transversal and inside the parallel lines.

$\angle 5$ and $\angle 4$ are congruent. They are alternate interior angles.
$\angle 7$ and $\angle 2$ are congruent. They are alternate interior angles.

Alternate exterior angles are also congruent. They are on opposite sides of the transversal and above and below the parallel lines.

$\angle 1$ and $\angle 8$ are congruent. They are alternate exterior angles.
$\angle 3$ and $\angle 6$ are congruent. They are alternate exterior angles.

Look at the diagram below. For each pair of angles, state whether they are corresponding (C), alternate interior (I), alternate exterior (E), vertical (V), or supplementary angles (S).

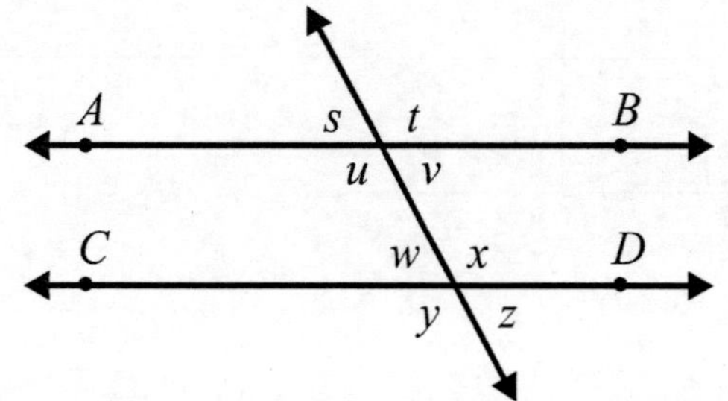

$\overline{AB}$ and $\overline{CD}$ are parallel.

1. $\angle u$, $\angle x$ _____
2. $\angle w$, $\angle s$ _____
3. $\angle t$, $\angle y$ _____
4. $\angle s$, $\angle t$ _____
5. $\angle w$, $\angle y$ _____
6. $\angle t$, $\angle x$ _____
7. $\angle w$, $\angle z$ _____
8. $\angle v$, $\angle w$ _____
9. $\angle v$, $\angle z$ _____
10. $\angle s$, $\angle z$ _____
11. $\angle t$, $\angle u$ _____
12. $\angle w$, $\angle x$ _____
13. $\angle w$, $\angle s$ _____
14. $\angle s$, $\angle v$ _____
15. $\angle x$, $\angle z$ _____

ANGLE RELATIONSHIPS

When two lines meet at a point, they form an angle. On the Nevada Math test, you will be asked to apply knowledge of angle relationships to real-world situations. You will also apply these skills in later classes in mathematics.

Example: Make a graphic relationship between a clock and a Cartesian plane. Draw and explain two congruent triangles using the digits on the clock.

1. Use the center of a clock as the origin of the Cartesian plane.
2. The x-axis aligns with the hours 3 and 9. The y-axis aligns with the hours 12 and 6.
3. The clock can be divided into four quadrants, each of 90 degrees on the Cartesian plane.

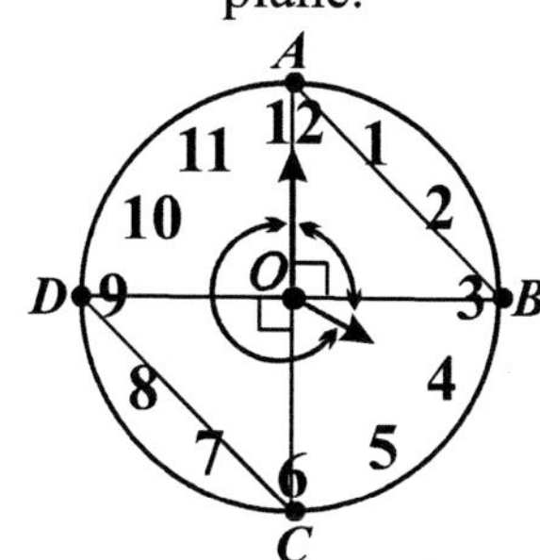

Point A = 12. Point B = 3. Point C = 6. Point D = 9. O is the center.

1. $\overline{OA} = \overline{OB} = \overline{OC} = \overline{OD}$. Points on the circumference are equal distance from the center.
2. $\angle AOB = \angle COD$. First and third quadrants of the Cartesian plane.
3. $\Delta AOB \cong \Delta COD$.

1. Find the measure of the two angles formed by the hands of the clock, a) at 7:00 AM and b) at 2:00 PM.

2. A grandfather clock that is supposed to strike a bell at every hour, does not work properly when the hands of the clock form angles in which one angle formed is twice the measure of the opposite angle. Is there a time when this event would happen? If so, find it.

3. Given: $\overline{PQ} \mid\mid \overline{MN}$
$\overline{AB}$ intersects $\overline{PQ}$ at R, and $\overline{MN}$ at S
$\angle ARP = 5x$ and $\angle ASN = 13x$
Find the measures of all angles in degrees.

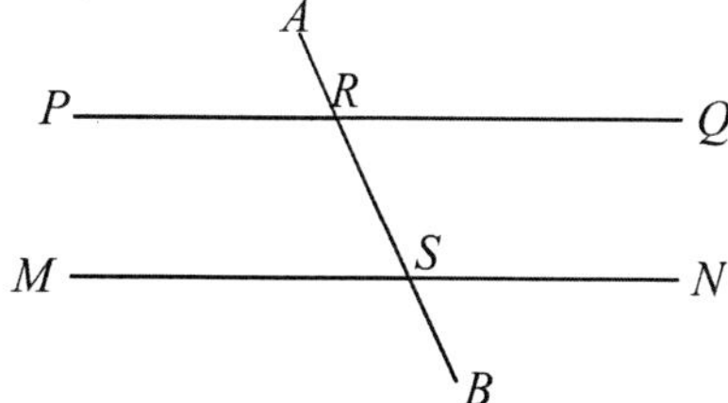

4. If two parallel lines are cut by a transversal, find the measures of:
   a. Two alternate exterior angles represented by $x + 14$ and $2x - 35$.
   b. Two corresponding angles represented by $4x + 10$ and $5x - 5$.
   c. Two exterior angles on the same side of the transversal whose values are $3x$ and $7x$.

5. Given $\overline{MN}$ and measures of angles as shown in the figures, find the value of the remaining angles.

   a. $\angle B = 62°$ and $\angle G = 103°$

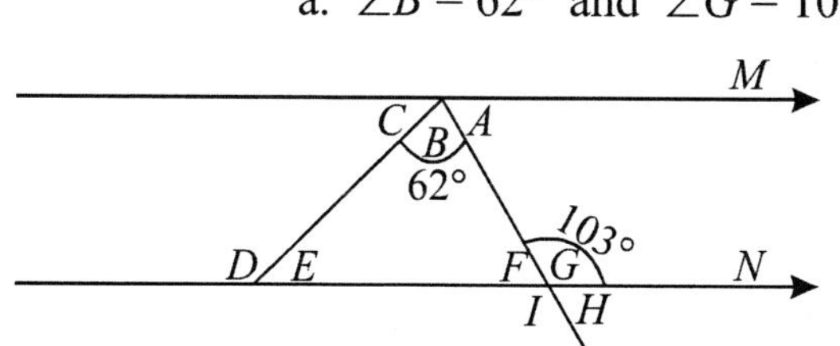

   b. $\angle C = 76°$ and $\angle F = 34°$

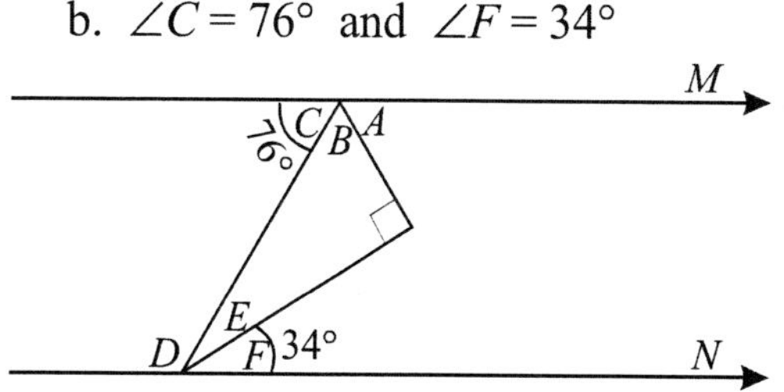

SUM OF INTERIOR ANGLES OF A POLYGON

Given a polygon, you can find the sum of the measures of the interior angles using the following formula:

Sum of the measures of the interior angles = $180°(n - 2)$
where n is the number of sides of the polygon

EXAMPLE: Find the sum of the measures of the interior angles of the following polygon:

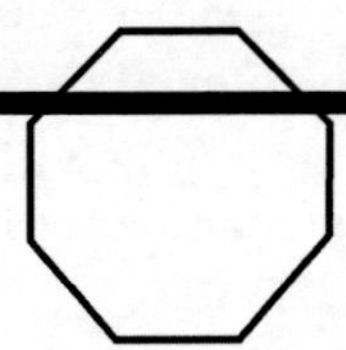

The figure has 8 sides. Using the formula we have $180°(8 - 2) = 180°(6) = 1080°$.

Using the formula, $180°(n - 2)$, find the sum of the interior angles of the following figures.

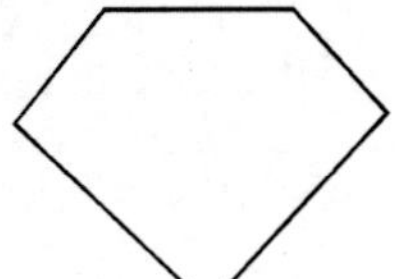
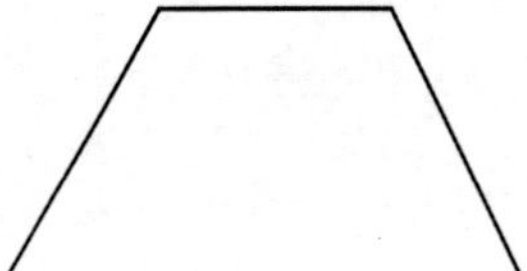
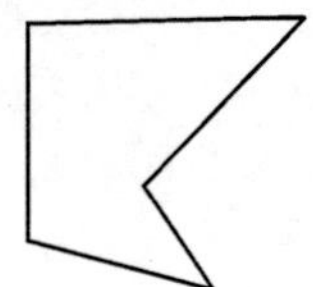
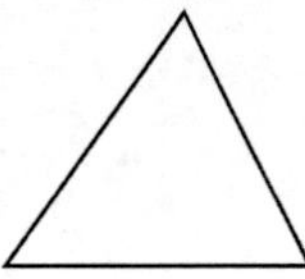

1. __________ 4. __________ 7. __________ 10. __________

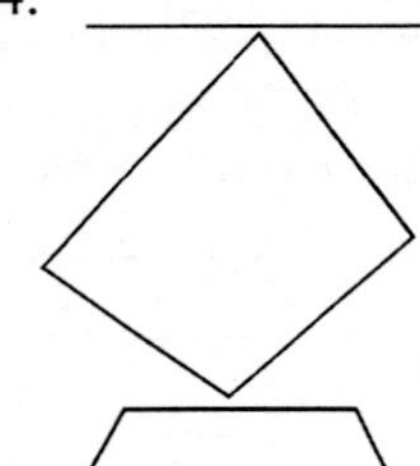
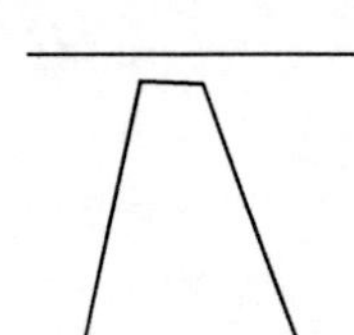

2. __________ 5. __________ 8. __________ 11. __________

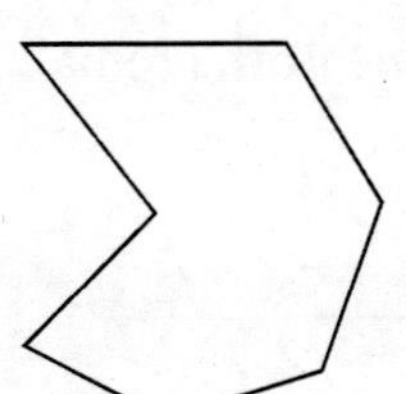
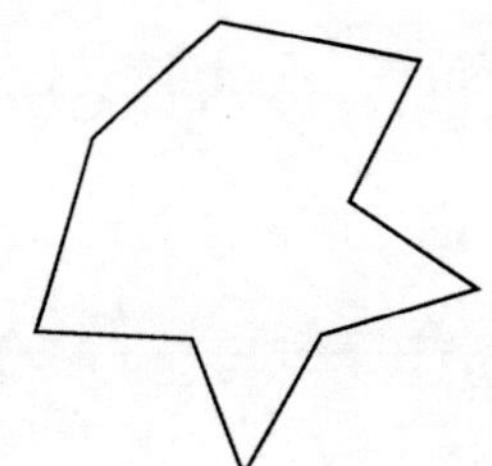
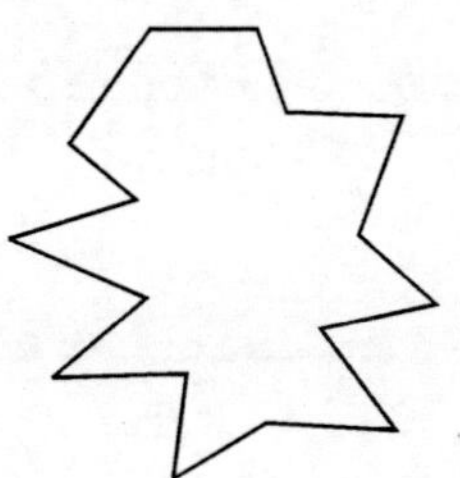

3. __________ 6. __________ 9. __________ 12. __________

EXTERIOR ANGLES OF A POLYGON

The sum of the exterior angles of a polygon is 360°. In the regular hexagon below, each side has been extended to form an exterior angle. We know there are 6 exterior angles and together they add to 360 so we have the equation $6a = 360$. If we divide both sides of the equation by 6 we have $a = 60$. Each exterior angle equals 60 degrees.

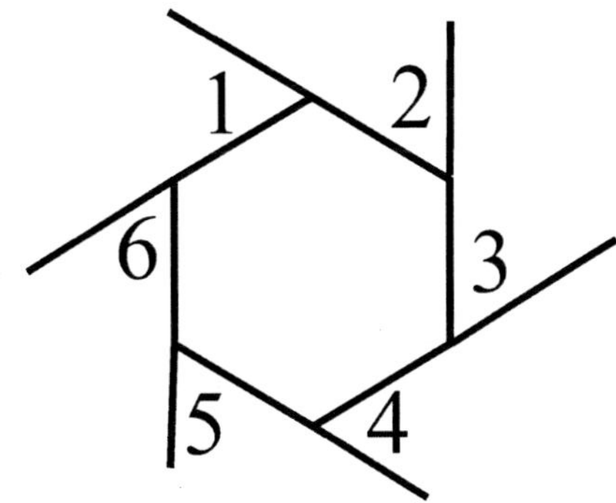

Find the measure of the exterior angles of the regular polygons below.

1.

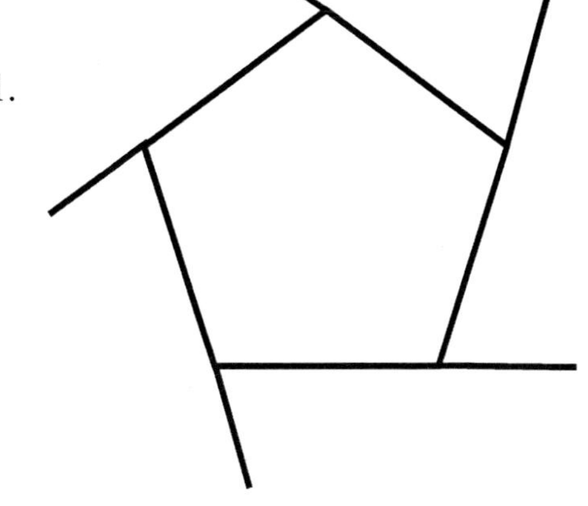

3.

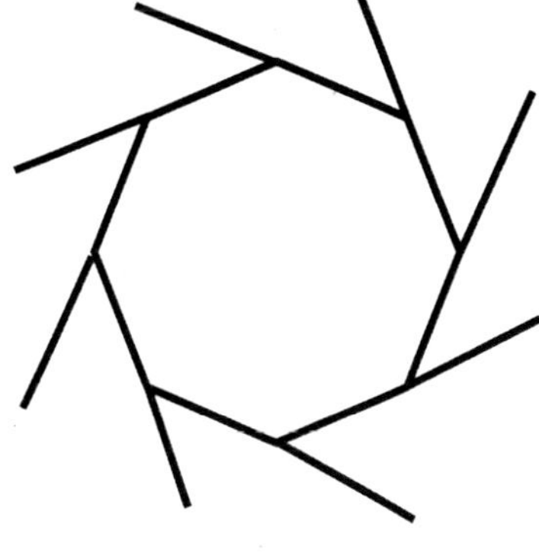

2.

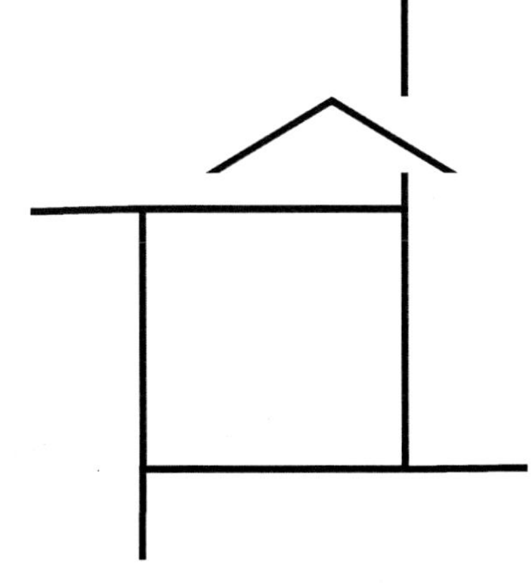

4.

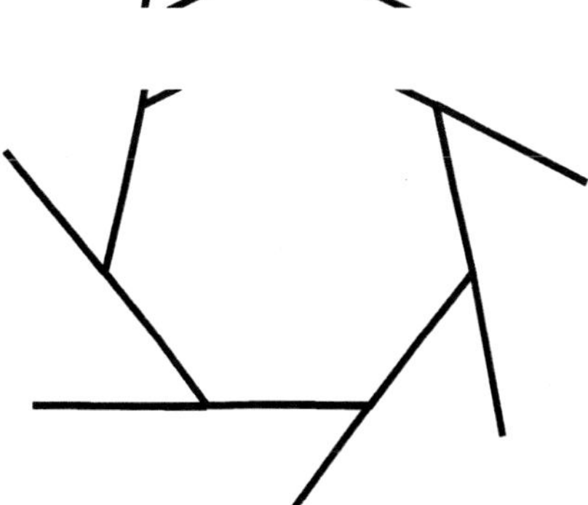

ANGLE APPLICATIONS

Using angles rules, workers in diverse fields can develop new innovations, complete creative projects, and maintain uniform designs.

EXAMPLE 1: An architect examined the drawing below. Using the ancient Incan building system of interlocking similar trapezoidal blocks, the architect hopes to create a wall which is resistant to earthquakes. What is the degree angle of the region labeled x?

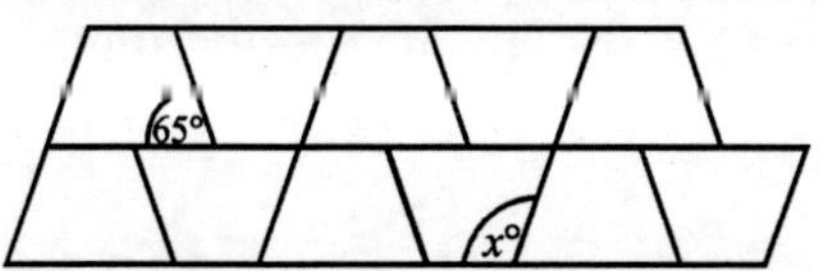

Solution: Using the rules of supplementary angles, we find that the corresponding angle is 180° – 65° = 115°. Using the rules of alternate interior angles, we find the 115° angle to be the same as $m\angle x$.

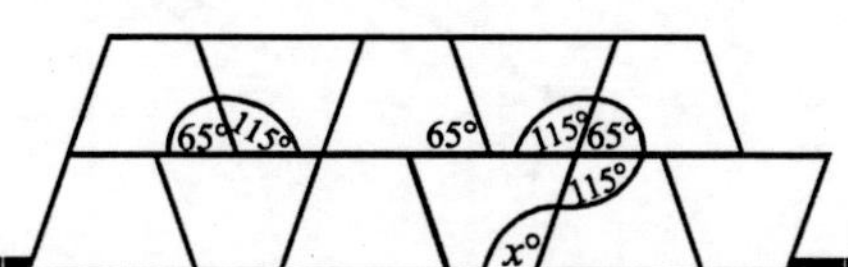

Examine the problems below. Using the rules you have learned in this chapter, solve for x.

1. Mr. Ramirez was installing a new slide at the city park. In order to be certain the slide is welded properly, he will need to know the angle where the supporting beams meet the slide. Knowing the sum of the angle measures of a triangle, which of the following statements is true?

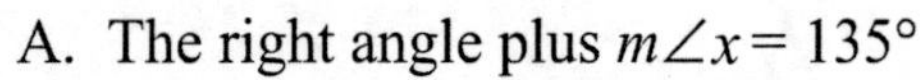

   A. The right angle plus $m\angle x = 135°$
   B. The right angle minus $m\angle x = 30°$
   C. $30° + m\angle x = 45°$
   D. 30° + the right angle = 110°

2. Danny and Eddie were trying to come up with the ideal model for a High-flying kite. Knowing the sum of the angle measures of a triangle and the laws of congruent shapes, which of the following statements is correct?

   A. $\angle d = \angle b$
   B. $\angle a + \angle c = 90°$
   C. $\angle c = 45°$
   D. $\angle d - \angle a = 15°$

3. Examine the quilt pattern on the right. Using the rule of vertical angles and knowing the sum of the angle measures of a triangle, which of the following statements is correct?

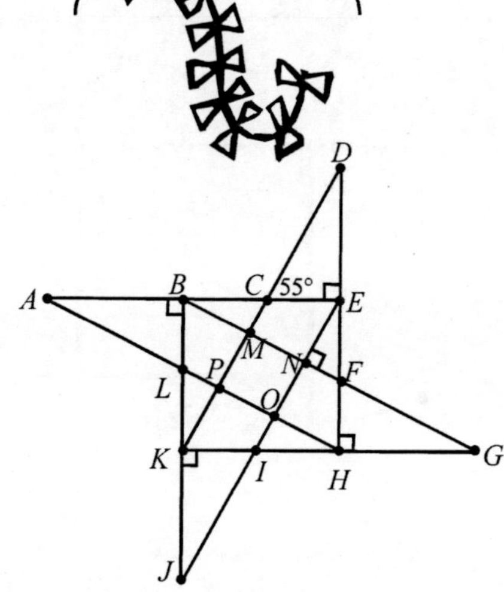

   A. $m\angle LAB + m\angle LPK = 135°$
   B. $m\angle KHL + m\angle HLB = 90°$
   C. $m\angle KHF - m\angle NEF = 25°$
   D. $m\angle MCE - m\angle IOH = 35°$

CHAPTER 13 REVIEW

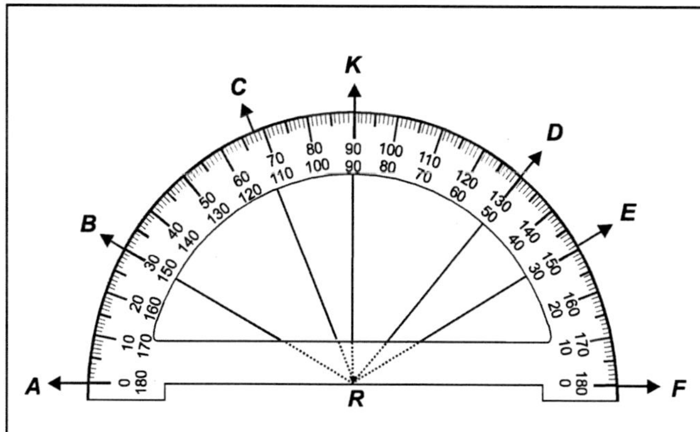

1. What is the measure of $\angle DRA$?

____

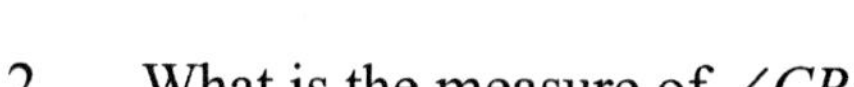

2. What is the measure of $\angle CRF$?

____

3. What is the measure of $\angle ARB$?

____

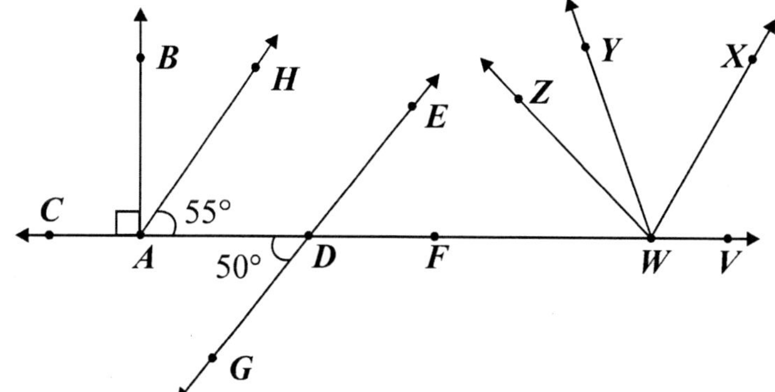

4. Which angle is a supplementary angle to $\angle EDF$?

____

5. What is the measure of angle $\angle GDF$?

____

6. Which 2 angles are right angles?

_______ and _______

7. What is the measure of $\angle EDF$?

____

8. Which angle is adjacent to $\angle BAD$?

____

9. Which angle is a complementary angle to $\angle HAD$?

____

10. What is the measure of $\angle HAB$?

____

11. What is the measure of $\angle CAD$?

____

12. What kind of angle is $\angle FDA$?

____

13. What kind of angle is $\angle GDA$?

____

14. Which angles are adjacent to $\angle EDA$?

_______ and _______

15. Measure $\angle VWX$ with a protractor.

____

16. Measure $\angle FWY$ with a protractor.

____

17. Measure $\angle VWY$ with a protractor.

____

Look at the diagram below. For each pair of angles, state whether they are corresponding (C), alternate interior (I), alternate exterior (E), vertical (V), or supplementary angles (S).

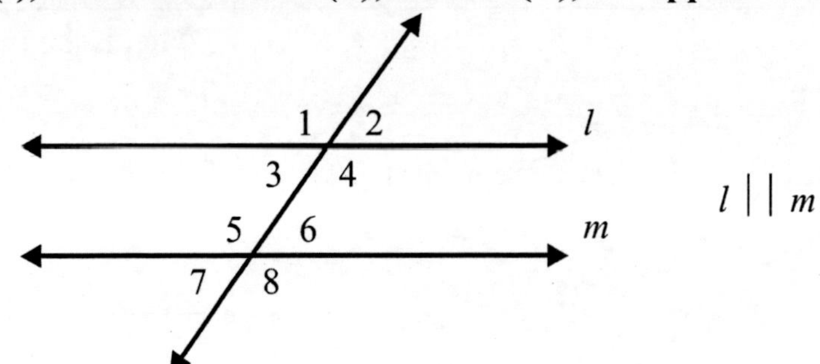

18. $\angle 1$ and $\angle 4$	______	25. $\angle 1$ and $\angle 2$	______
19. $\angle 2$ and $\angle 6$	______	26. $\angle 4$ and $\angle 5$	______
20. $\angle 1$ and $\angle 3$	______	27. $\angle 6$ and $\angle 8$	______
21. $\angle 5$ and $\angle 8$	______	28. $\angle 3$ and $\angle 6$	______
22. $\angle 5$ and $\angle 7$	______	29. $\angle 4$ and $\angle 8$	______
23. $\angle 6$ and $\angle 5$	______	30. $\angle 1$ and $\angle 5$	______
24. $\angle 2$ and $\angle 7$	______	31. $\angle 2$ and $\angle 3$	______

32. The second hand on a clock moves from the 6 to the 9. How many degrees has the second hand traveled?

33. If you constructed a pie chart for J & K's budget, what would be the central angle measure for office expenses?

34. If you constructed a pie chart for J & K's budget, what would be the central angle measure for advertising?

J & K Manufacturing Company 2-million-dollar spending budget	
Cost of Goods Sold	\$750,000
Advertising	\$500,000
Payroll and Benefits	\$350,000
Warehouse Supplies	\$160,000
Office Expenses	\$150,000
Utilities	\$56,000
Other	\$34,000

35. Lem wanted to cut a rusted stop sign into quarters in his shop class. Which of the following statements is factual?

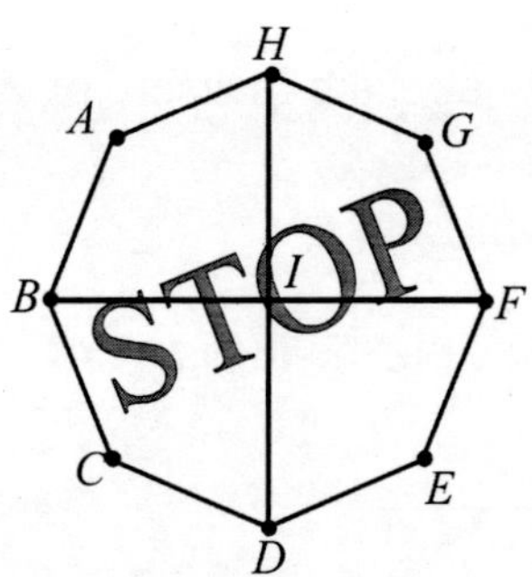

A. $m\angle ABF = 67.5°$
B. $m\angle CBF + m\angle BAH = 135°$
C. $m\angle HGF = 67.5°$
D. $m\angle BAH - m\angle HIF = 35°$

36. A railroad track cuts through two parallel roads. If angle 1 and angle 2 are congruent, which of the following statements is factual?

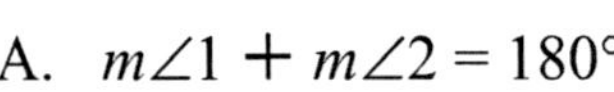

A. $m\angle 1 + m\angle 2 = 180°$
B. $m\angle 1 - m\angle 2 = 0$
C. $m\angle 1 + m\angle 2 = 90°$
D. $m\angle 1 - m\angle 2 = 90°$

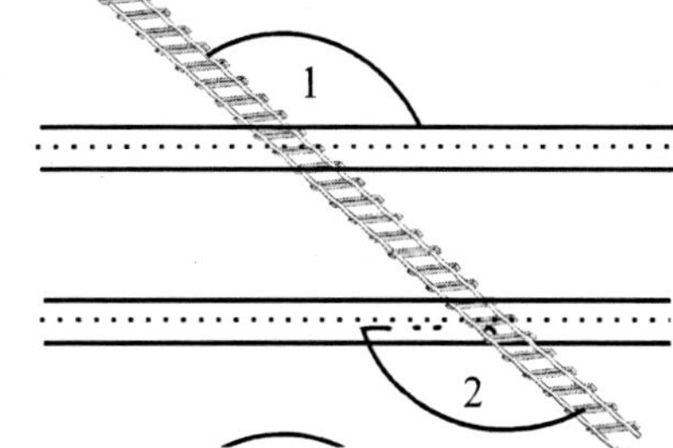

37. Peter is building a stained-glass window for his bathroom. If the blue triangle is isosceles and is also congruent to the yellow triangle, which of the following statements is factual?

A. $m\angle 1 + m\angle 2 = 180°$
B. $m\angle 1 - m\angle 2 = 90°$
C. $m\angle 1 - m\angle 2 = 30°$
D. $m\angle 1 + m\angle 2 = 90°$

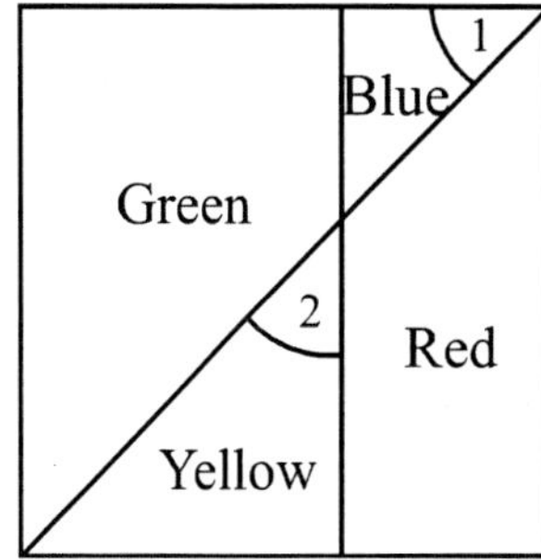

38. What is the sum of the measures of the interior angles in the figure below?

____

39. If you were given a 10 sided regular polygon, what would be the measure of the exterior angles?

____

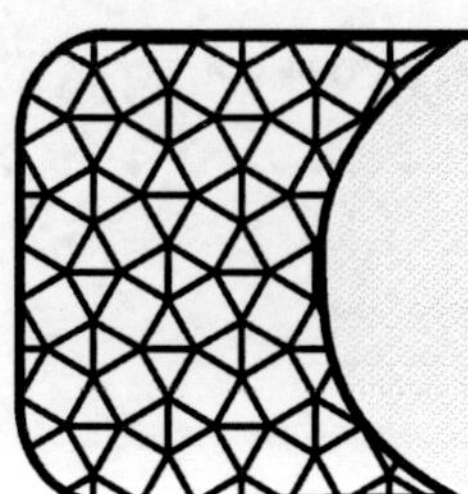

CHAPTER 14: TRIANGLES

Standards 2.12.2, 4.12.2 and 4.12.7

INTERIOR ANGLES OF A TRIANGLE

The three interior angles of a triangle always add up to 180°.

EXAMPLE 1:

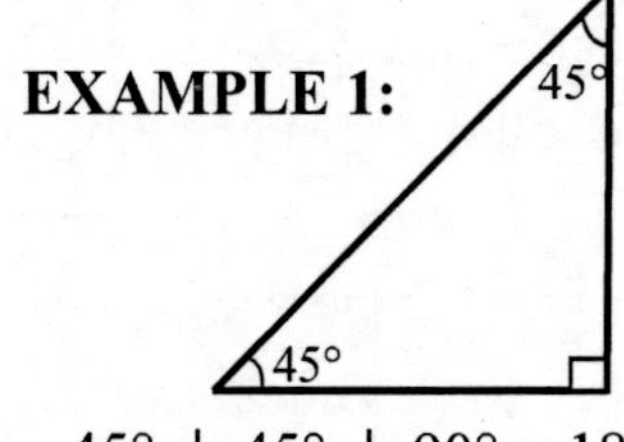

$45° + 45° + 90° = 180°$

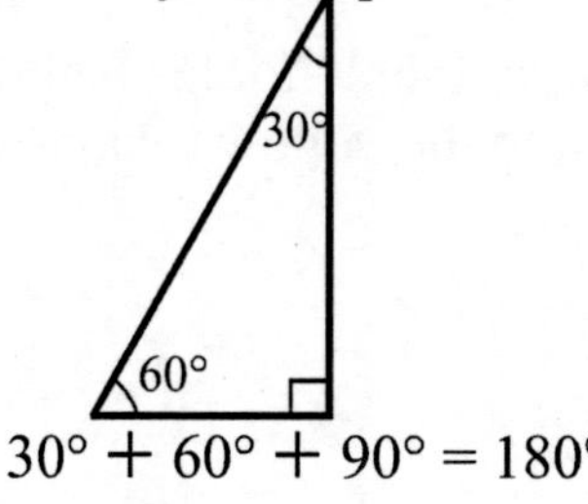

$30° + 60° + 90° = 180°$

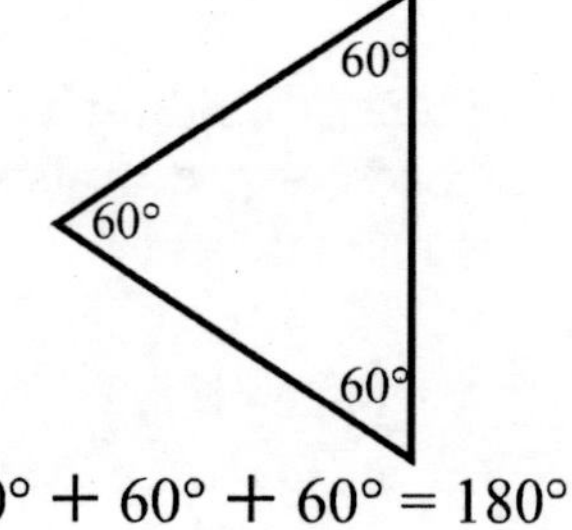

$60° + 60° + 60° = 180°$

EXAMPLE 2: Find the missing angle in the triangle.

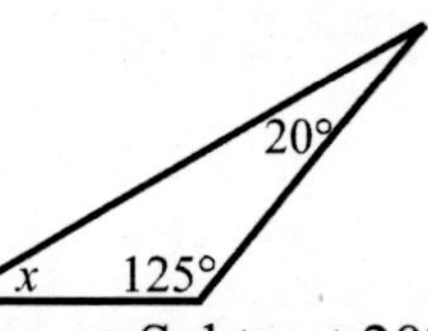

Solution:

$$\begin{array}{rl} 20° + 125° + x = 180° & \\ -20° -125° \quad -20° -125° & \\ \hline x = 180° - 20° - 125° & \\ x = 35°. & \end{array}$$

Subtract 20° and 125° from both sides to get x by itself.

The missing angle is 35°.

Find the missing angle in the triangles.

1. 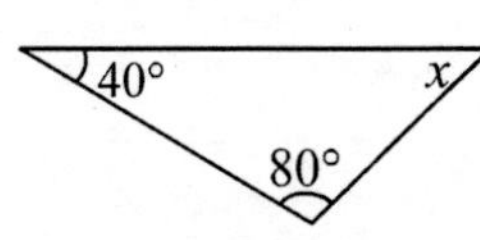

4. 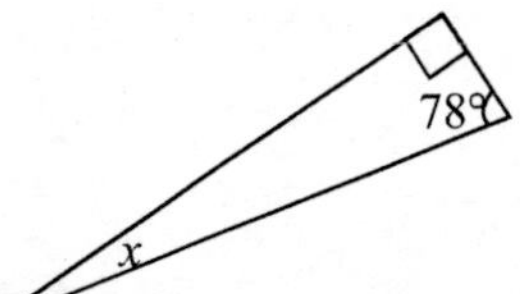

7. 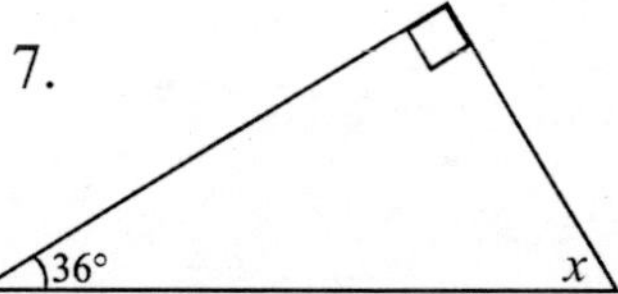

2. 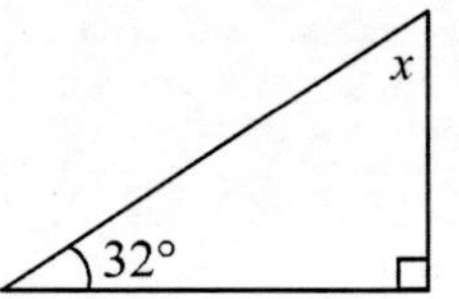

5. 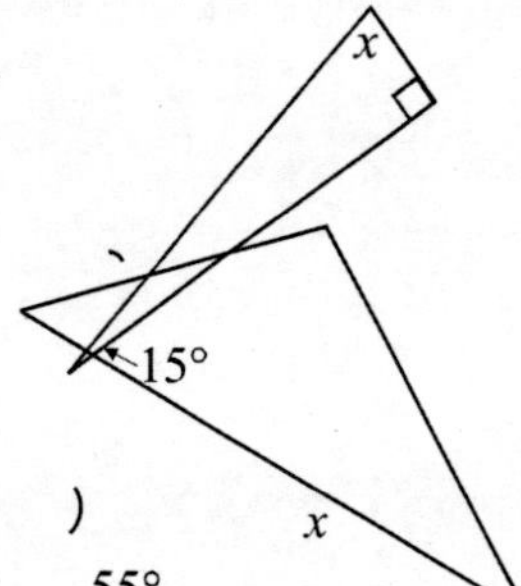

8. 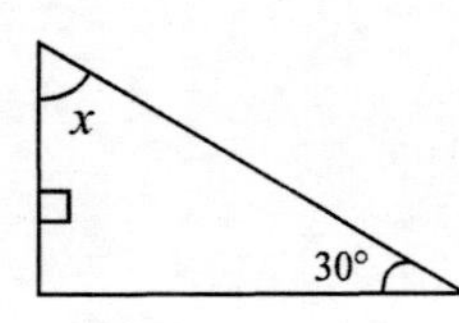

3. 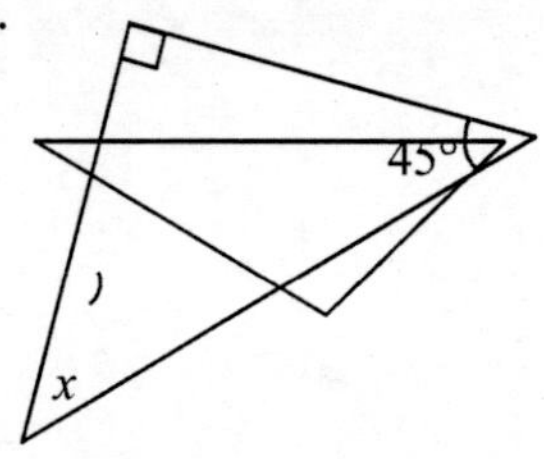

6. 

9. 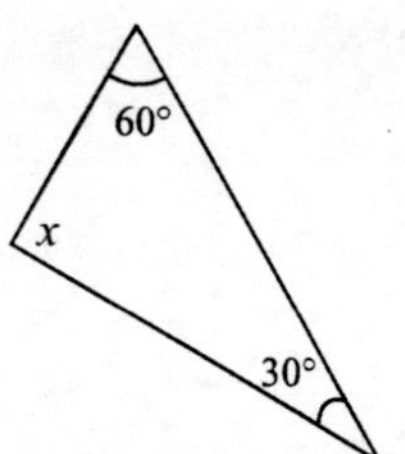

SOLVING PROPORTIONS

Two **ratios (fractions)** that are **equal** to each other are called **proportions. For example,** $\frac{1}{4} = \frac{2}{8}$.
Read the following example to see how to find a number missing from a proportion.

EXAMPLE: $\frac{5}{15} = \frac{8}{x}$

Step 1: To find x, you first multiply the two numbers that are diagonal to each other. **15 × 8 = 120** $\frac{5}{\boxed{15}} = \frac{\boxed{8}}{x}$

Step 2 Then, divide the product (120) by the other number in the proportion (5).
120 ÷ 5 = 24

Therefore, $\frac{5}{15} = \frac{8}{24}$ $\mathbf{x = 24}$

Practice finding the number missing from the following proportions. First, multiply the two numbers that are diagonal from each other. Then divide by the other number.

1. $\frac{2}{5} = \frac{6}{x}$
2. $\frac{9}{3} = \frac{x}{5}$
3. $\frac{x}{12} = \frac{3}{4}$
4. $\frac{7}{x} = \frac{3}{9}$
5. $\frac{12}{x} = \frac{2}{5}$
6. $\frac{12}{x} = \frac{4}{3}$
7. $\frac{27}{3} = \frac{x}{2}$
8. $\frac{1}{x} = \frac{3}{12}$
9. $\frac{15}{2} = \frac{x}{4}$
10. $\frac{7}{14} = \frac{x}{6}$
11. $\frac{5}{6} = \frac{10}{x}$
12. $\frac{4}{x} = \frac{3}{6}$
13. $\frac{x}{5} = \frac{9}{15}$
14. $\frac{9}{18} = \frac{x}{2}$
15. $\frac{5}{7} = \frac{35}{x}$
16. $\frac{x}{2} = \frac{8}{4}$
17. $\frac{15}{20} = \frac{x}{8}$
18. $\frac{x}{40} = \frac{5}{100}$
19. $\frac{4}{7} = \frac{x}{28}$
20. $\frac{7}{6} = \frac{42}{x}$
21. $\frac{x}{8} = \frac{1}{4}$

SIMILAR TRIANGLES

Two triangles are similar if the measurements of the three angles in both triangles are the same. If the three angles are the same, then their corresponding sides are proportional.

CORRESPONDING SIDES - The triangles below are similar. Therefore, the two shortest sides from each triangle, ***c*** and ***f***, are corresponding. The two longest sides from each triangle, ***a*** and ***d***, are corresponding. The two medium length sides, ***b*** and ***e***, are corresponding.

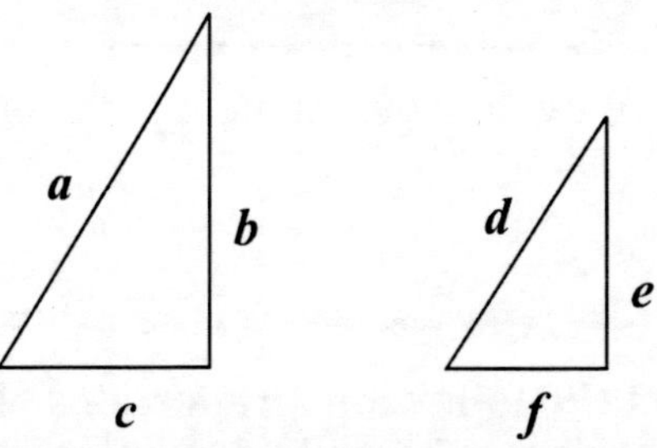

PROPORTIONAL - The corresponding sides of similar triangles are proportional to each other. This means if we know all the measurements of one triangle, and we only know one measurement of the other triangle, we can figure out the measurements of the other two sides with proportion problems. The two triangles below are similar.

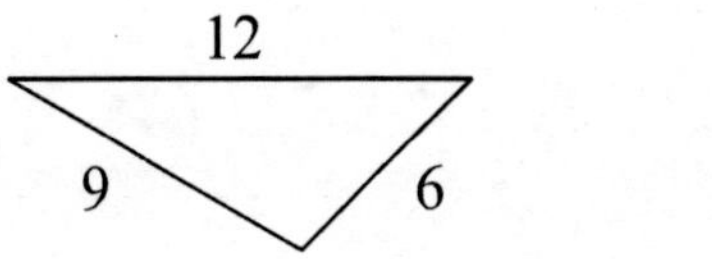

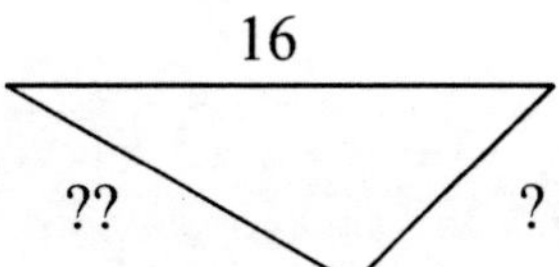

Note: **To set up the proportion correctly, it is important to keep the measurements of each triangle on opposite sides of the equal sign.**

To find the short side:

Step 1: Set up the proportion.

$\frac{\textit{long side}}{\textit{short side}} \quad \frac{12}{6} = \frac{16}{?}$

Step 2: Solve the proportion as you did on the previous page.

$16 \times 6 = 96$
$96 \div 12 = 8$

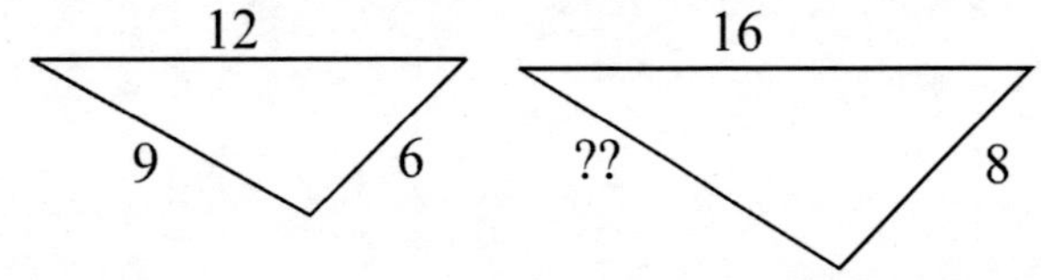

To find the medium length side:

Step 1: Set up the proportion.

$\frac{\textit{long side}}{\textit{medium}} \quad \frac{12}{9} = \frac{16}{??}$

Step 2: Solve the proportion as you Did on the previous page.

$16 \times 9 = 144$
$144 \div 12 = 12$

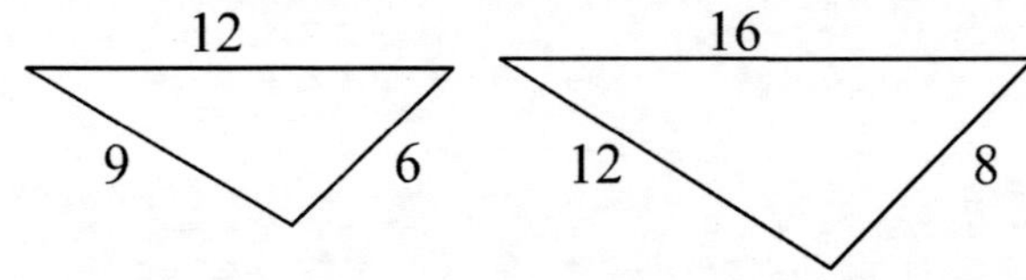

MORE SIMILAR TRIANGLES

Find the missing side from the following similar triangles.

1.

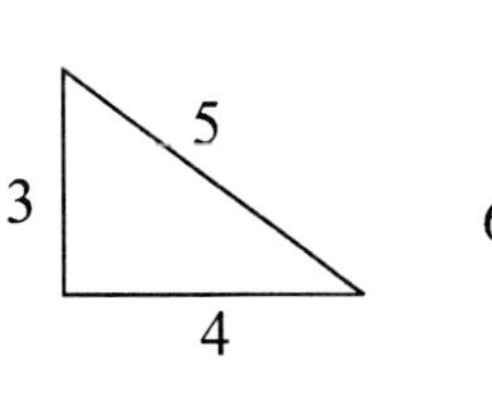

5
3
4
?
6

5.

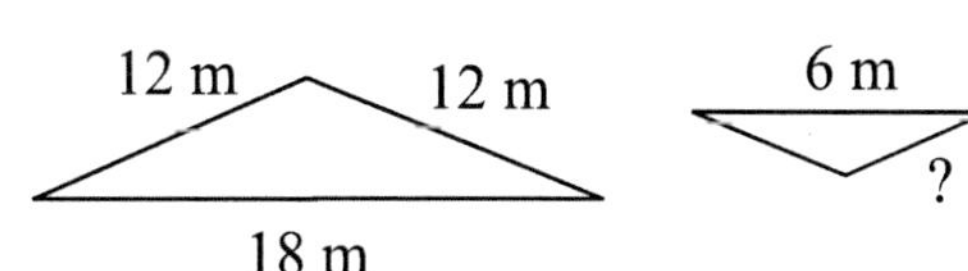

12 m
12 m
18 m
6 m
?

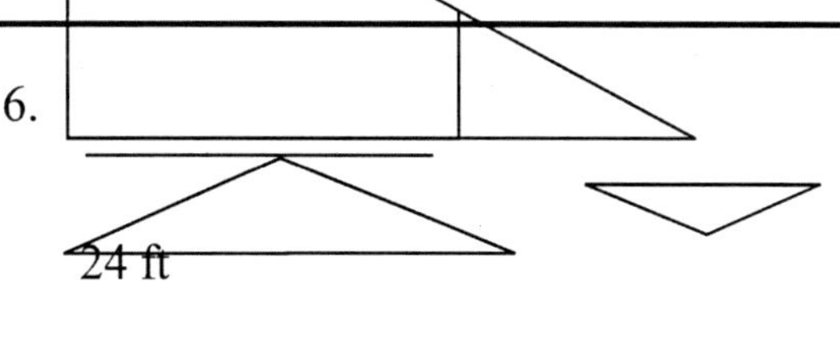

2.

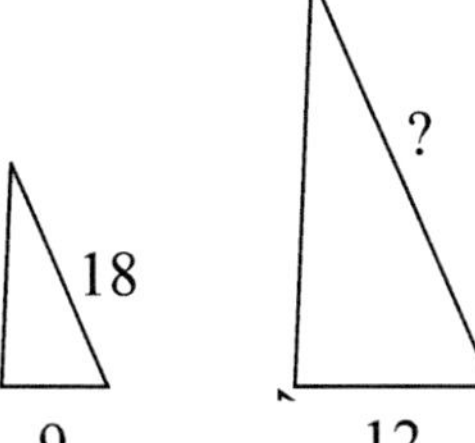

?
18
9
12

6.

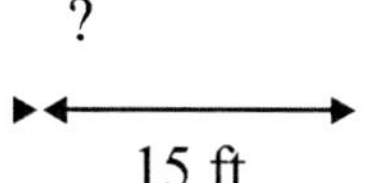

24 ft
?
25 ft
15 ft

3.

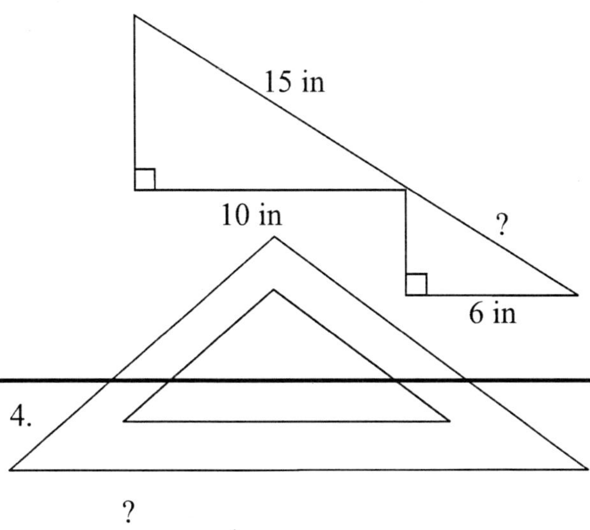

15 in
10 in
?
6 in

4.

?
4
5
10

7.

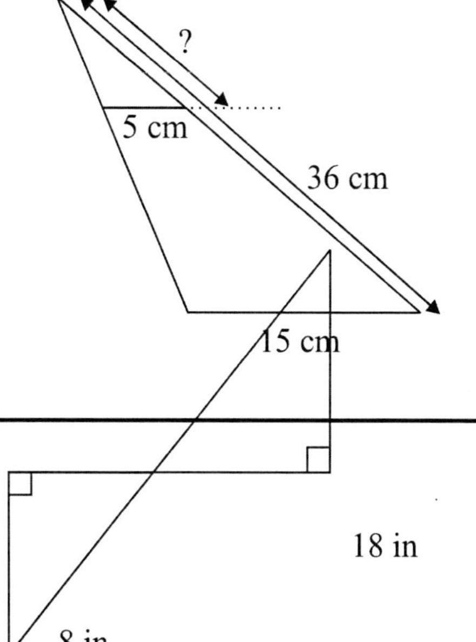

?
5 cm
36 cm
15 cm

8.

18 in
8 in
?
12 in

USING PROPORTIONS WITH PARALLEL LINES

Proportions can be found using parallel lines and transversals. When three parallel lines intersect two transversals, proportionate line segments are formed.

EXAMPLE 1: In the figure below, line a is parallel to line b and lines y and z are transversals. Find the measure of x.

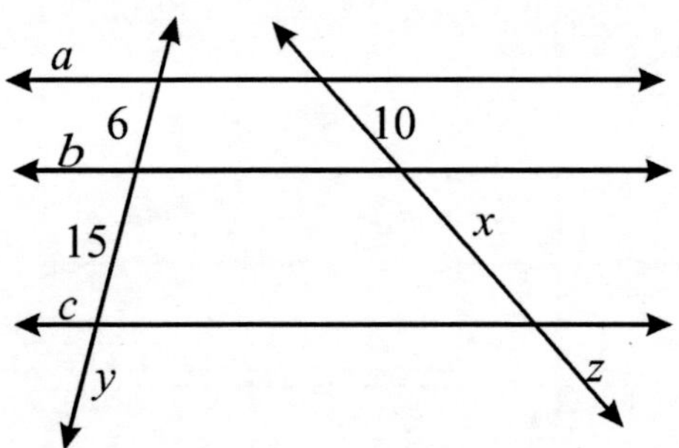

In the figure above, two trapezoids are formed. There is one small trapezoid with line a as its top; and a larger trapezoid with line b as its top. These trapezoids are similar to each other, and therefore, their sides are proportionate.

Step 1: Set up the proportions.

$$\frac{\text{shorter (left) side of the small trapezoid}}{\text{longer (right) side of the small trapezoid}} = \frac{\text{shorter side of the large trapezoid}}{\text{longer side of the large trapezoid}}$$

Step 2: Plug the values into the proportion and solve for x.

$\frac{6}{15} = \frac{10}{x}$ Cross multiply.

$6x = 150$

$x = 25$ The length of the longer side of the trapezoid is 25.

SIMILAR FIGURES

The measures of corresponding sides of similar figures can also be found by setting up a proportion.

EXAMPLE 1: The following rectangles are similar.

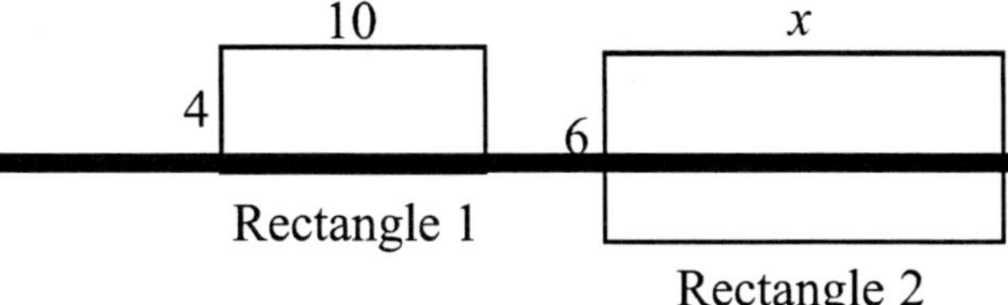

Rectangle 1

Rectangle 2

Step 1: Set up the proportion.

$$\frac{\text{Short side rectangle 1}}{\text{Long side rectangle 1}} = \frac{\text{Short side rectangle 2}}{\text{Long side rectangle 2}}$$

Step 2: $\frac{4}{10} = \frac{6}{x}$ Cross multiply.

$4x = 60$

$x = 15$

All of the pairs of figures below are similar. Find the missing side for each pair.

1.

2.

4.

5.

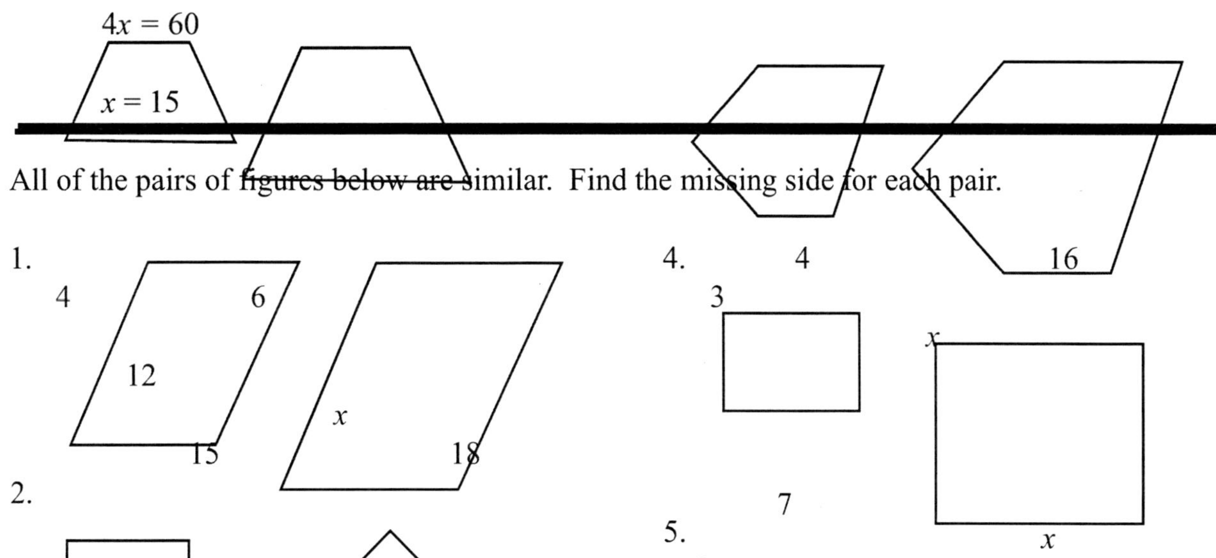

3.

6.

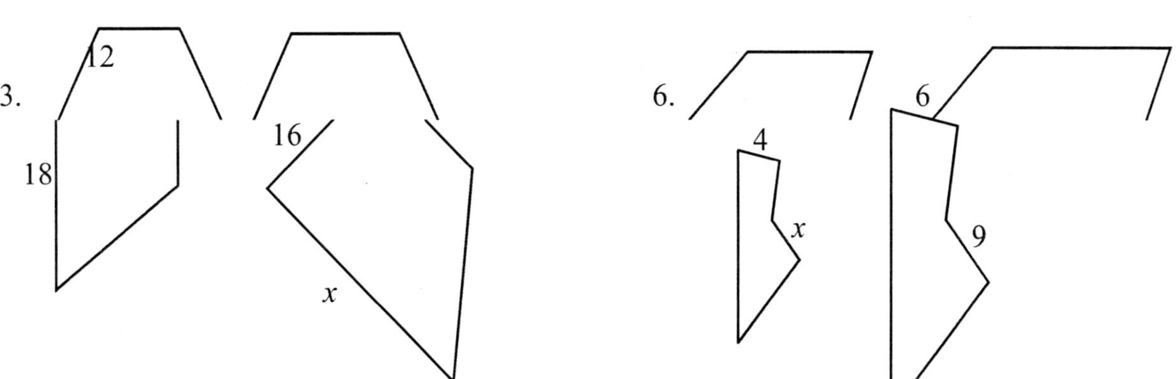

PYTHAGOREAN THEOREM

Pythagoras was a Greek mathematician and philosopher who lived around 600 B.C. He started a math club among Greek aristocrats called the Pythagoreans. Pythagoras formulated the **Pythagorean Theorem** which states that in a **right triangle**, the sum of the squares of the legs of the triangle are equal to the square of the hypotenuse. Most often you will see this formula written as $a^2 + b^2 = c^2$. **This relationship is only true for right triangles.**

EXAMPLE: Find the length of side *c*.

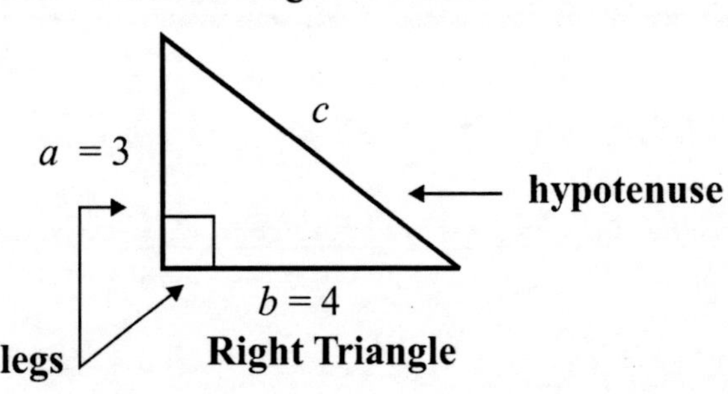

Formula: $a^2 + b^2 = c^2$

$3^2 + 4^2 = c^2$

$9 + 16 = c^2$

$25 = c^2$

$\sqrt{25} = \sqrt{c^2}$

$5 = c$

Find the hypotenuse of the following triangles. Leave answers in radical form. The first one has been done for you.

1.

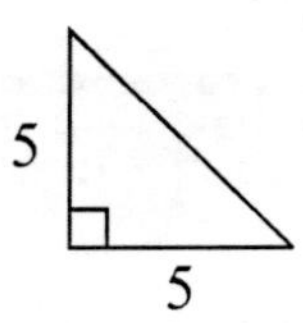

5

5

$c = \underline{5\sqrt{2}}$

2.

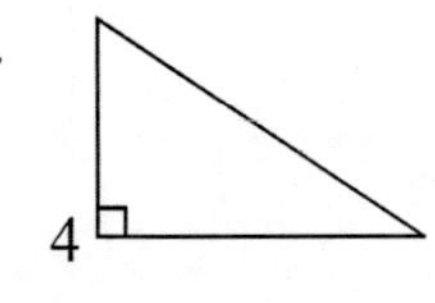

4

8

$c =$ ______

3.

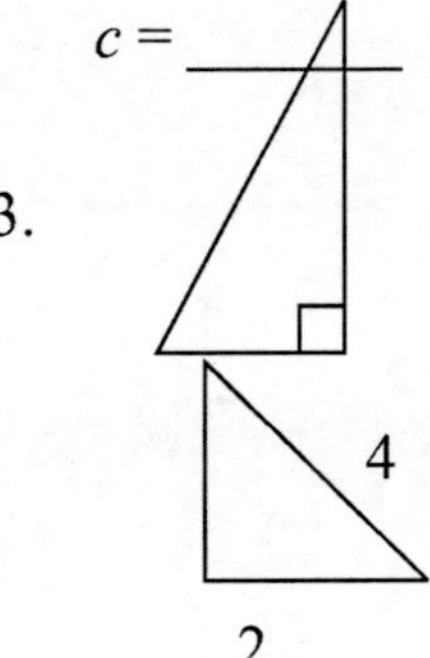

4

2

$c =$ ______

4.

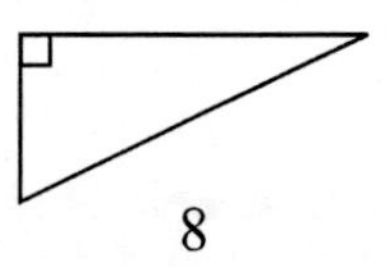

8

1

$c =$ ______

5.

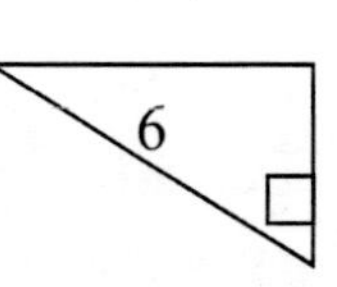

6

3

$c =$ ______

6.

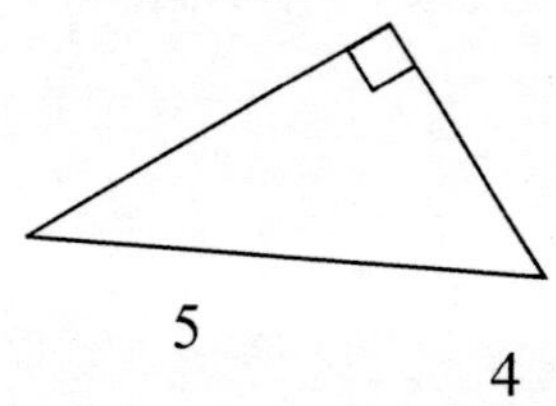

5

4

$c =$ ______

7.

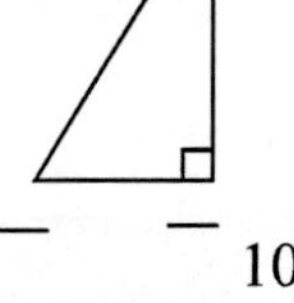

10

3

$c =$ ______

8.

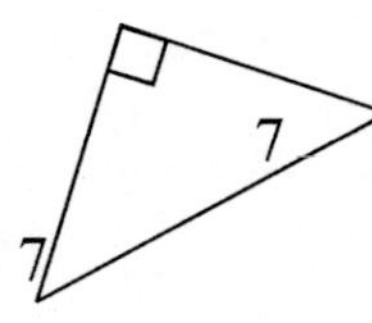

7

7

$c =$ ______

9.

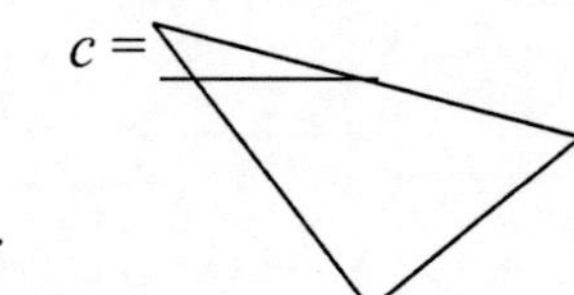

4

3

$c =$ ______

FINDING THE MISSING LEG OF A RIGHT TRIANGLE

In the right triangle shown, the measurement of the hypotenuse is known as well as one of the legs. To find the measurement of the other leg, use the Pythagorean Theorem by filling in the known measurements, and then solve for the unknown side.

In the formula, $a^2 + b^2 = c^2$, a and b are the legs, and c is always the hypotenuse.
$9^2 + b^2 = 41^2$. Now solve for b algebraically.

$81 + b^2 = 1681$
$b^2 = 1681 - 81$
$b^2 = 1600$
$\sqrt{b^2} = \sqrt{1600}$
$b = 40$

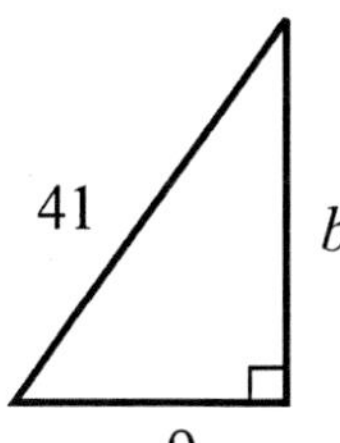

Practice finding the measure of the missing leg in each right triangle below. Simplify square roots.

1. 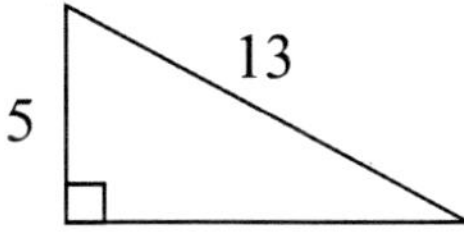

____

2. 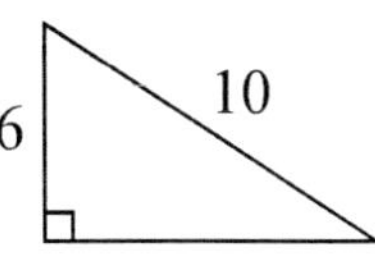

____

3. 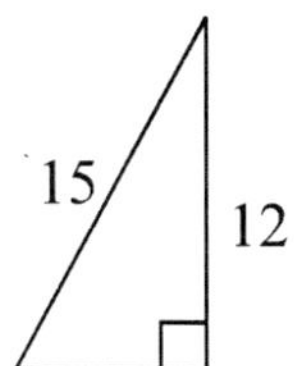

____

4. 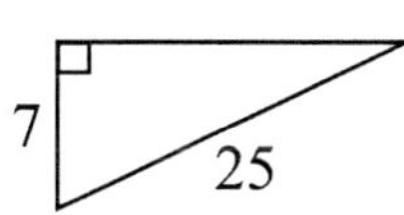

____

5. 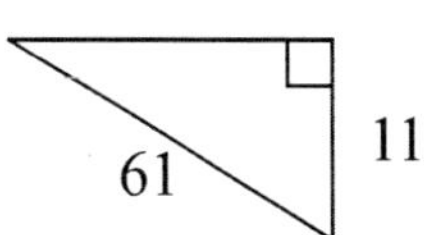

____

6. 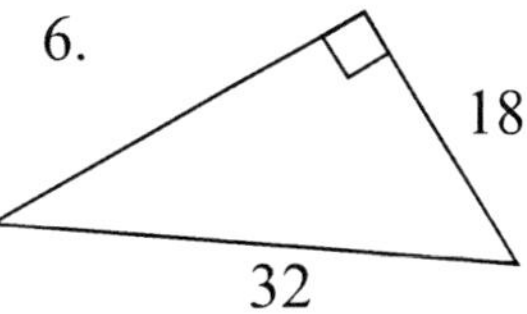

____

7. 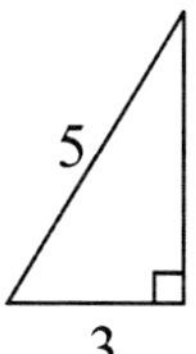

____

8. 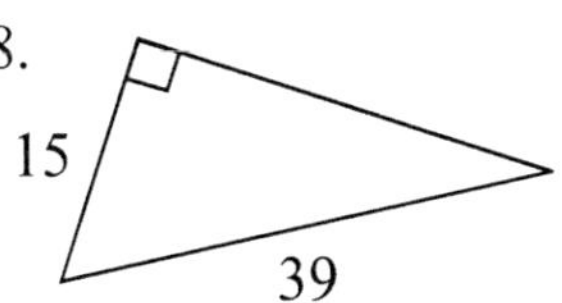

____

9. 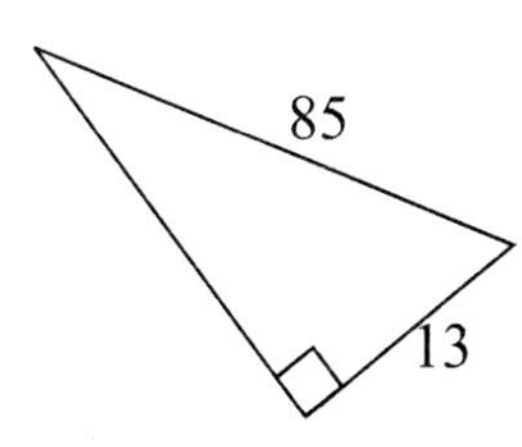

____

APPLICATIONS OF THE PYTHAGOREAN THEOREM

The Pythagorean Theorem can be used to determine the distance between two points in some situations. Recall that the formula is written $a^2 + b^2 = c^2$.

EXAMPLE: Find the distance between point B and point A given that the length of each square is 1 inch long and 1 inch wide.

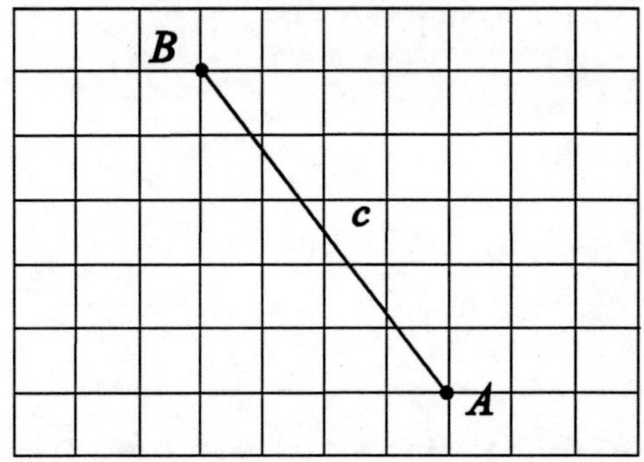

Step 1: Draw a straight line between the two points. We will call this side c.

Step 2: Draw two more lines, one from point B and one from point A. These lines should make a 90° angle. The two new lines will be labeled a and b. Now we can use the Pythagorean Theorem to find the distance from Point B to Point A.

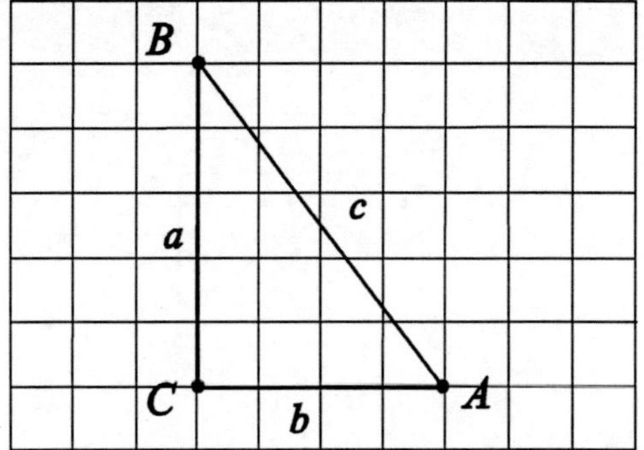

Step 3: Find the length of a and c by counting the number of squares each line has. We find that $a = 5$ inches and $b = 4$ inches. Now, substitute the values found into the Pythagorean Theorem.

$$a^2 + b^2 = c^2$$
$$5^2 + 4^2 = c^2$$
$$25 + 16 = c^2$$
$$41 = c^2$$
$$\sqrt{41} = \sqrt{c^2}$$
$$\sqrt{41} = c$$

Use the Pythagorean theorem to find the distances asked.

Below is a diagram of the mall. Use the grid to help answer questions 1 and 2. Each square is 25 feet × 25 feet.

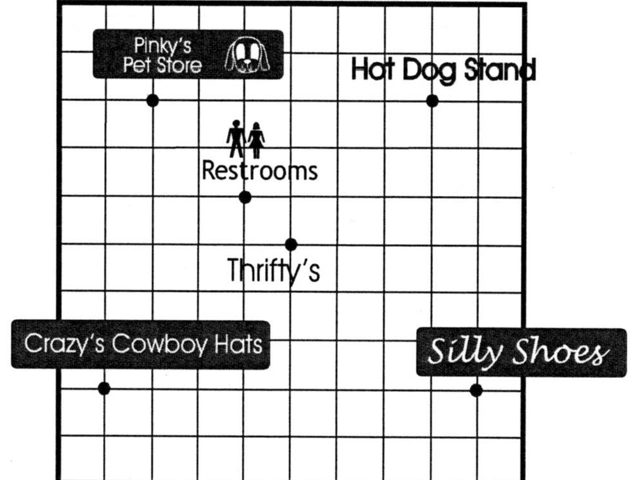

1. Marty walks from Pinky's Pet Store to the restrooms to wash his hands. How far did he walk?

2. Betty needs to meet her friend at Silly Shoes, but she wants to get a hot dog first. If Betty is at Thrifty's, how far will she walk to meet her friend?

Below is a diagram of a football field. Use the grid on the football field to help find the answers to questions 3 and 4. Each square is 10 yards × 10 yards.

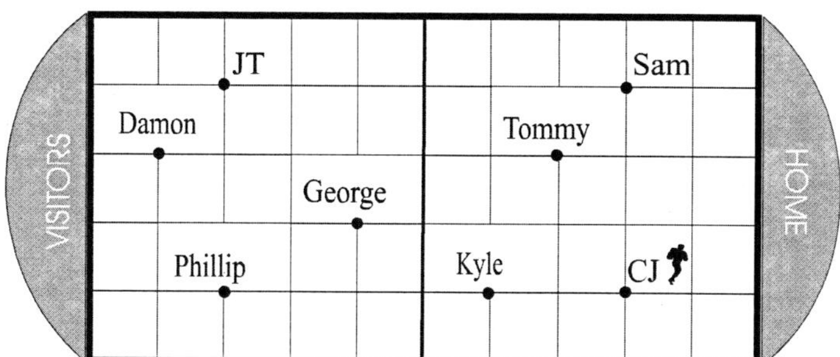

3. George must throw the football to a teammate before he is tackled. If CJ is the only person open, how far must George be able to throw the ball?

4. Damon has the football and is about to make a touchdown. If Phillip tries to stop him, how far must he run to reach Damon?

SPECIAL RIGHT TRIANGLES

Two right triangles are special right triangles if they have fixed ratios among their sides.

45-45-90 Triangles

In a 45-45-90 triangle, the two sides opposite the 45° angles will always be equal. The length of the hypotenuse is $\sqrt{2}$ times the length of one of the sides opposite a 45° angle.

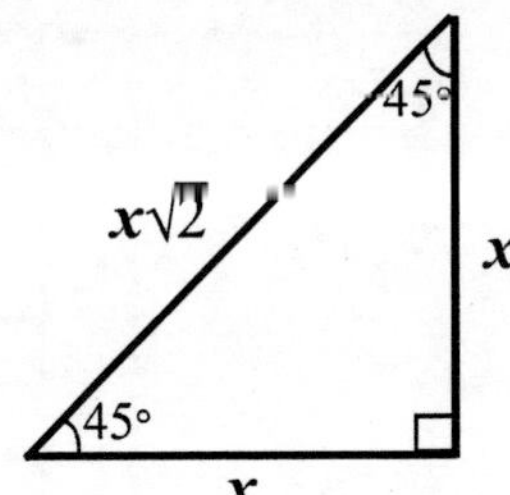

EXAMPLE 1: What are the lengths of sides a and b?

Step 1: The two sides opposite the 45° angles are equal. Therefore, side $b = 3$.

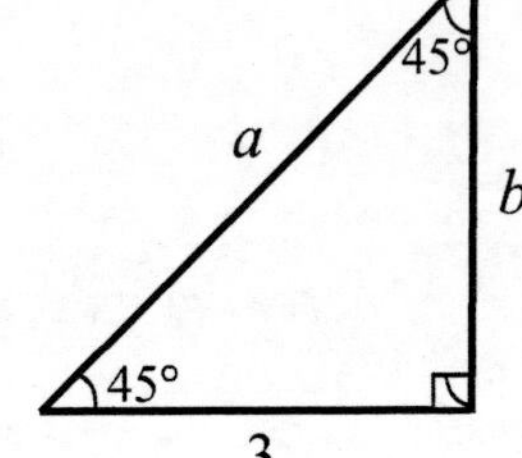

Step 2: The hypotenuse is $\sqrt{2}$ × the length of a side opposite a 45° angle.
Therefore, $a = 3 \times \sqrt{2}$
Simplify: $a = 3\sqrt{2}$

30-60-90 Triangles

In a 30-60-90 triangle, the side opposite the 30° angle is the shortest leg. The side opposite the 60° angle is $\sqrt{3}$ times as long as the shortest leg, and the hypotenuse is twice as long as the shortest leg.

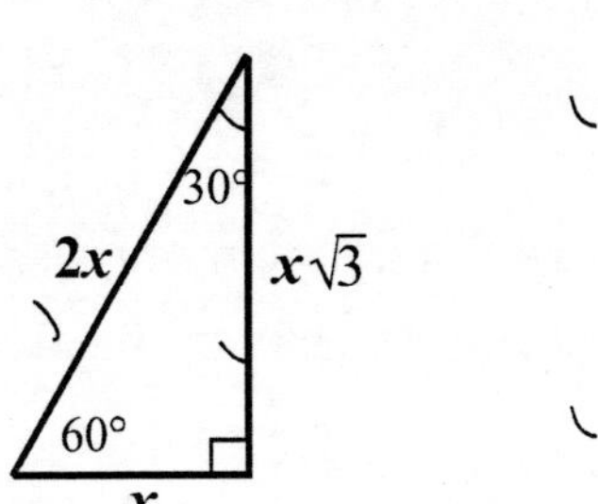

EXAMPLE 2: What are the lengths of sides a and b?

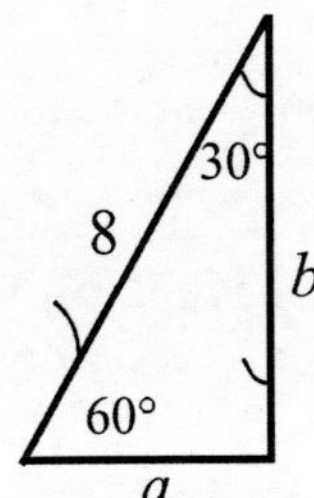

Step 1: The hypotenuse is 2 times the side opposite the 30° angle. Write the above sentence using algebra and then solve.

$$8 = 2a$$
$$\frac{8}{2} = \frac{2a}{2}$$
$$4 = a$$

Step 2: Now that it is known that the shortest leg has a length of 4, the side opposite the 60° angle can be calculated easily.

$$b = a \times \sqrt{3}$$
$$b = 4 \times \sqrt{3}$$
$$b = 4\sqrt{3}$$

Find the missing leg of each of the special right triangles. Simplify your answers.

1. 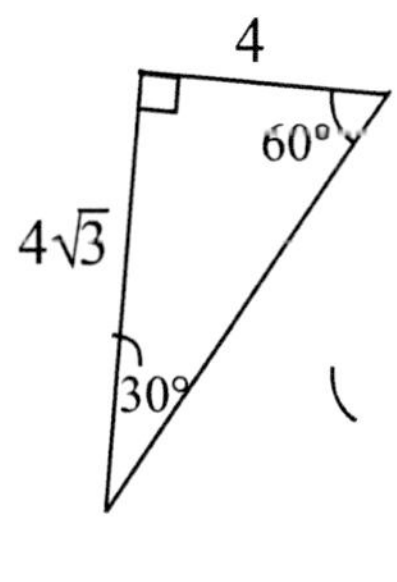

____

3. 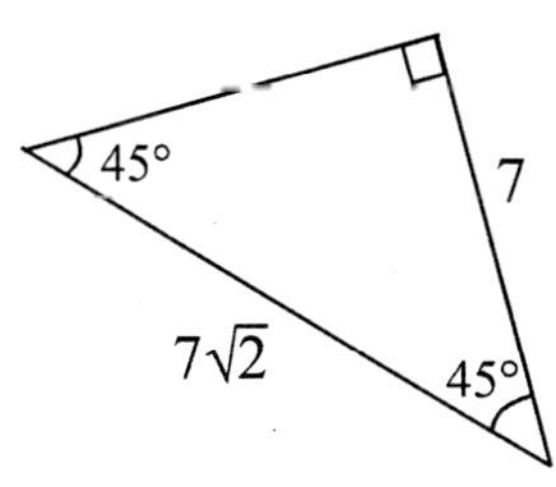

____

5. 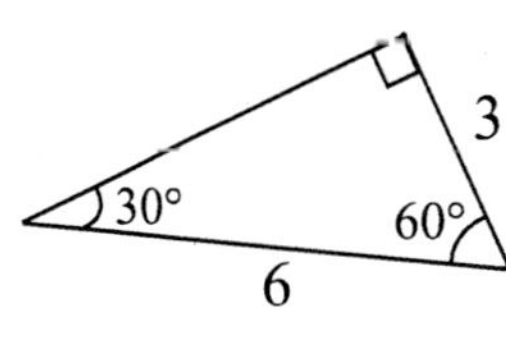

____

2. 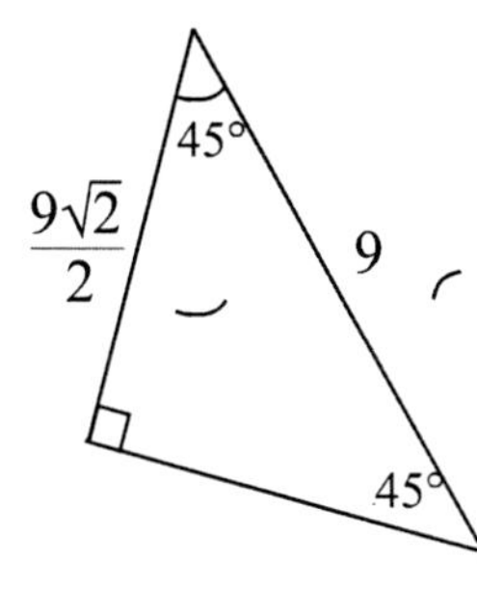

____

4. 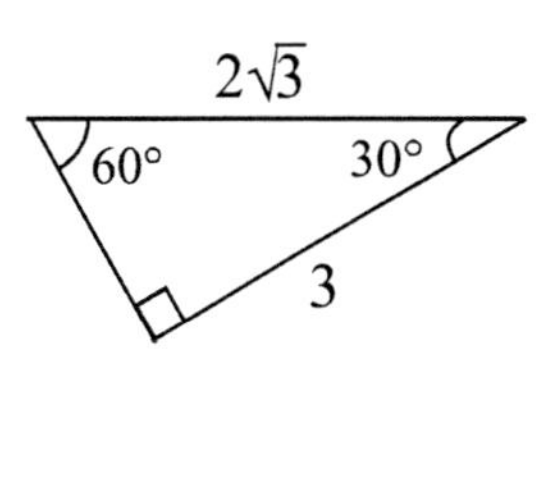

____

6. 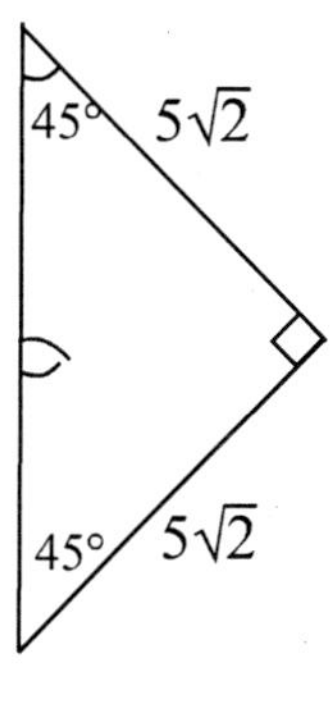

____

Find the lengths of sides *a* and *b* in each of the special right triangles.

7. 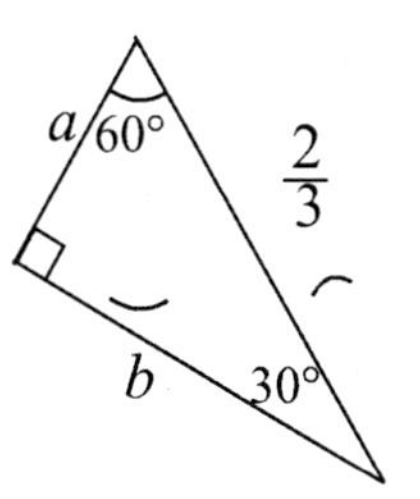

$a =$ _____ $b =$ _____

9. 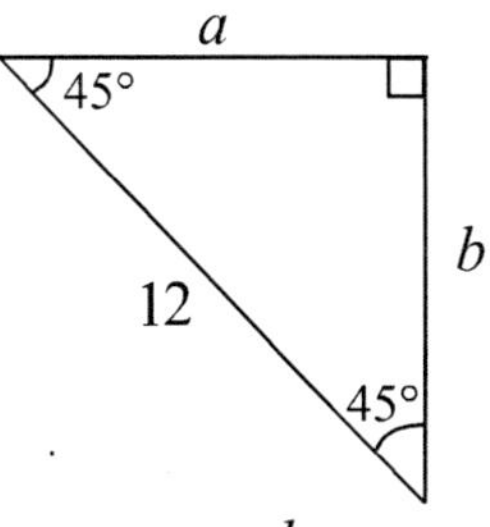

$a =$ _____ $b =$ _____

11. 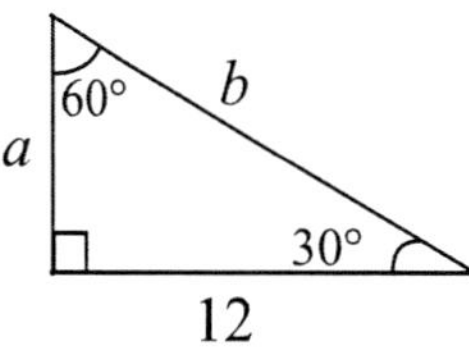

$a =$ _____ $b =$ _____

8. 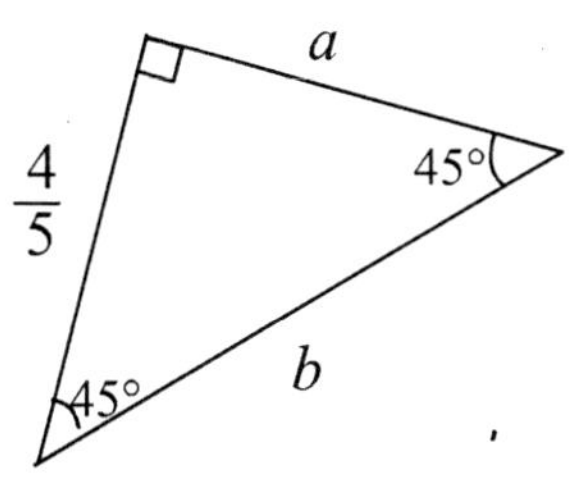

$a =$ _____ $b =$ _____

10. 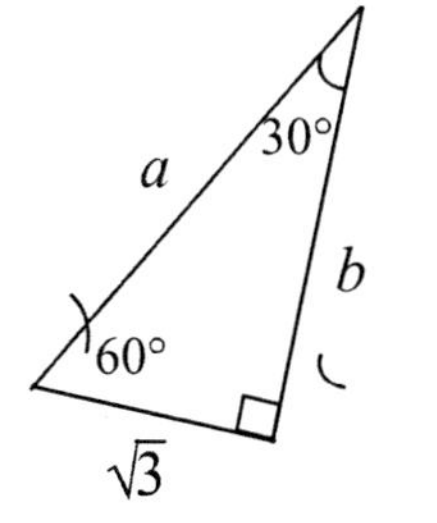

$a =$ _____ $b =$ _____

12. 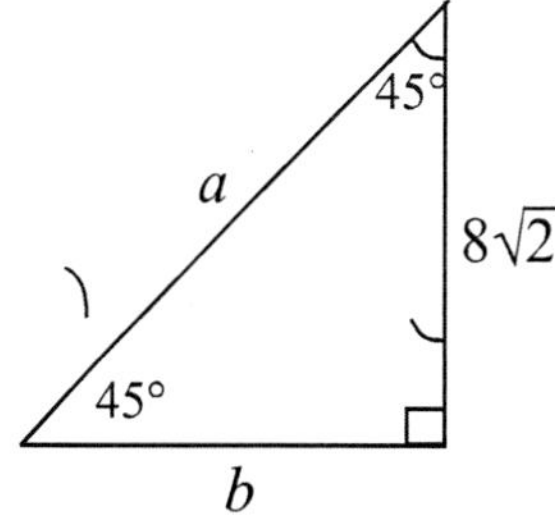

$a =$ _____ $b =$ _____

BASIC TRIGONOMETRIC RATIOS

Trigonometry is a mathematical topic that applies the relationships between sides and angles in right triangles. Recall that a right triangle has one 90° angle and two acute angles. Consider the right triangle shown below. Note that the angles are labeled with capital letters. The sides are labeled with lowercase letters that correspond to the angles opposite them.

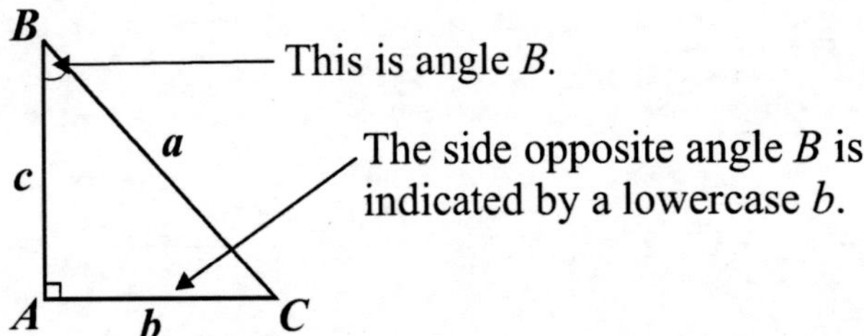

Trigonometric ratios are ratios of the measures of two sides of a right triangle and are related to the acute angles of a right triangle, not the right angle. The value of a trigonometric ratio is dependent only on the size of the acute angle and is not affected by the lengths of the sides of the triangle.

We will consider the three basic trigonometric ratios in this section: **sine, cosine, and tangent**. Definitions and descriptions of the sine, cosine, and tangent functions are presented below.

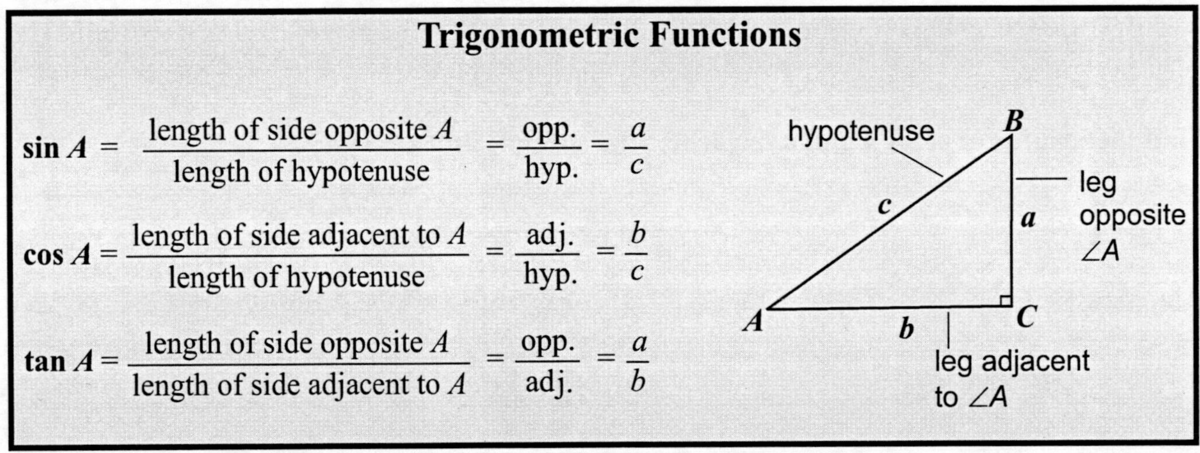

$$\sin A = \frac{\text{length of side opposite } A}{\text{length of hypotenuse}} = \frac{\text{opp.}}{\text{hyp.}} = \frac{a}{c}$$

$$\cos A = \frac{\text{length of side adjacent to } A}{\text{length of hypotenuse}} = \frac{\text{adj.}}{\text{hyp.}} = \frac{b}{c}$$

$$\tan A = \frac{\text{length of side opposite } A}{\text{length of side adjacent to } A} = \frac{\text{opp.}}{\text{adj.}} = \frac{a}{b}$$

EXAMPLE 1: For right triangle ABC, find $\sin A$, $\cos A$, $\tan A$, $\sin C$, $\cos C$, and $\tan C$.

$$\sin A = \frac{\text{opp.}}{\text{hyp.}} = \frac{3}{5} = 0.6 \qquad \sin C = \frac{\text{opp.}}{\text{hyp.}} = \frac{4}{5} = 0.8$$

$$\cos A = \frac{\text{adj.}}{\text{hyp.}} = \frac{4}{5} = 0.8 \qquad \cos C = \frac{\text{adj.}}{\text{hyp.}} = \frac{3}{5} = 0.6$$

$$\tan A = \frac{\text{opp.}}{\text{adj.}} = \frac{3}{4} = 0.75 \qquad \tan C = \frac{\text{opp.}}{\text{adj.}} = \frac{4}{3} = 1.\overline{3}$$

Find sin A, cos A, tan A, sin B, cos B, and tan B in each of the following right triangles. Express answers as fractions.

1.

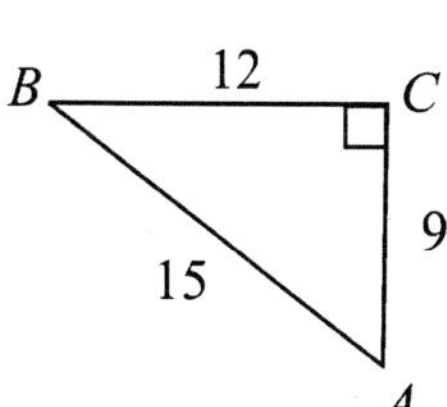

2.

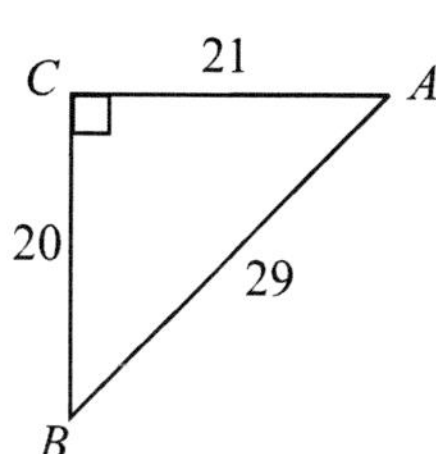

3.

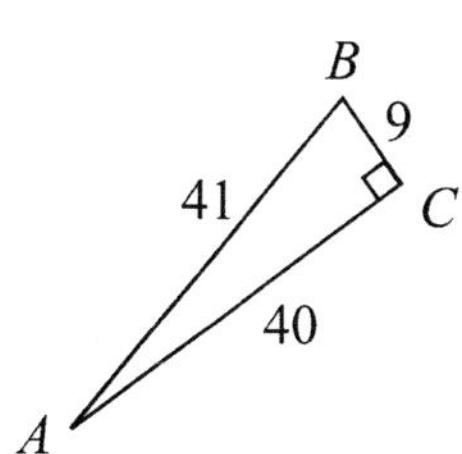

4.

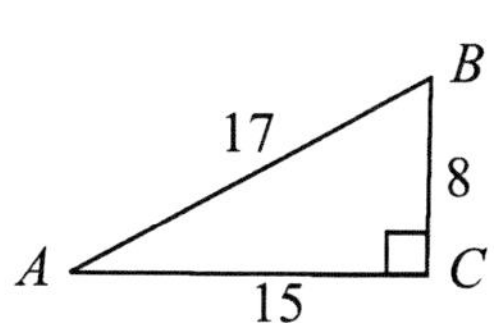

5.

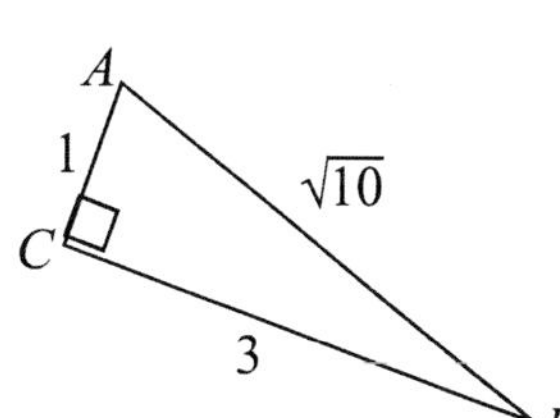

6.

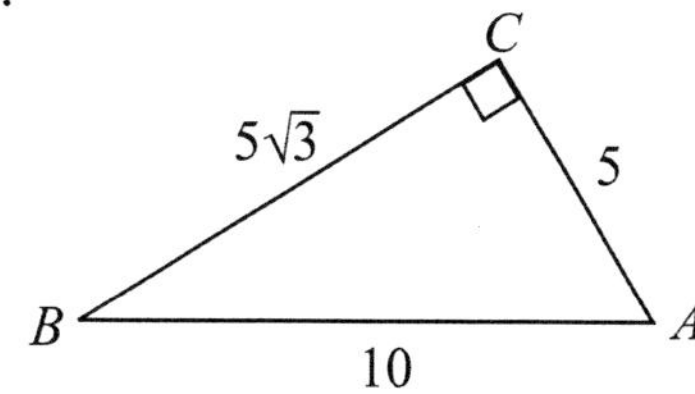

7.

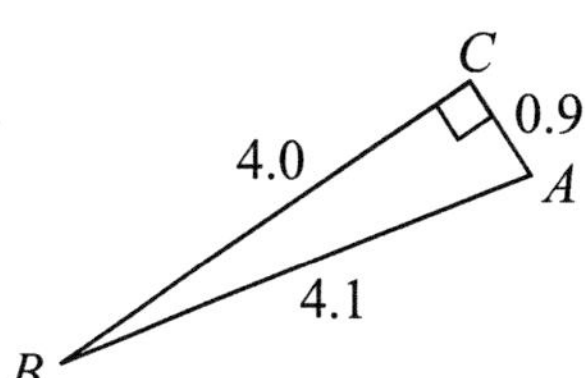

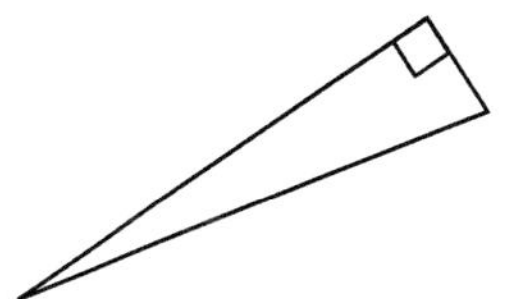

8.

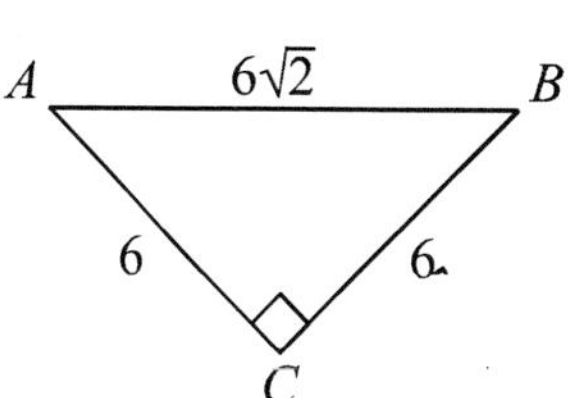

9.

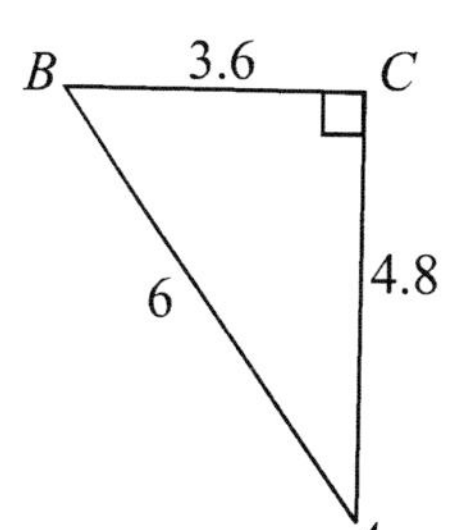

CHAPTER 14 REVIEW

1. Find the missing angle.

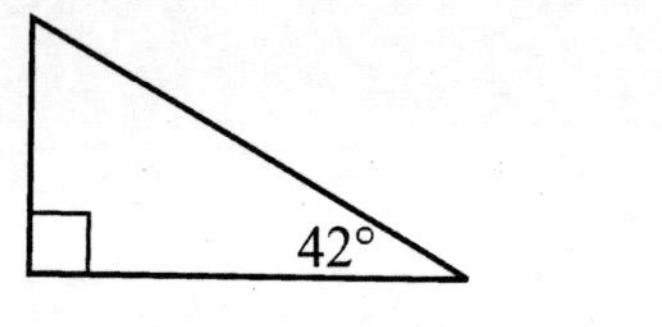

2. What is the length of line segment $\overline{WY}$?

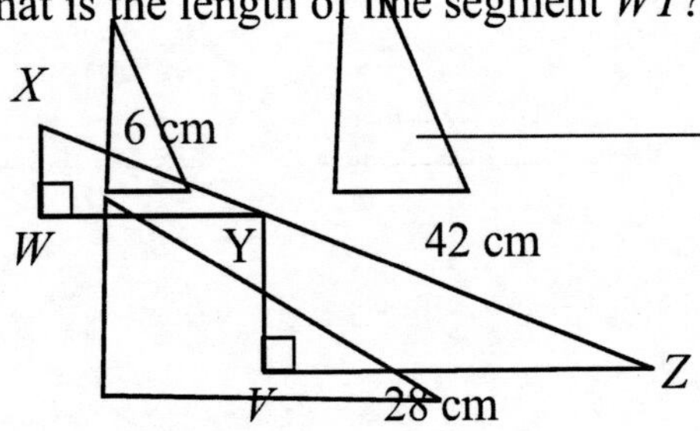

3. Find the missing side.

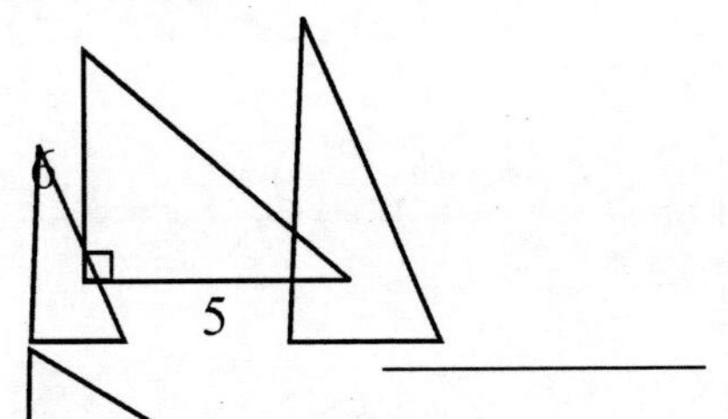

4. Find the measure of the missing leg of the right triangle below.

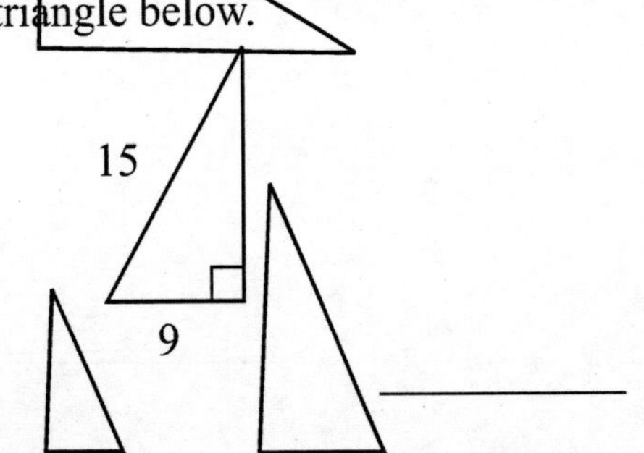

5. The following two triangles are similar. Find the length of the missing side.

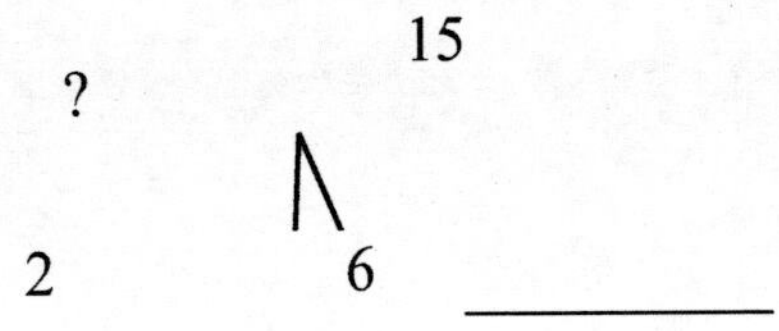

For questions 6 and 7, find the missing angle and sides.

6.

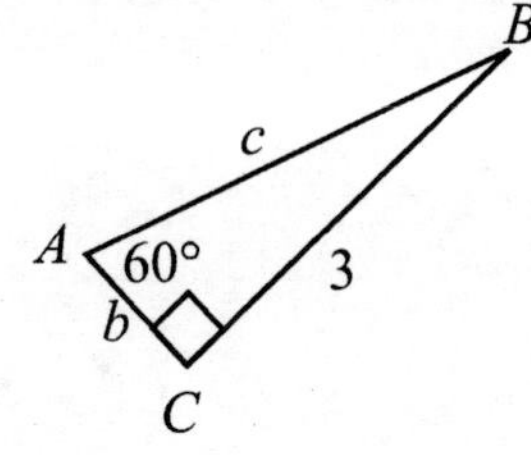

7.

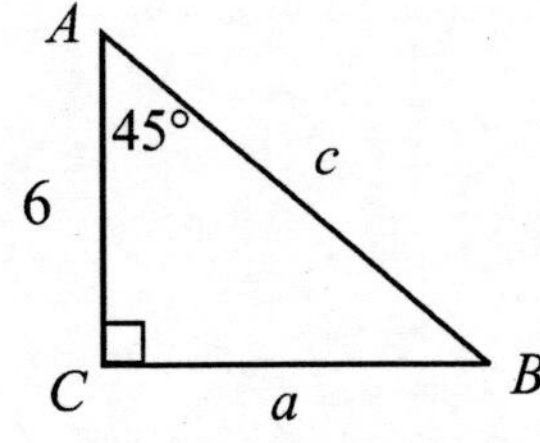

8. Find the missing side of the similar trapezoids below.

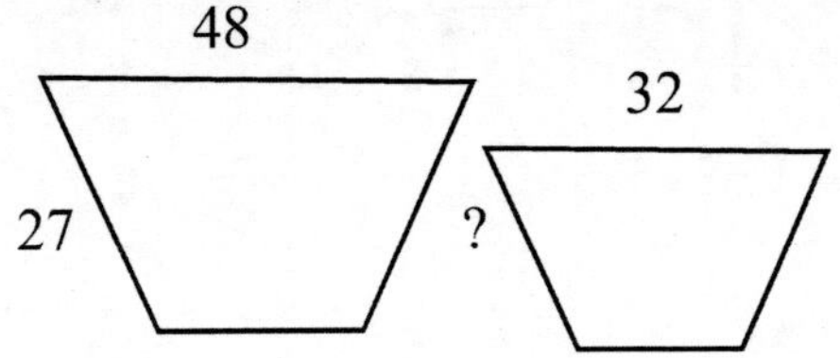

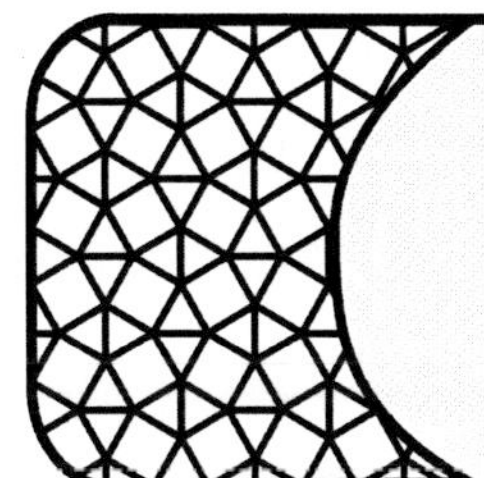

CHAPTER 15: PLANE GEOMETRY

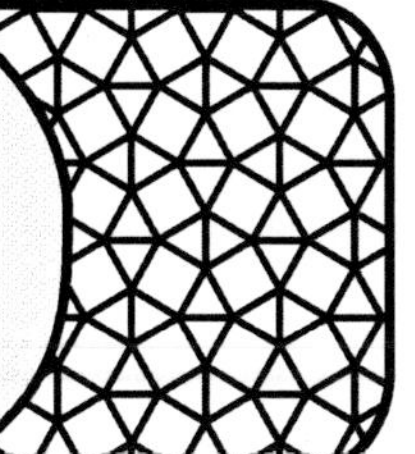

Standards 3.12.5 and 4.12.1

PERIMETER

The **perimeter** is the distance around a polygon. To find the perimeter, add the lengths of the sides.

EXAMPLES:

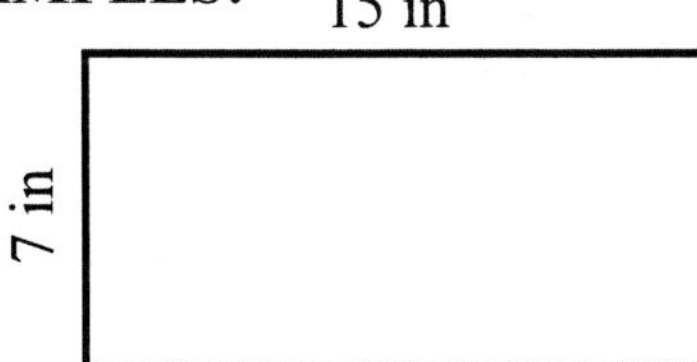

$P = 7 + 15 + 7 + 15$
$P = 44$ in

4 cm
6 cm
5 cm

$P = 4 + 6 + 5$
$P = 15$ cm

15 ft
8 ft
20 ft
10 ft
12 ft

$P = 8 + 15 + 20 + 12 + 10$
$P = 65$ ft

Find the perimeter of the following polygons.

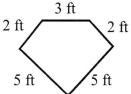

1. 8 in, 5 in ________

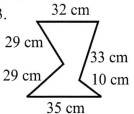

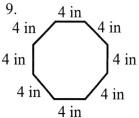

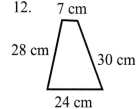

2. 3 ft, 2 ft, 2 ft, 5 ft, 5 ft ________

3. 32 cm, 29 cm, 33 cm, 29 cm, 10 cm, 35 cm ________

4. 13 cm, 15 cm, 10 cm ________

5. 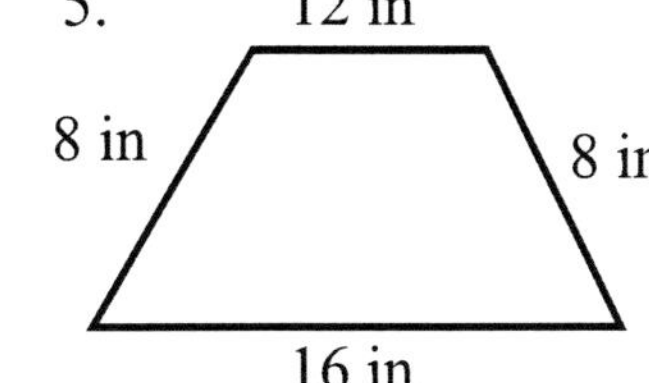

____

6. 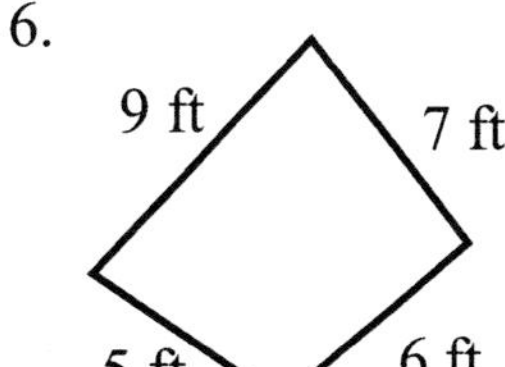

____

7. 6 ft, 6 ft ________

8. 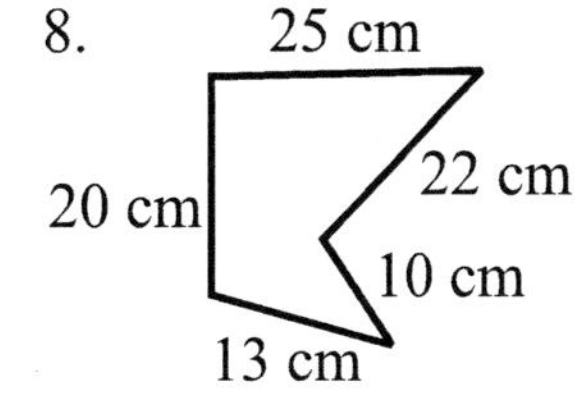

____

9. 4 in, 4 in, 4 in, 4 in, 4 in, 4 in, 4 in, 4 in ________

10. 8 cm, 1 cm ________

11. 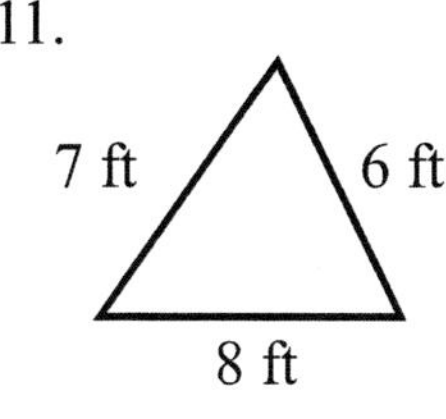

____

12. 7 cm, 28 cm, 30 cm, 24 cm ________

AREA OF SQUARES AND RECTANGLES

Area - area is always expressed in square units, such as in^2, cm^2, ft^2, and m^2.

The area (A) of squares and rectangles equals length (ℓ) times width (w). $A = \ell w$

EXAMPLE:

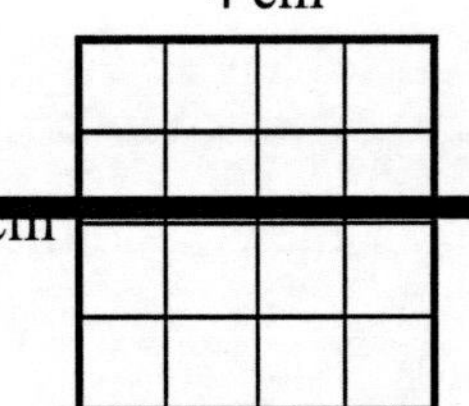

$A = \ell w$
$A = 4 \times 4$
$A = 16$ cm^2

If a square has an area of 16 cm^2, it means that it will take 16 squares that are 1 cm on each side to cover the area of a square that is 4 cm on each side.

Find the area of the following squares and rectangles, using the formula $A = \ell w$.

1.
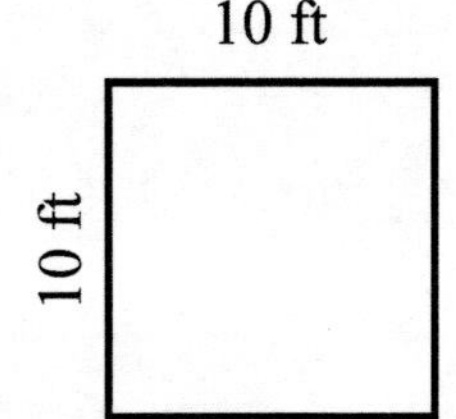

____

2.
5 cm
2 cm

____

3.
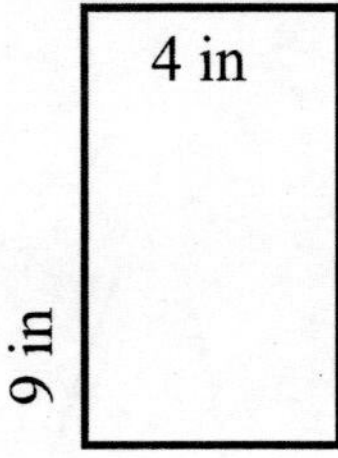

____

4.
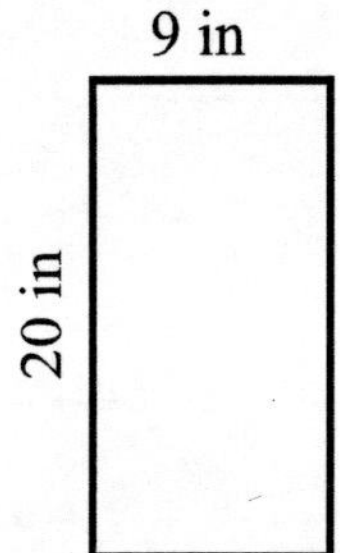

____

5.
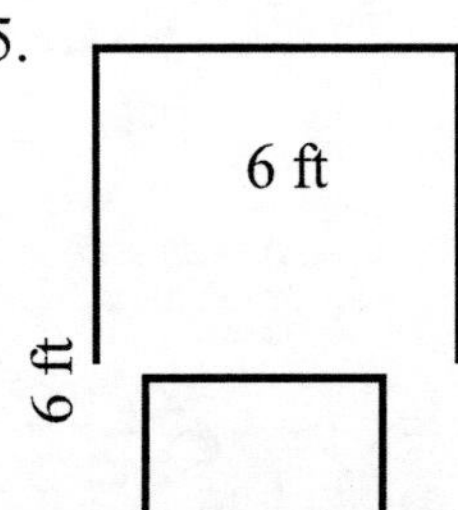

____

6.
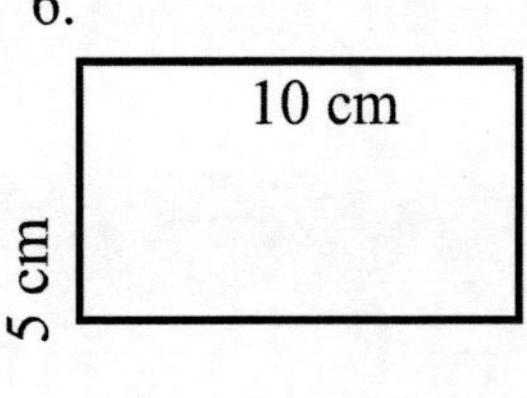

____

7.
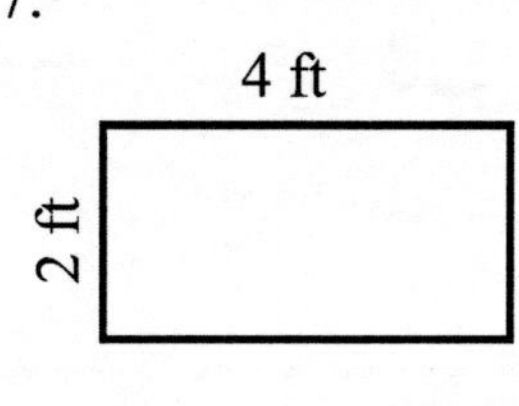

____

8.
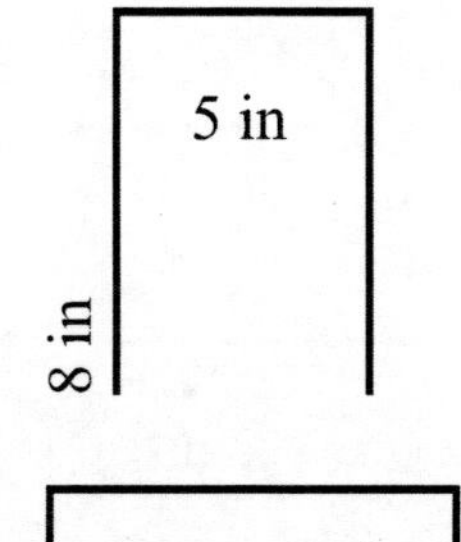

____

9.
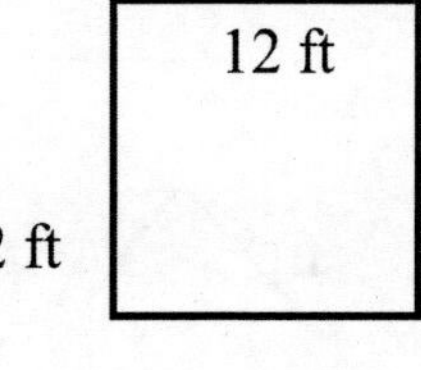

____

10.
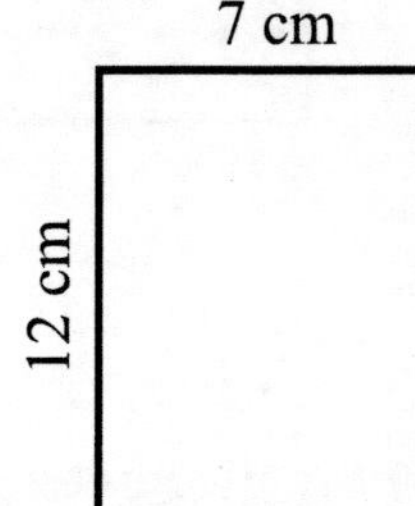

____

11.
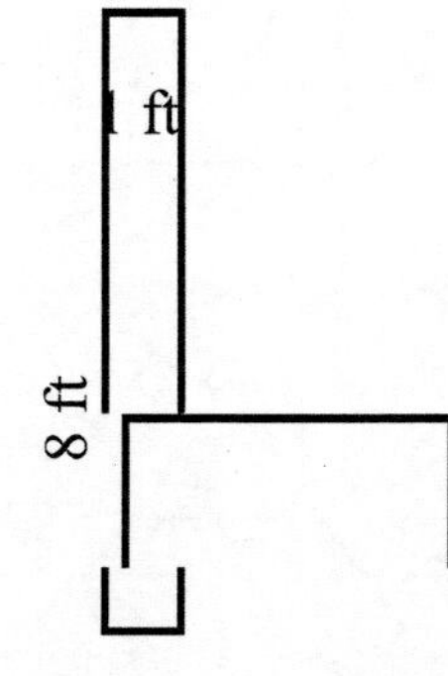

____

12.
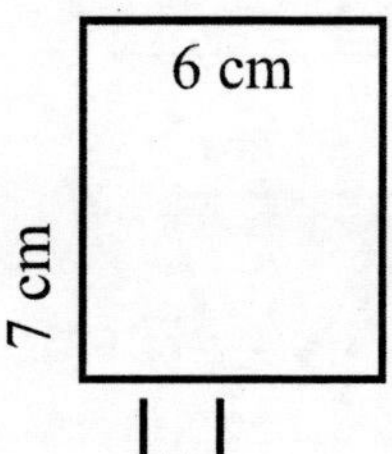

____

AREA OF TRIANGLES

EXAMPLE: Find the area of the following triangle.

The formula for the area of a triangle is as follows:

$A = \frac{1}{2} \times b \times h$

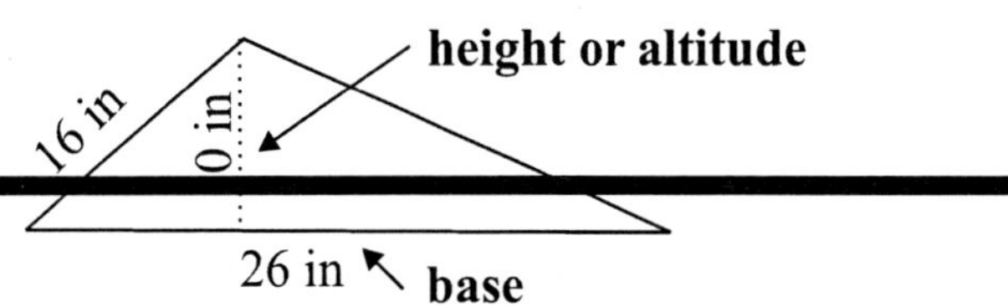

A = **area**
b = **base**
h = **height or altitude**

Step 1: Insert measurements from the triangle into the formula: $A = \frac{1}{2} \times 26 \times 10$

Step 2: Cancel and multiply. $A = \frac{1}{\cancel{2}_1} \times \frac{\cancel{26}^{13}}{1} \times \frac{10}{1} = 130 \text{ in}^2$

Note: **Area is always expressed in square units such as in^2, ft^2, cm^2, or m^2.**

Find the area of the following triangles..

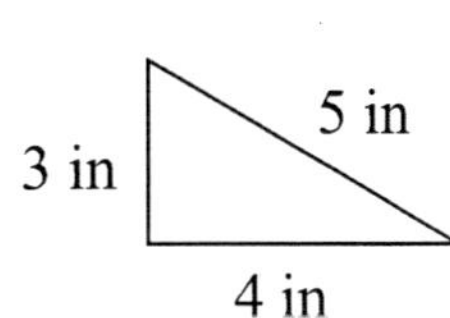

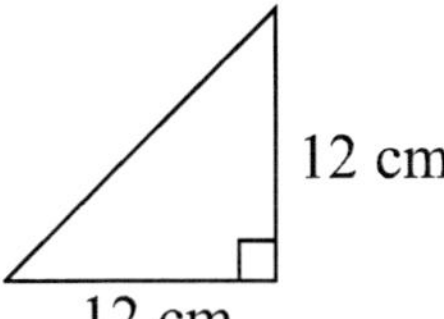

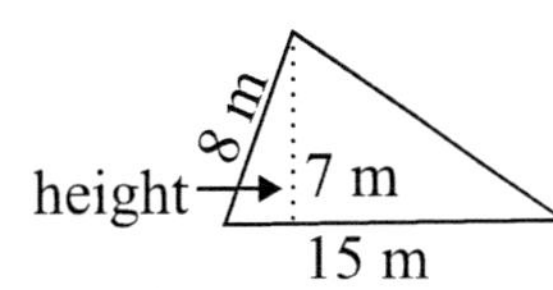

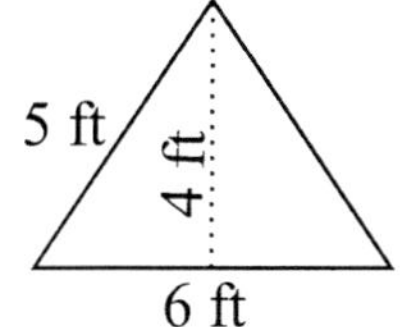

1. __________ in^2 4. __________ cm^2 7. __________ m^2 10. __________ ft^2

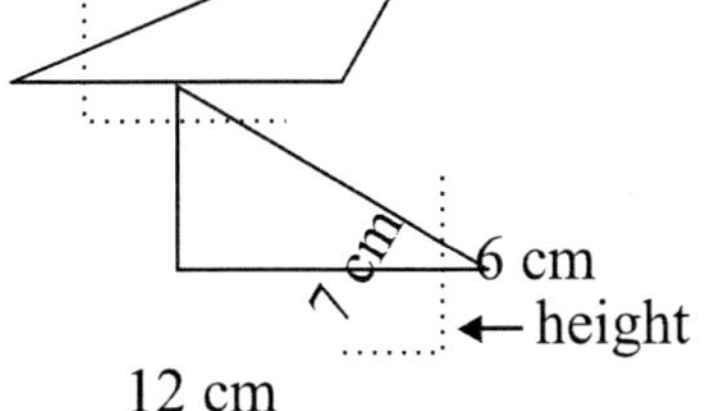

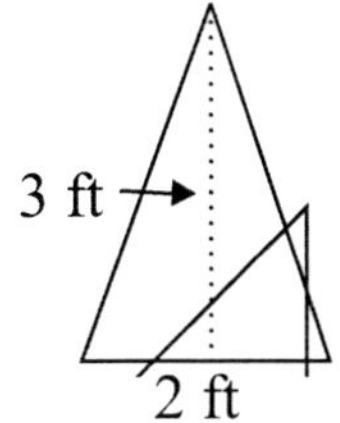

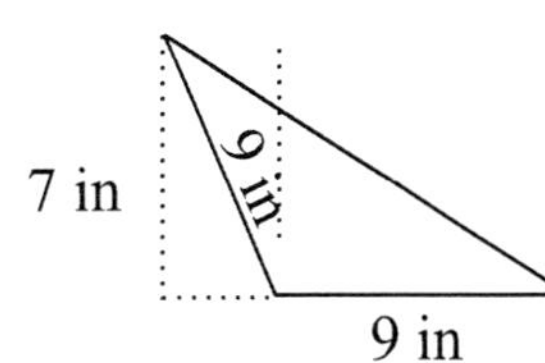

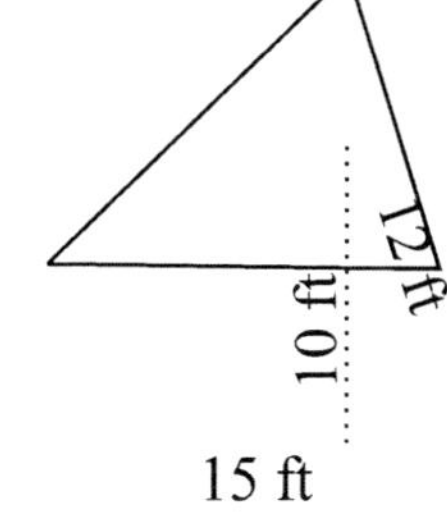

2. __________ cm^2 5. __________ ft^2 8. __________ in^2 11. __________ ft^2

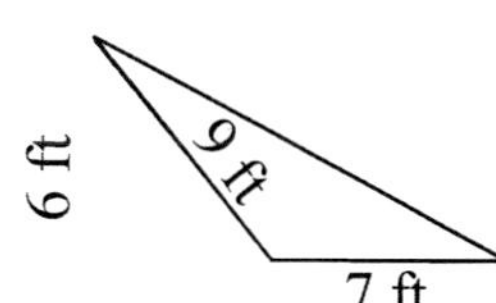

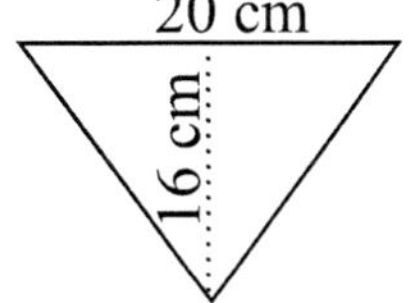

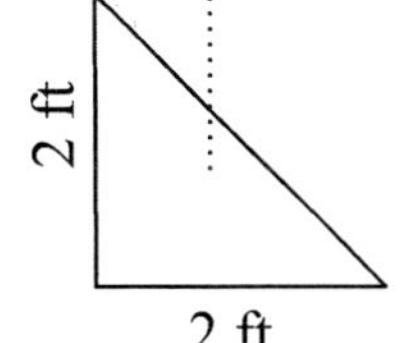

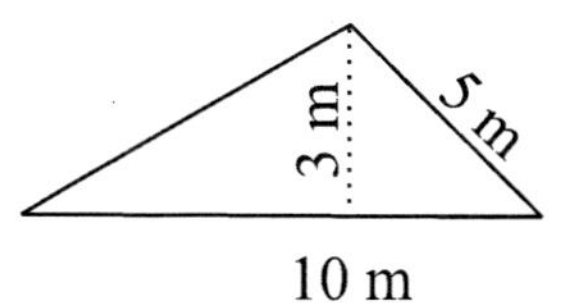

3. __________ ft^2 6. __________ cm^2 9. __________ ft^2 12. __________ m^2

AREA OF TRAPEZOIDS AND PARALLELOGRAMS

EXAMPLE 1: Find the area of the following parallelogram.

The formula for the area of a parallelogram is: $A = bh$.

A = **area**
b = **base**
h = **height**

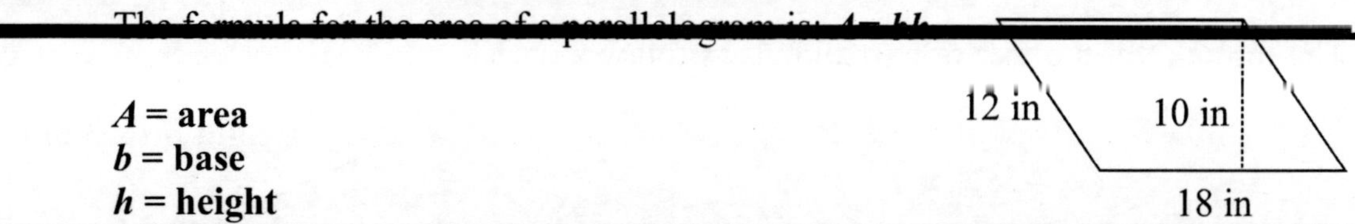

Step 1: Insert measurements from the parallelogram into the formula: $A = 18 \times 10$.

Step 2: Multiply. $18 \times 10 = 180 \text{ in}^2$

EXAMPLE 2: Find the area of the following trapezoid.

The formula for the area of a trapezoid is $A = \frac{1}{2} h (b_1 + b_2)$. A trapezoid has two bases that are parallel to each other. When you add the length of the two bases together and then multiply by $\frac{1}{2}$, you find their average length.

A = **area**
b = **base**
h = **height**

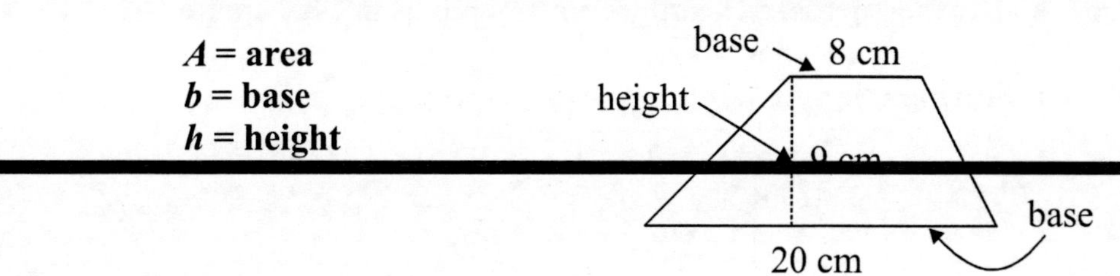

Insert measurements from the trapezoid into the formula and solve -
$\frac{1}{2} \times 9 \cdot (8 + 20) = 126 \text{ cm}^2$

Find the area of the following parallelograms and trapezoids.

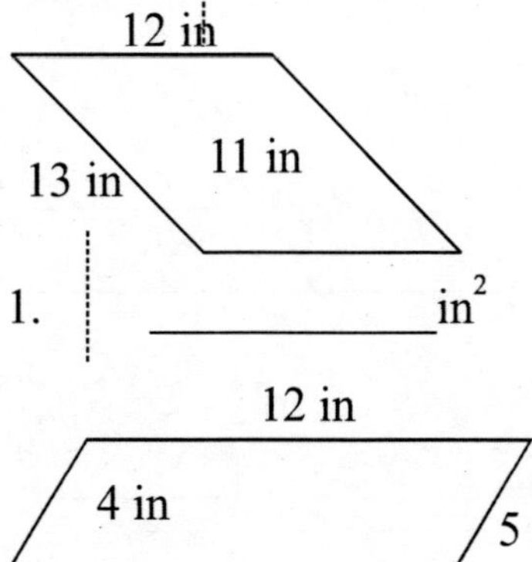

1. ____________ in^2

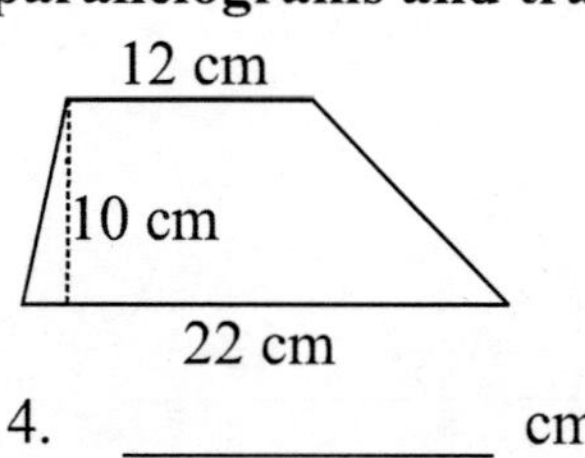

4. ____________ cm^2

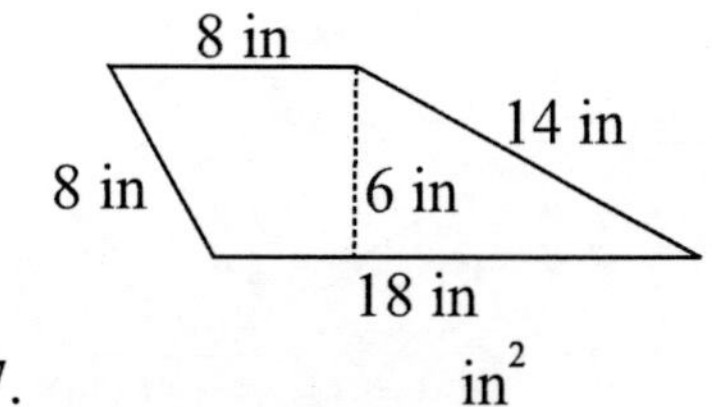

7. ____________ in^2

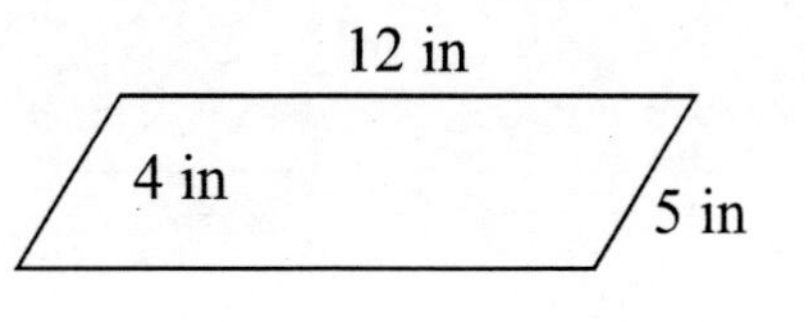

2. ____________ in^2

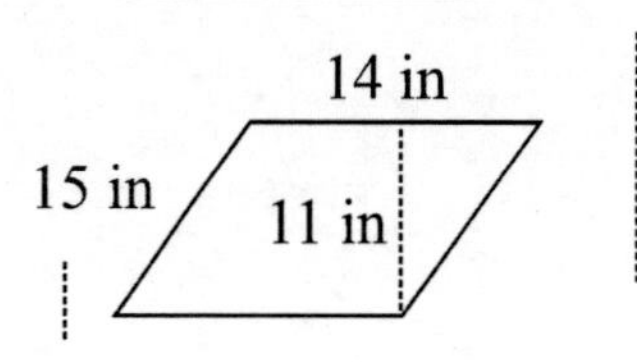

5. ____________ in^2

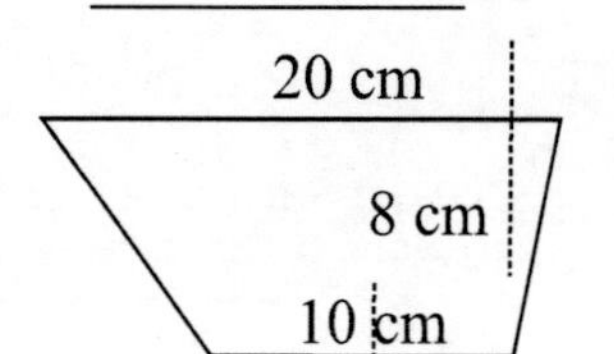

8. ____________ cm^2

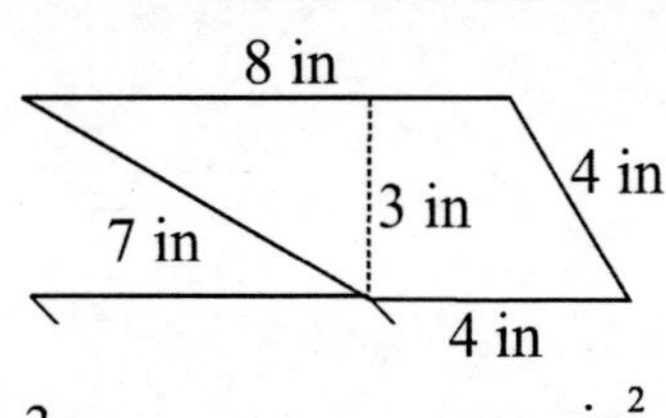

3. ____________ in^2

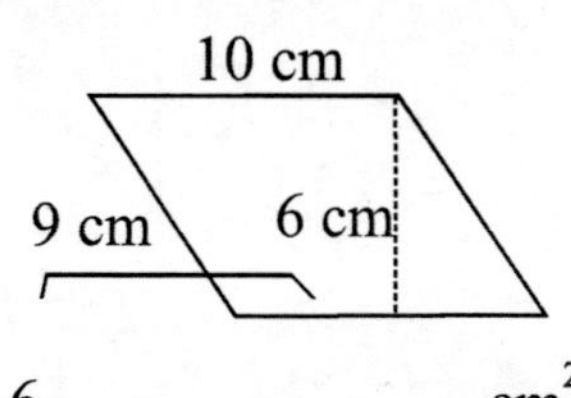

6. ____________ cm^2

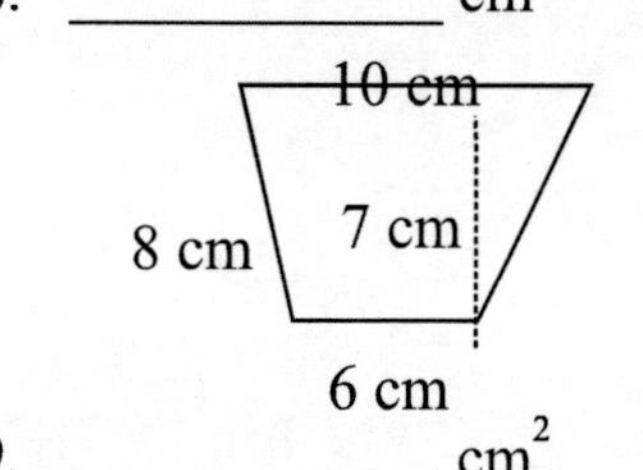

9. ____________ cm^2

PARTS OF A CIRCLE

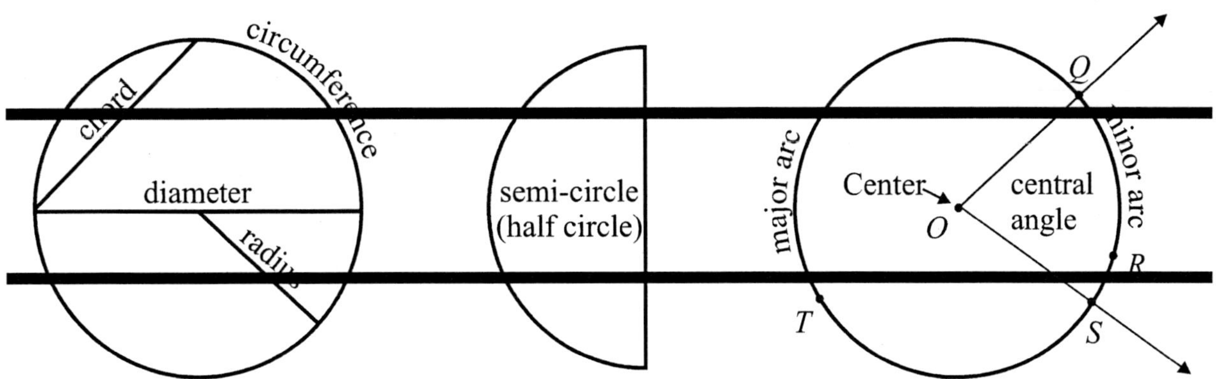

A **central angle** of a circle has the center of the circle as its vertex. The rays of a central angle each contain a radius of the circle. $\angle QOS$ is a central angle.

The points Q and S separate the circle into **arcs**. The arc lies on the circle itself. It does not include any points inside or outside the circle. $\widehat{QRS}$, or $\widehat{QS}$, is a **minor arc** because it is less than a semicircle. A minor arc can be named by 2 or 3 points. $\widehat{QTS}$ is a **major arc** because it is more than a semicircle. A major arc must be named by 3 points.

An **inscribed angle** is an angle whose vertex lies on the circle and whose sides contain **chords** of the circle. $\angle ABC$ is an inscribed angle.

These angles are not inscribed.

Refer to the figure on the right, and answer the following questions.

1. Identify the 2 line segments that are chords of the circle but not diameters. __________ __________
2. Identify the largest major arc of the circle that contains point S. __________
3. Identify the center of the circle. __________
4. Identify the inscribed angle. __________
5. Identify the central angle. __________
6. Identify the line segment that is a diameter of the circle. __________

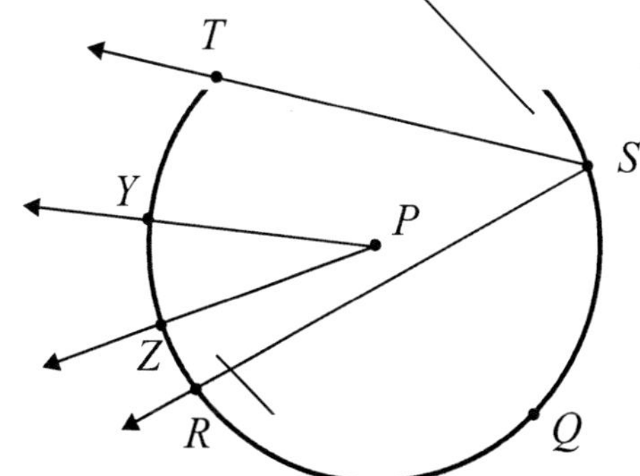

CIRCUMFERENCE

Circumference, C - the distance around the outside of a circle
Diameter, d - a line segment passing through the center of a circle from one side to the other
Radius, r - a line segment from the center of a circle to the edge of the circle
Pi, π - the ratio of the circumference of a circle to its diameter **$\pi = 3.14$** or **$\pi = \frac{22}{7}$**

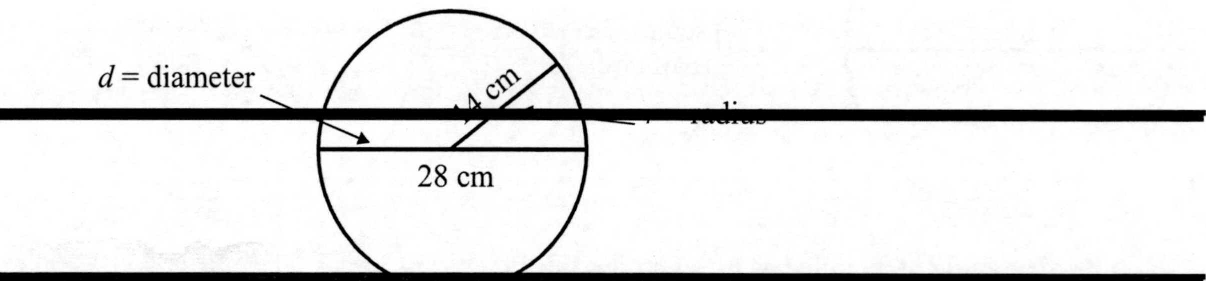

The formula for the circumference of a circle is $C = 2\pi r$ or $C = \pi d$. (The formulas are equal because the diameter is equal to twice the radius, $d = 2r$.)

EXAMPLE:	**EXAMPLE:**
Find the circumference of the circle above.	**Find the circumference of the circle above.**
$C = \pi d$ Use = 3.14	$C = 2\pi r$
$C = 3.14 \times 28$	$C = 2 \times 3.14 \times 14$
$C = 87.92$ cm	$C = 87.92$ cm

Use the formulas given above to find the circumference of the following circles. Use $\pi = 3.14$.

1. 8 in — $C =$ __________
2. 14 ft — $C =$ __________
3. 2 cm — $C =$ __________
4. 6 m — $C =$ __________
5. 8 ft — $C =$ __________

Use the formulas given above to find the circumference of the following circles. Use $\pi = \frac{22}{7}$.

6. 3 ft — $C =$ __________
7. 12 in — $C =$ __________
8. 6 m — $C =$ __________
9. 5 cm — $C =$ __________
10. 16 in — $C =$ __________

AREA OF A CIRCLE

The formula for the area of a circle is $A = \pi r^2$. The area is how many square units of measure would fit inside a circle.

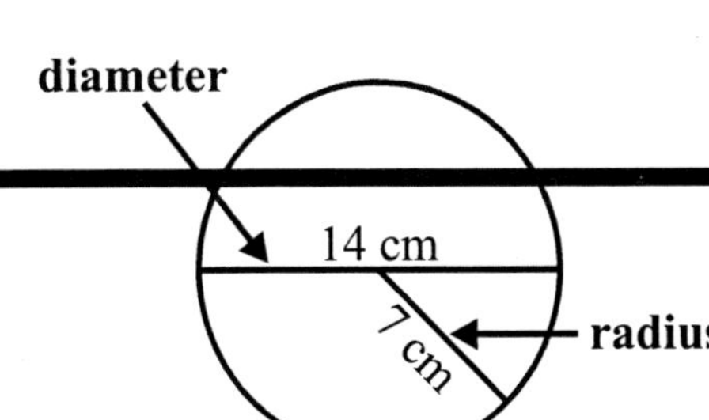

$\pi = \frac{22}{7}$ or $\pi = 3.14$

EXAMPLE: Find the area of the circle, using both values for π.

Let $\pi = \frac{22}{7}$	Let $\pi = 3.14$
$A = \pi r^2$	$A = \pi r^2$
$A = \frac{22}{7} \times \frac{7^2}{1}$	$A = 3.14 \times 7^2$
$A = \frac{22}{1} \times \frac{49}{7}$ → $A = \frac{22}{\cancel{7}_1} \times \frac{\cancel{49}^7}{1} = 154 \text{ cm}^2$	$A = 3.14 \times 49 = 153.86 \text{ cm}^2$

Find the area of the following circles. Remember to include units.

	$\pi = 3.14$	$\pi = \frac{22}{7}$
1. 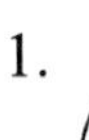  5 in	$A =$ ______	$A =$ ______
2. 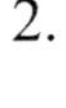 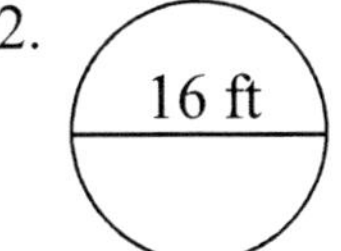 16 ft	$A =$ ______	$A =$ ______
3. 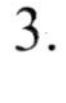 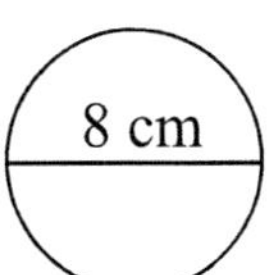 8 cm	$A =$ ______	$A =$ ______
4. 3 m	$A =$ ______	$A =$ ______

Fill in the chart below. Include appropriate Units.

	Radius	Diameter	Area $\pi = 3.14$	Area $\pi = \frac{22}{7}$
5.	9 ft			
6.		4 in		
7.	8 cm			
8.		20 ft		
9.	14 m			
10.		18 cm		
11.	12 ft			
12.		6 in		

TWO-STEP AREA PROBLEMS

Solving the problems below will require two steps. You will need to find the area of two figures, and then either add or subtract the two areas to find the answer. **Carefully read the EXAMPLES below.**

EXAMPLE 1:

Find the area of the living room below.

Figure 1

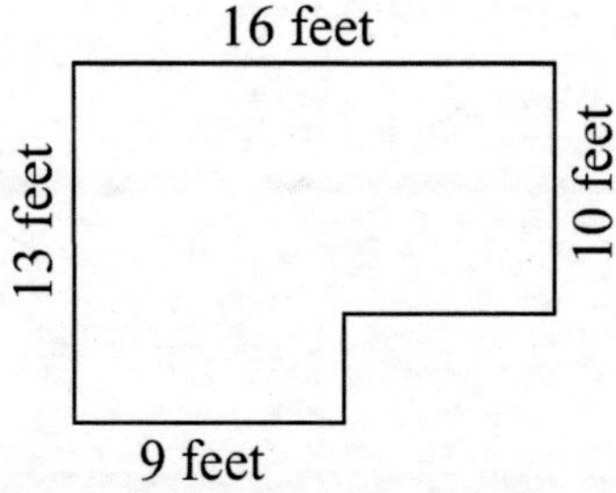

Step 1: Complete the rectangle as in Figure 2, and compute the area as if it were a complete rectangle.

Figure 2

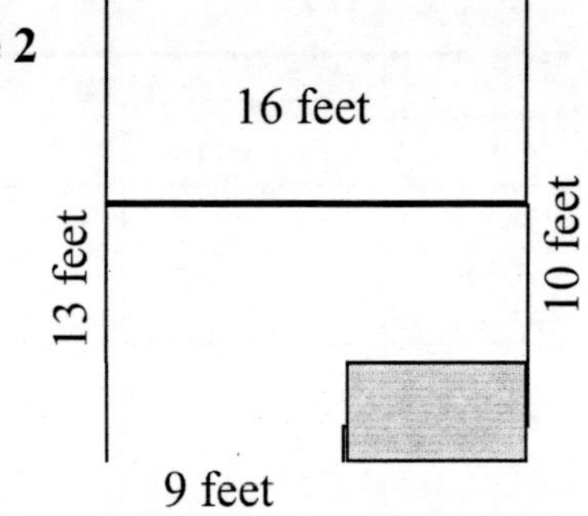

$A = \text{length} \times \text{width}$
$A = 16 \times 13$
$A = 208 \text{ ft}^2$

Step 2: Figure the area of the shaded part.

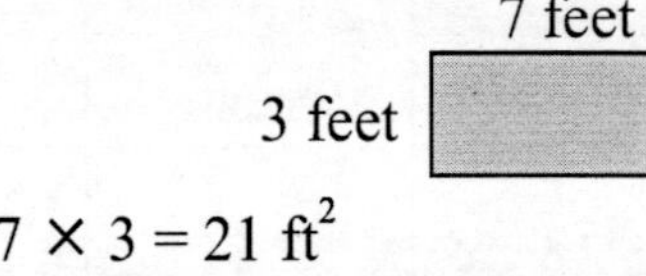

$7 \times 3 = 21 \text{ ft}^2$

Step 3: Subtract the area of the shaded part from the area of the complete rectangle.

$208 - 21 = 187 \text{ ft}^2$

EXAMPLE 2:

Find the area of the shaded sidewalk.

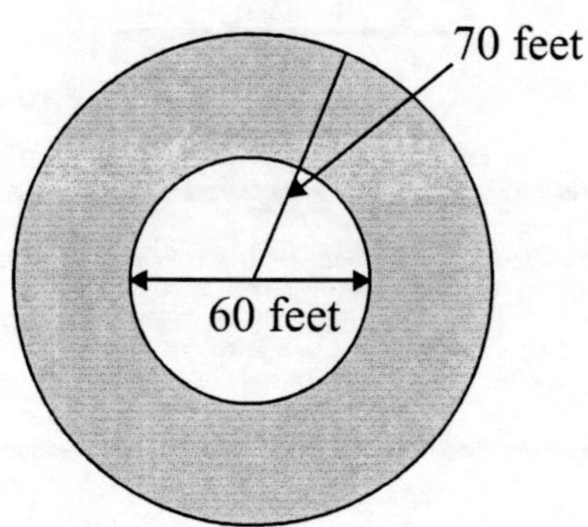

Step 1: Find the area of the outside circle.

$\pi = 3.14$
$A = 3.14 \times 70 \times 70$
$A = 15{,}386 \text{ ft}^2$

Step 2: Find the area of the inside circle.

$\pi = 3.14$
$A = 3.14 \times 30 \times 30$
$A = 2{,}826 \text{ ft}^2$

Step 3: Subtract the area of the inside circle from the area of the outside circle.

$15{,}386 - 2{,}826 = 12{,}560 \text{ ft}^2$

Find the area of the following figures.

1.

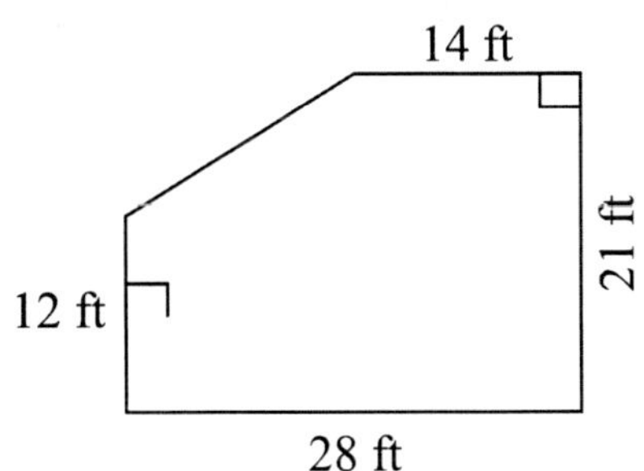

2.

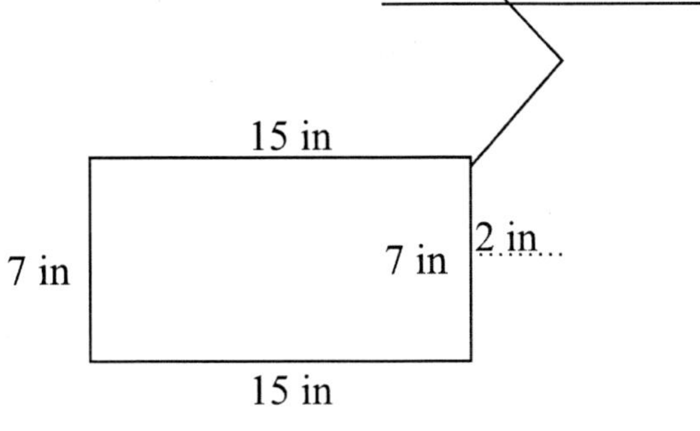

3. What is the area of the shaded circle? Use $\pi = 3.14$, and round the answer to the nearest whole number.

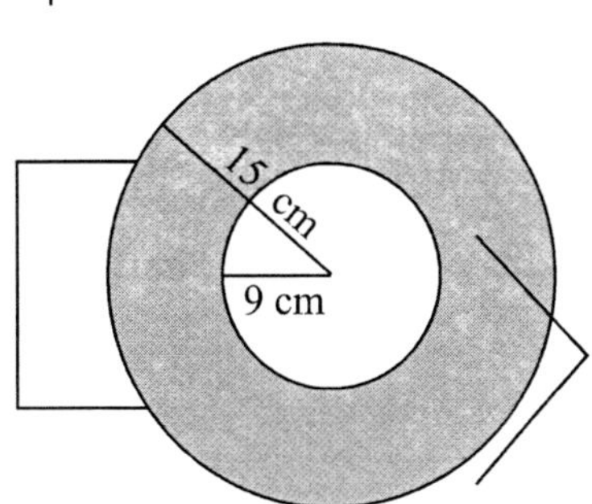

4.

1 ft

5 ft

4 ft

18 ft

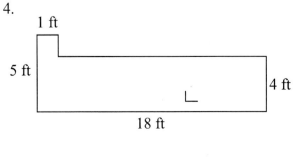

5. What is the area of the rectangle that is shaded? Use $\pi = 3.14$ and round to the nearest whole number.

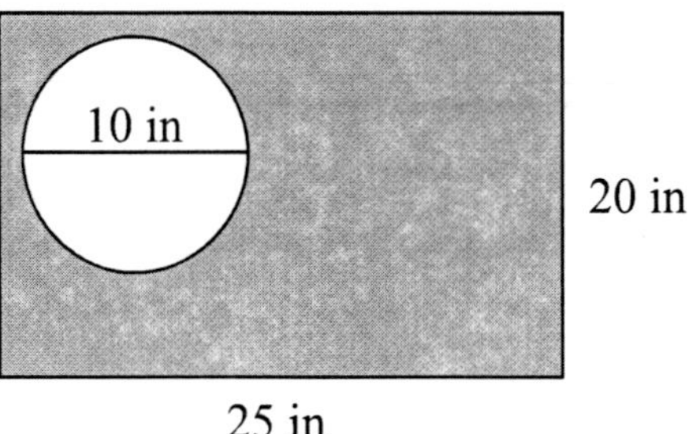

6. What is the area of the shaded part?

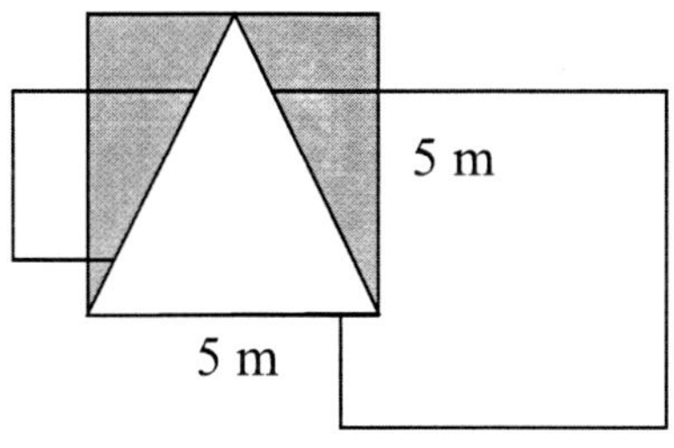

7. What is the area of the shaded part?

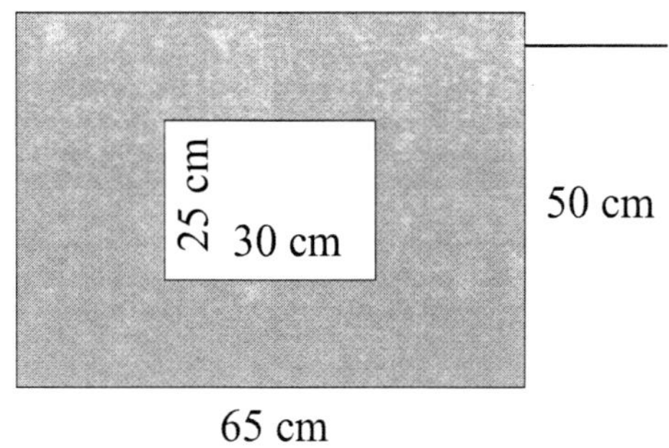

8. 6 m

24 m

12 m

12 m

CHAPTER 15 REVIEW

1. What is the length of line segment $\overline{WY}$? __________

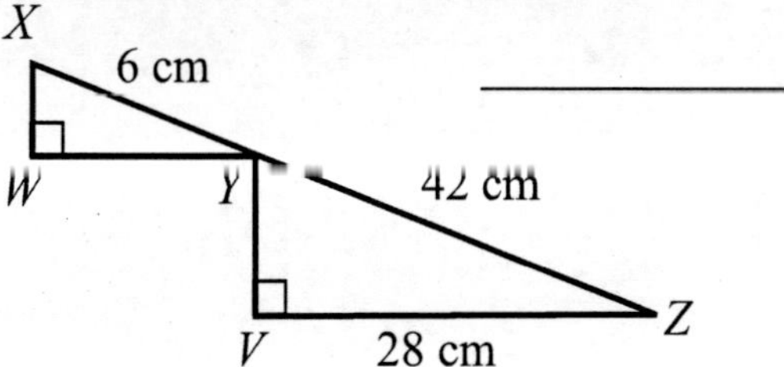

2. Find the missing side. Round your answer to two decimal places.

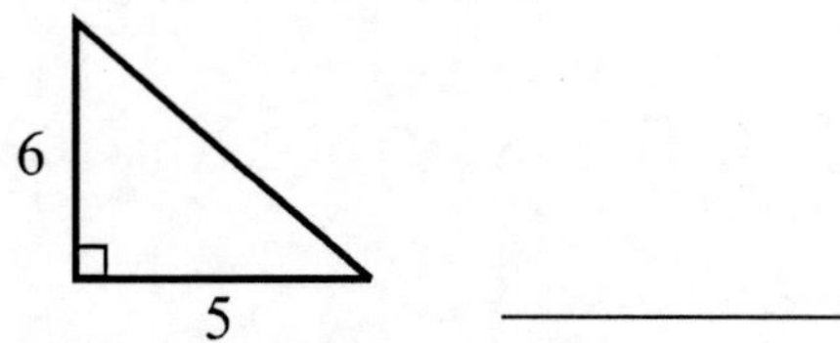

____

3. Find the area of the shaded region of the figure below.

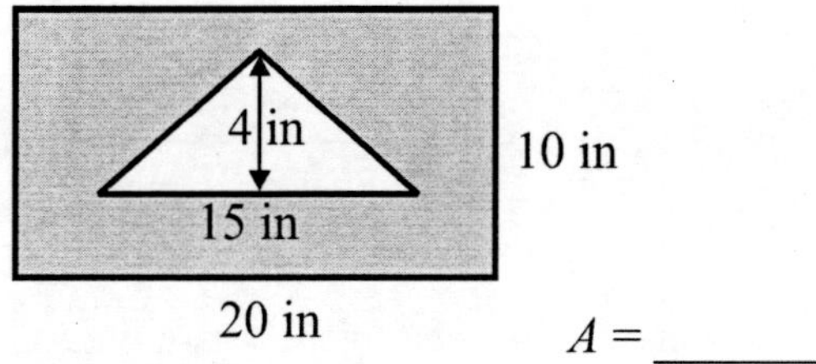

$A =$ ________

4. Calculate the perimeter of the following figure.

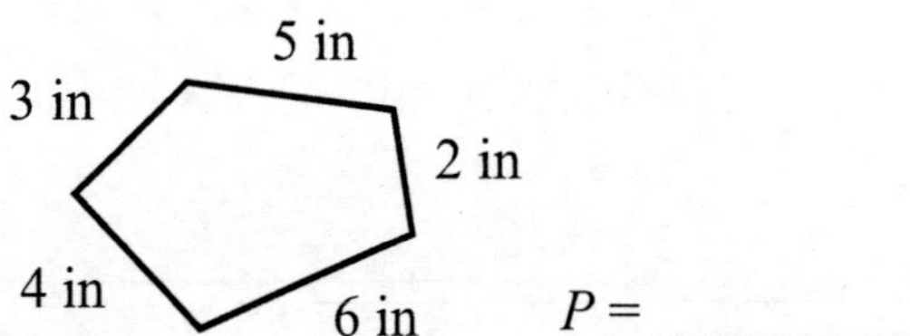

$P =$ ________

Calculate the perimeter and area of the following figures.

5.

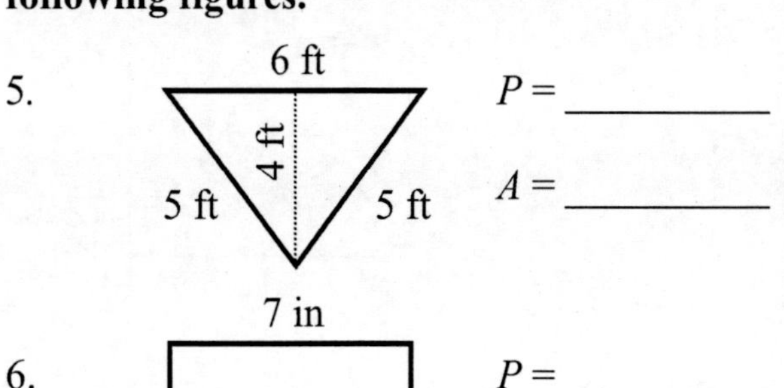

$P =$ ________

$A =$ ________

6.

$P =$ ________

$A =$ ________

Calculate the circumference and the area of the following circles.

7.

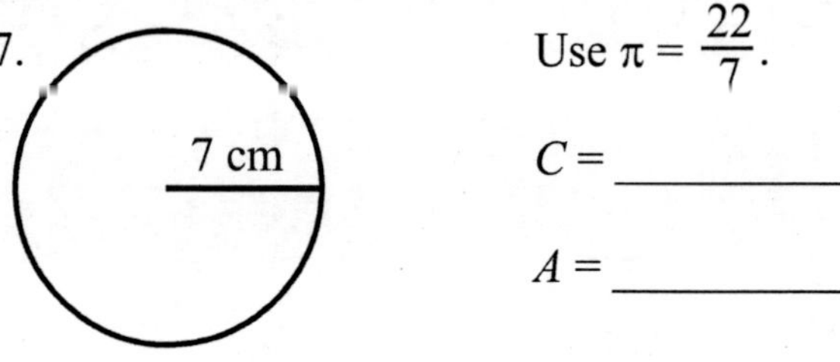

Use $\pi = \frac{22}{7}$.

$C =$ ________

$A =$ ________

8.

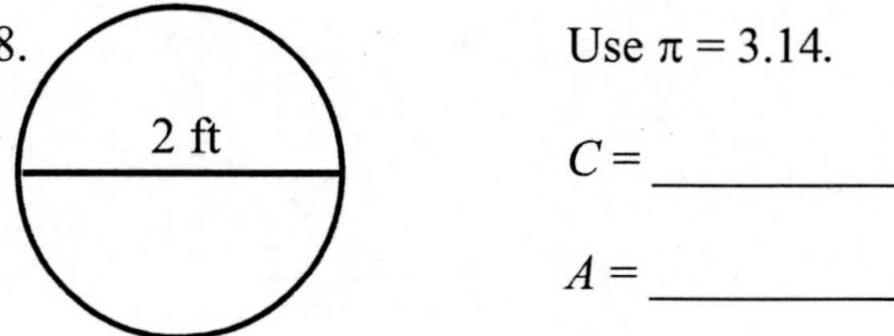

Use $\pi = 3.14$.

$C =$ ________

$A =$ ________

9. Use $\pi = 3.14$ to find the area of the shaded part. Round your answer to the nearest whole number.

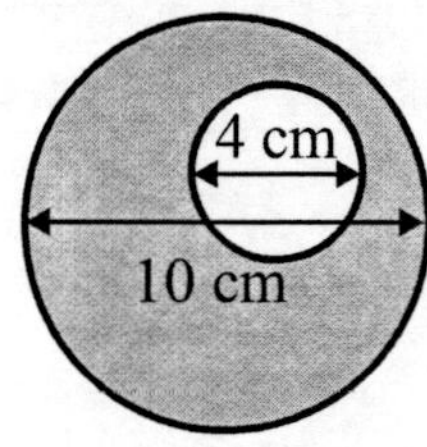

$A =$ ________

10. Use a ruler to measure the dimensions of the following figure in inches. Find the perimeter.

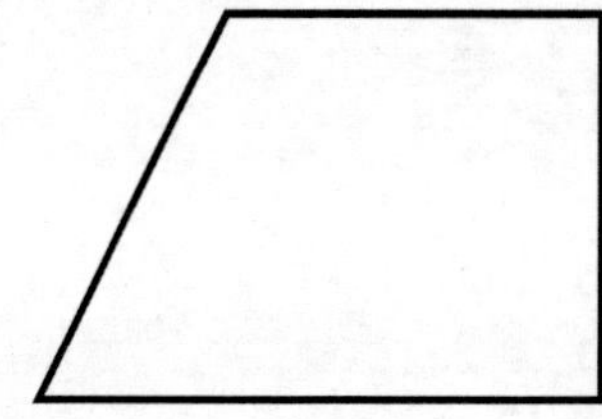

$P =$ ________

11. Use a ruler to measure the dimensions of the following figure in centimeters. Find the perimeter and area.

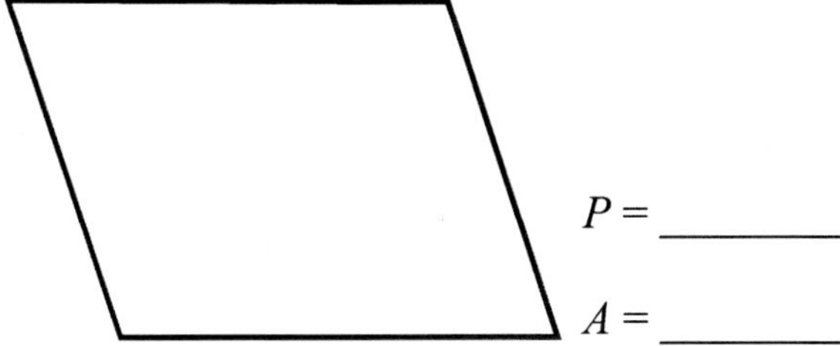

$P =$ ________

$A =$ ________

12. The shaded area below represents Grimes National Park. On the grid below, each square represents 10 square miles. Estimate the area of Grimes National Park.

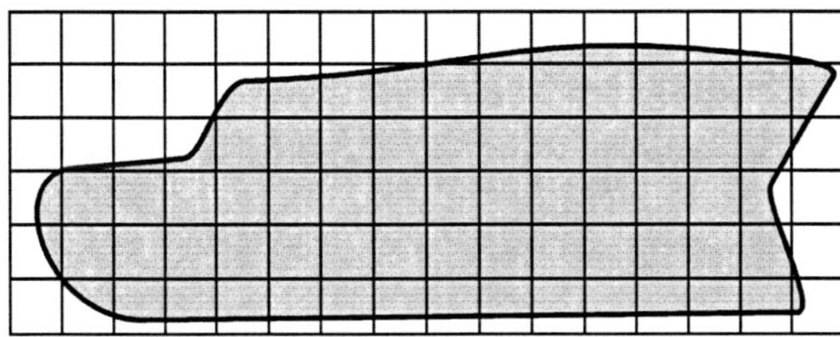

Area is about ________

13. The following two triangles are similar. Find the length of the missing side.

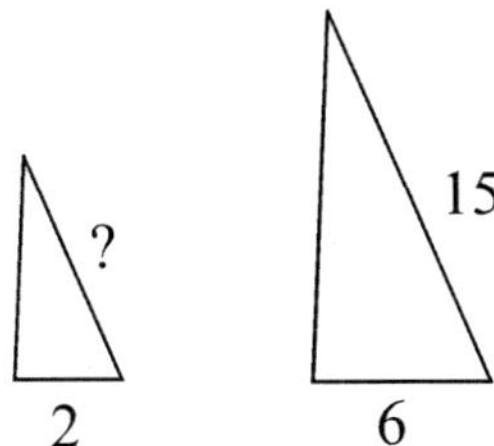

____

Find the area of the following figures.

14.

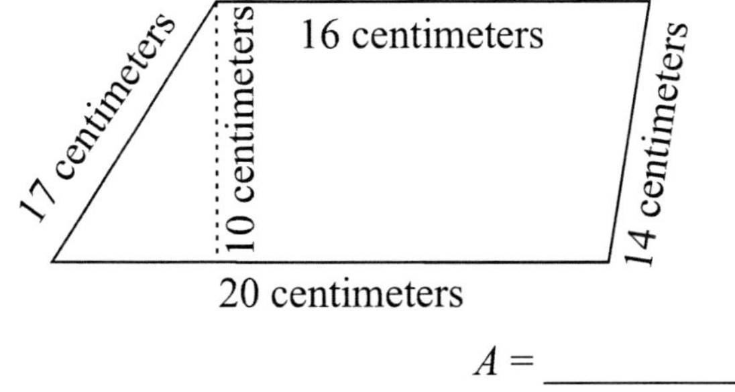

$A =$ ________

15.

18 inches

12 inches

9 inches

10 inches

10 inches

$A =$ ________

16.

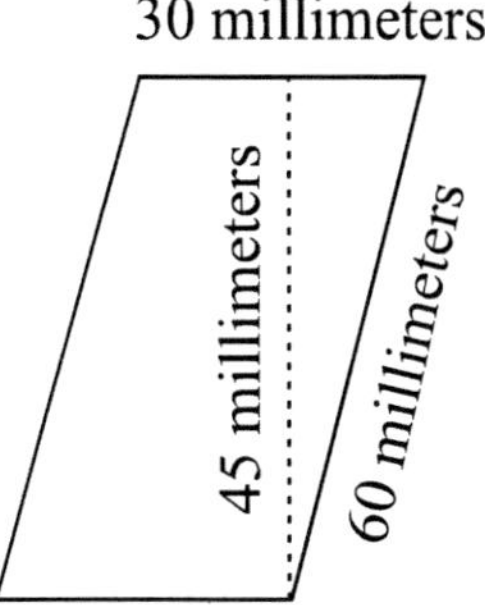

$A =$ ________

17.

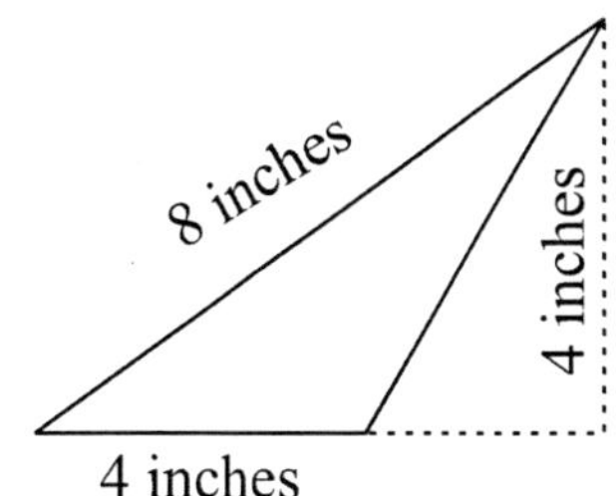

$A =$ ________

18. Find the missing side of the triangle below.

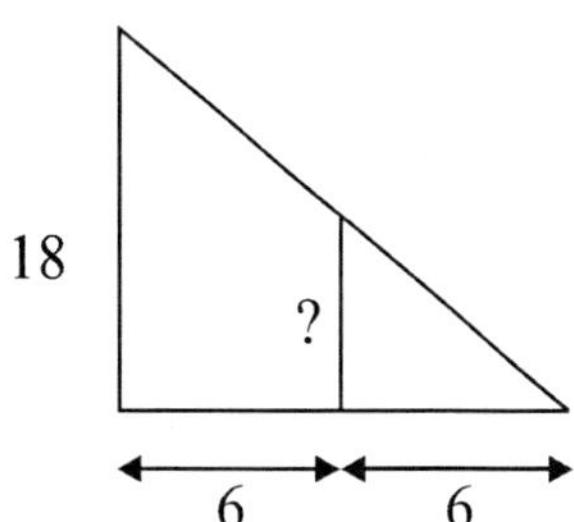

____

19. What is the area of a square which measures 8 inches on each side?

$A =$ ________

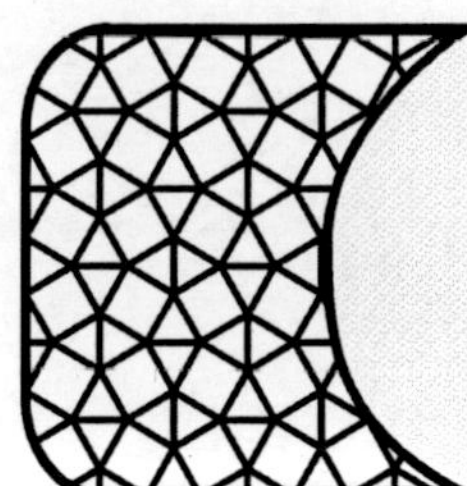
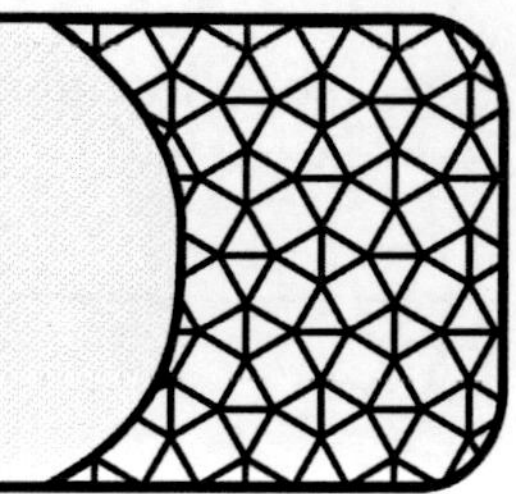

CHAPTER 16: SOLID GEOMETRY

Standard 3.12.5

In this chapter, you will learn about the following three-dimensional shapes.

SOLIDS

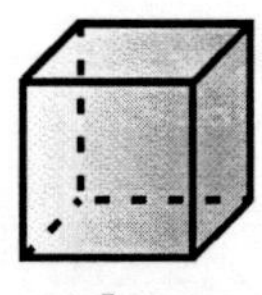
cube

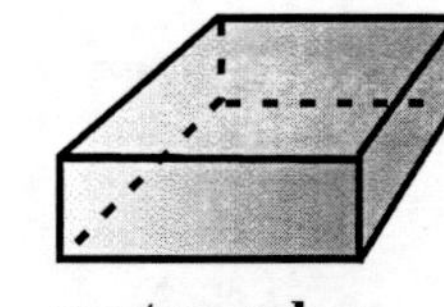
rectangular prism

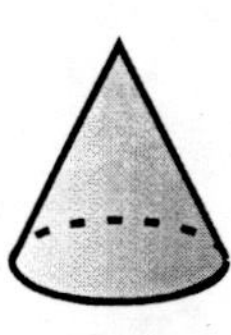
cone

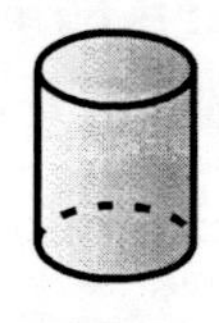
cylinder

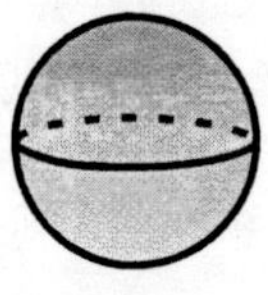
sphere

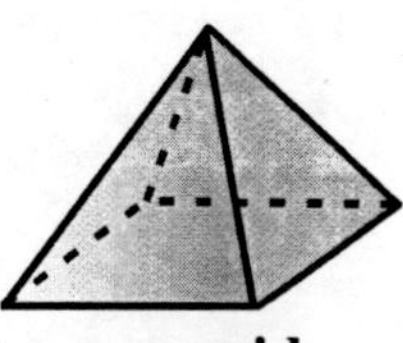
pyramid

UNDERSTANDING VOLUME

Volume - Measurement of volume is expressed in cubic units such as in^3, ft^3, m^3, cm^3, or mm^3. The volume of a solid is the number of cubic units that can be contained in the solid.

First, let's look at rectangular solids.

EXAMPLE:

How many 1 cubic centimeter cubes will it take to fill up the figure below?

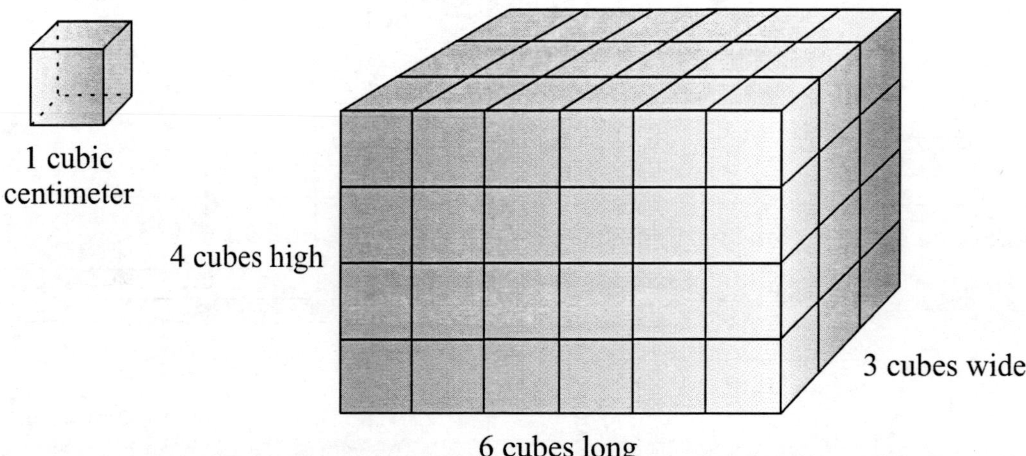

To find the volume, you need to multiply the length, times the width, times the height.

Volume of a rectangular solid = length × width × height ($V = l\,w\,h$).

$V = 6 \times 3 \times 4 = 72\ cm^3$

VOLUME OF CUBES

EXAMPLE: Find the volume of the cube pictured at the right.

Each side(s) of a cube has the same measure.

The formula for the volume of a cube is $V = s^3$ ($s \times s \times s$).

Step 1: Insert measurements from the figure into the formula.

Step 2: Multiply to solve. $5 \times 5 \times 5 = 125$ cm^3

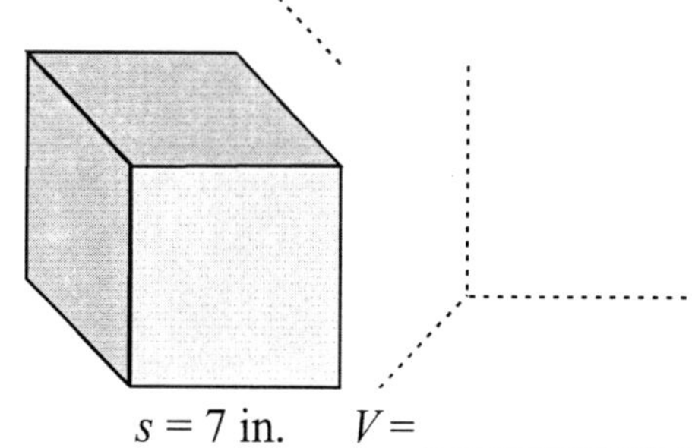

$s = 5$ cm

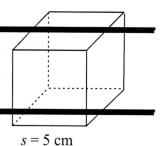

Note: **Volume is always expressed in cubic units such as in^3, ft^3, m^3, cm^3, or mm^3.**

Answer each of the following questions about cubes.

1. If a cube is 3 centimeters on each edge, what is the volume of the cube? ________

2. If the measure of the edge is doubled to 6 centimeters on each edge, what is the volume of the cube? ________

3. What if the edge of a 3 centimeter cube is tripled to become 9 centimeters on each edge? What will the volume be? ________

4. How many cubes with edges measuring 3 centimeters would you need to stack together to make a solid 12 centimeter cube? ________

5. What is the volume of a 2 centimeter cube? ________

6. Jerry built a 2 inch cube to hold his marble collection. He wants to build a cube with a volume 8 times larger. How much will each edge measure? ________

Find the volume of the following cubes.

7. $s = 7$ in. $V =$ ________

8. 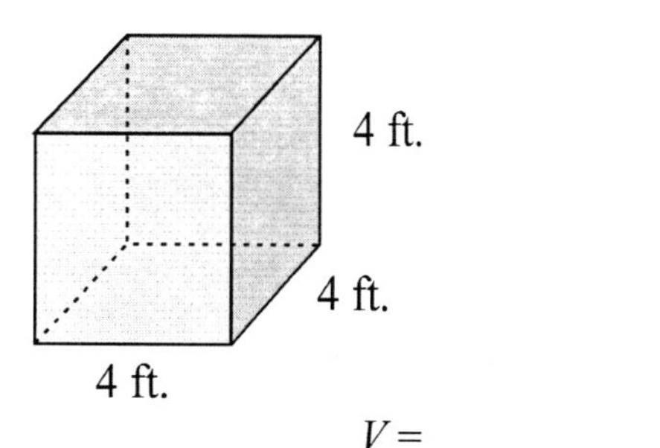

$V =$ ________

9. 12 inches = 1 foot

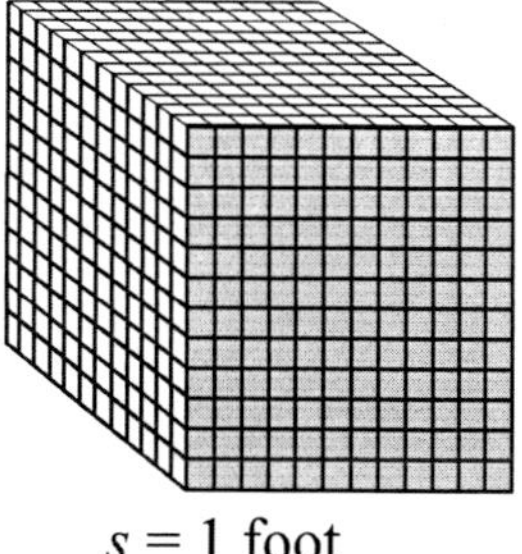

$s = 1$ foot

How many cubic inches are in a cubic foot? ________

VOLUME OF RECTANGULAR PRISMS

Find the volume of the following rectangular prisms (boxes) and cubes.

1.

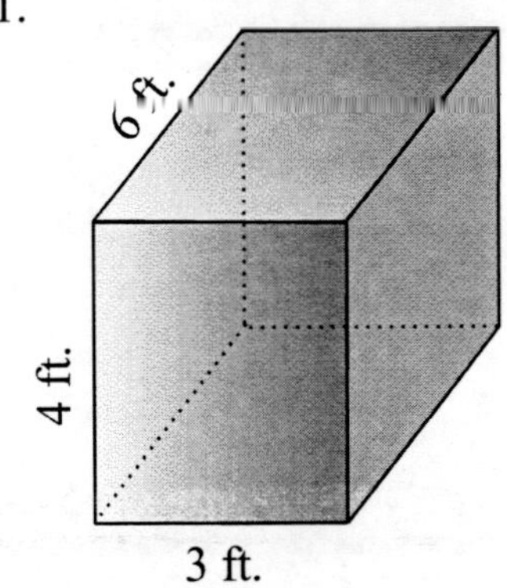

$V =$ ____________

2.

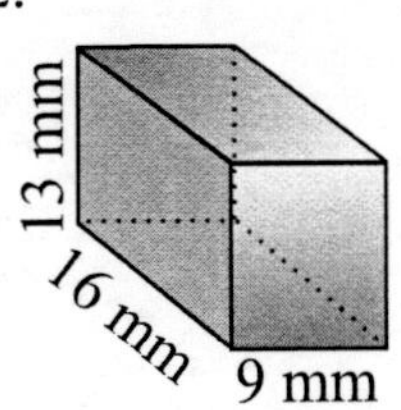

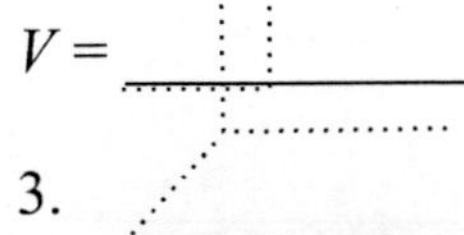

$V =$ ____________

3.

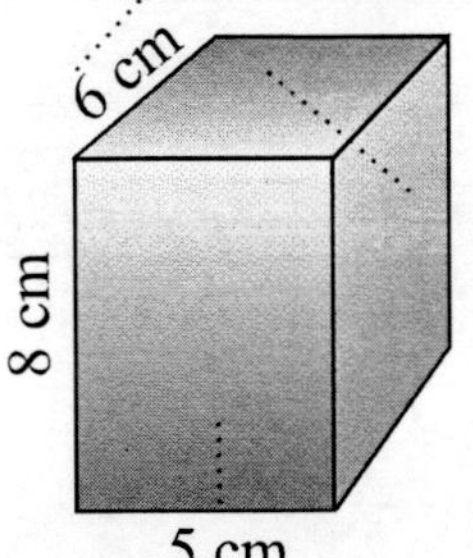

$V =$ ____________

4.

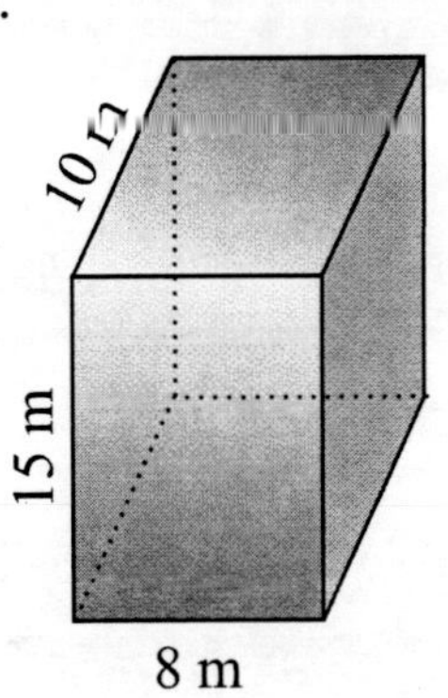

$V =$ ____________

5.

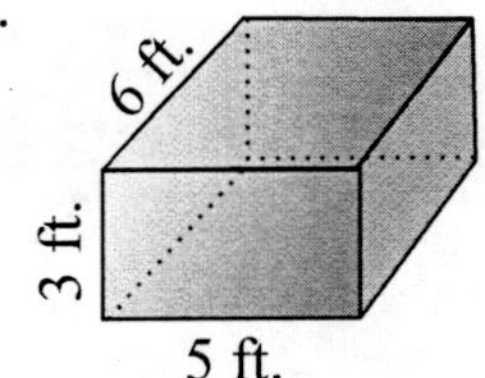

$V =$ ____________

6.

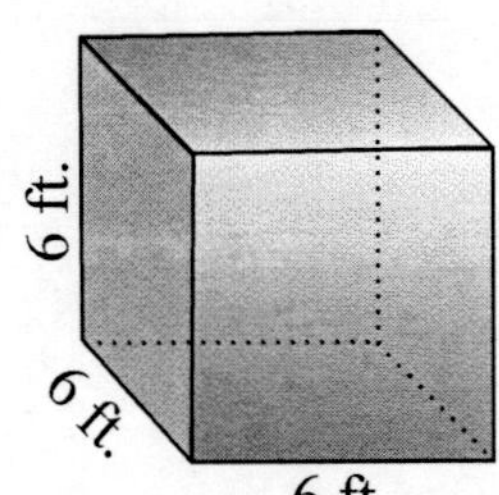

$V =$ ____________

7.

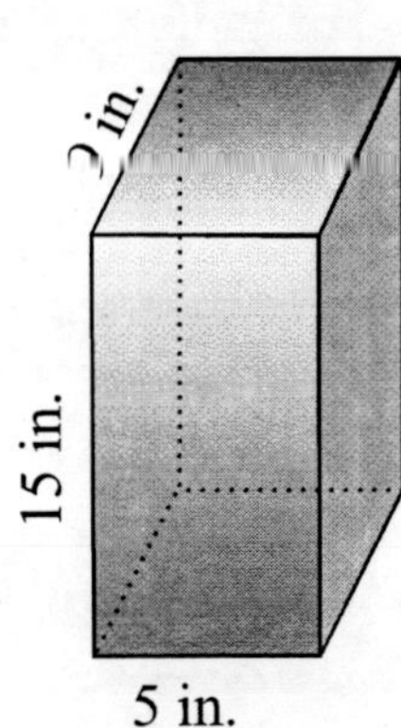

$V =$ ____________

8.

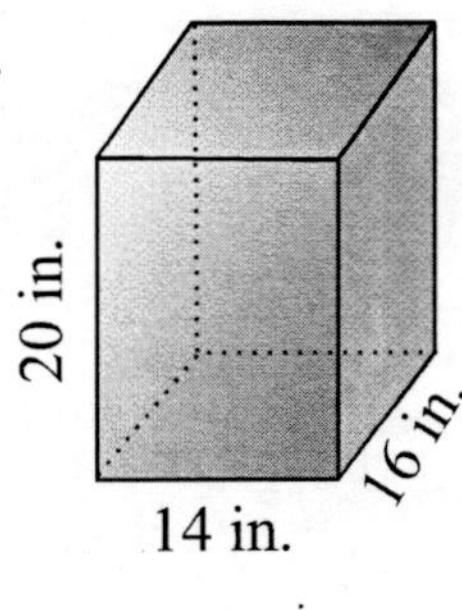

$V =$ ____________

9.

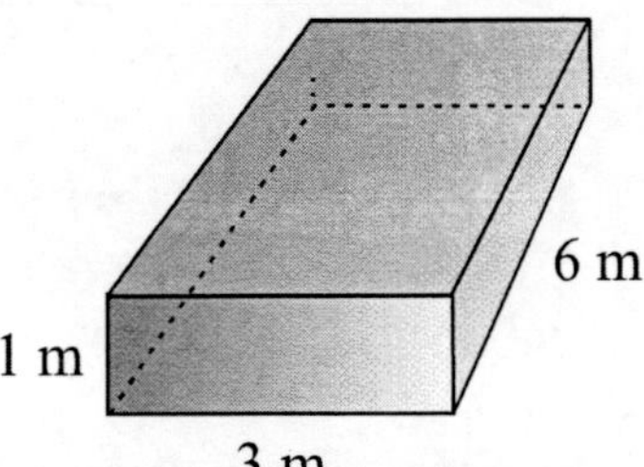

$V =$ ____________

VOLUME OF SPHERES, CONES, CYLINDERS, AND PYRAMIDS

To find the volume of a solid, substitute the measurements given for the solid in the correct formula and solve. Volumes are expressed in cubic units such as in^3, ft^3, m^3, cm^3, or mm^3.

Sphere
$V = \frac{4}{3}\pi r^3$

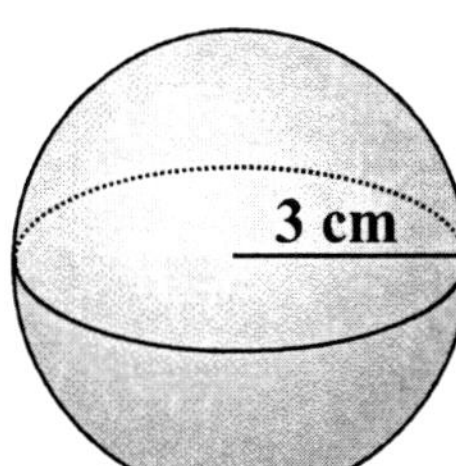

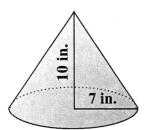

$V = \frac{4}{3}\pi r^3 \quad \pi = 3.14$
$V = \frac{4}{3} \times 3.14 \times 27$
$V = 113.04 \text{ cm}^3$

Cone
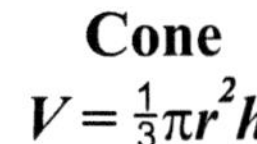
$V = \frac{1}{3}\pi r^2 h$

10 in.
7 in.

$V = \frac{1}{3}\pi r^2 h \quad \pi = 3.14$
$V = \frac{1}{3} \times 3.14 \times 49 \times 10$
$V = 512.87 \text{ in}^3$

Cylinder
 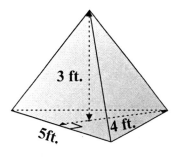
$V = \pi r^2 h$

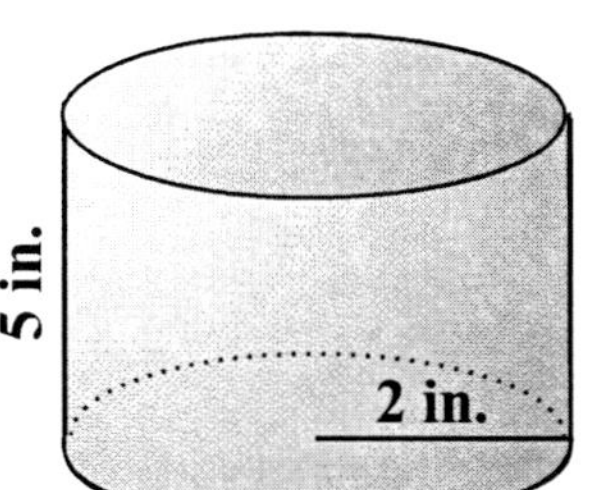

$V = \pi r^2 h \quad \pi = \frac{22}{7}$
$V = \frac{22}{7} \times 4 \times 5$
$V = 62\frac{6}{7} \text{ in}^3$

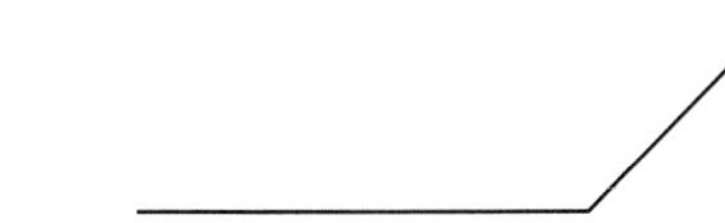

Pyramids

$V = \frac{1}{3}Bh \quad B =$ area of rectangular base

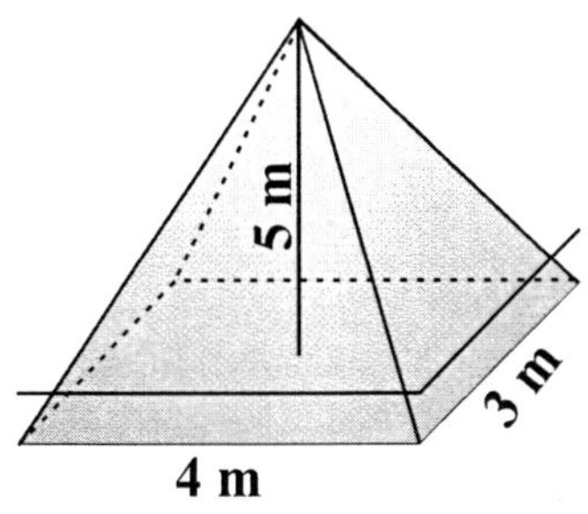

$V = \frac{1}{3}Bh \quad B = l \times w$
$V = \frac{1}{3} \times 4 \times 3 \times 5$
$V = 20 \text{ m}^3$

$V = \frac{1}{3}Bh \quad B =$ area of triangular base

3 ft.
4 ft.
5ft.

$B = \frac{1}{2} \times 5 \times 4 = 10 \text{ ft}^2$
$V = \frac{1}{3} \times 10 \times 3$
$V = 10 \text{ ft}^3$

Find the volume of the following shapes. Use π = 3.14.

1. 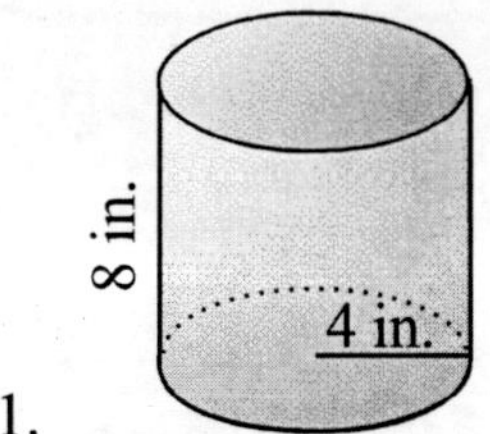

$V =$ ________

7. 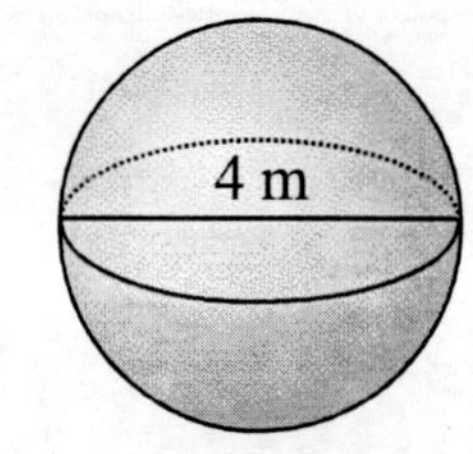

$V =$ ________

2. 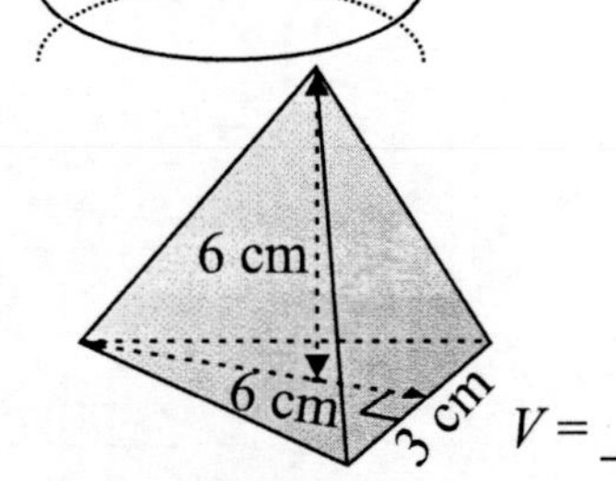

$V =$ ________

8. 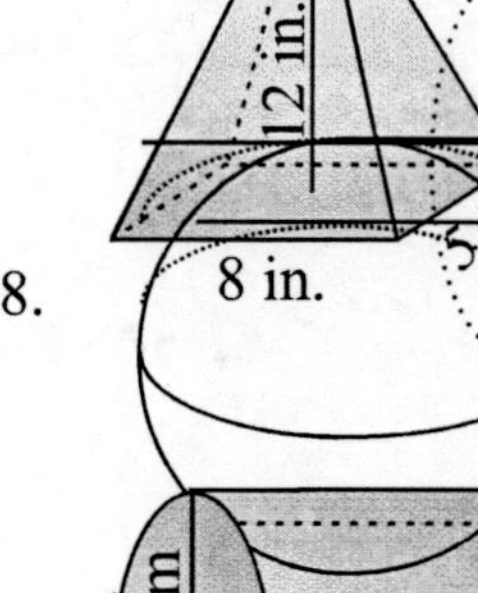

$V =$ ________

3. 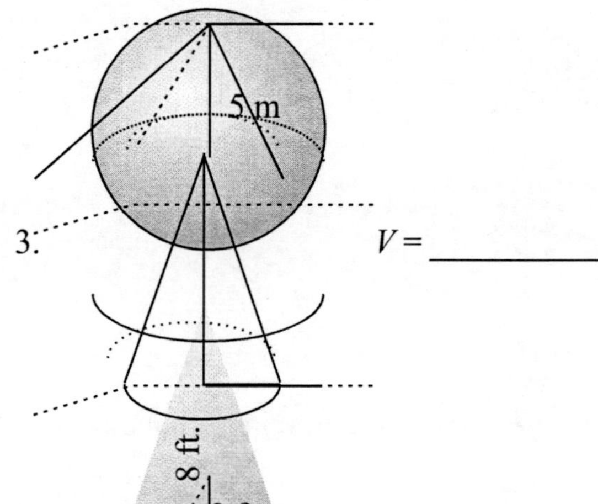

$V =$ ________

9. 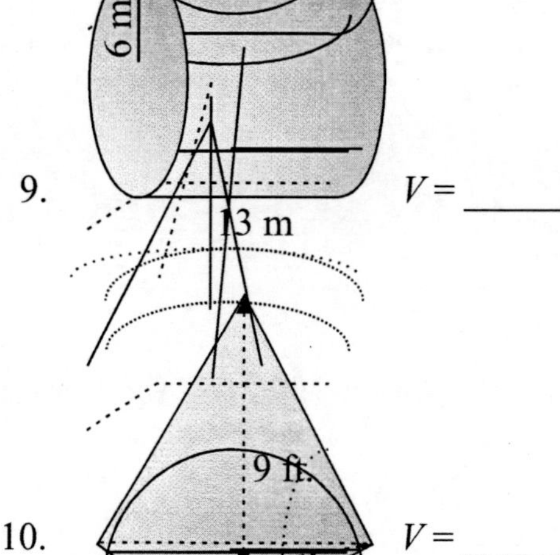

$V =$ ________

4. 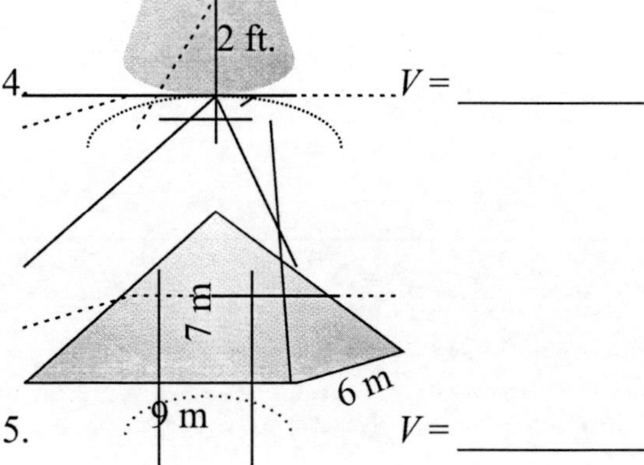

$V =$ ________

10.

$V =$ ________

5.

$V =$ ________

11. 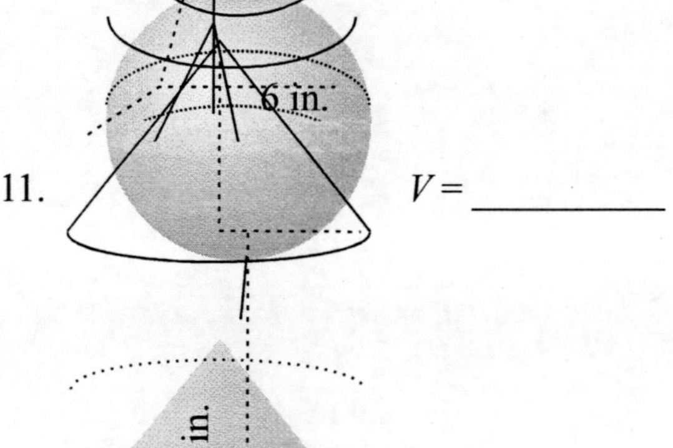

$V =$ ________

6. 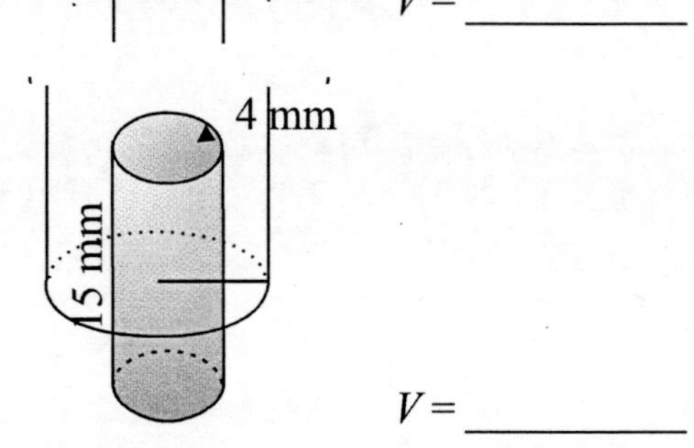

$V =$ ________

12.

$V =$ ________

GEOMETRIC RELATIONSHIPS OF SOLIDS

In the previous chapter, you looked at geometric relationships between 2-dimensional figures. Questions on the Nevada High School Proficiency Exam in Mathematics may also ask about the relationships between 3-dimensional figures. The formulas for finding the volumes of geometric solids are given below.

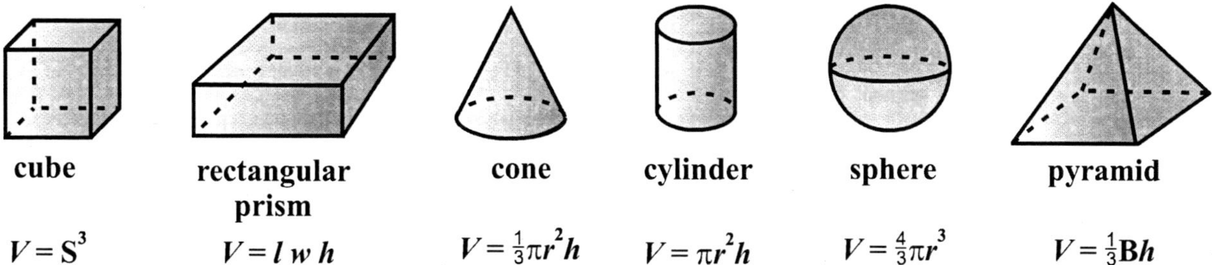

cube	**rectangular prism**	**cone**	**cylinder**	**sphere**	**pyramid**
$V = S^3$	$V = l\,w\,h$	$V = \frac{1}{3}\pi r^2 h$	$V = \pi r^2 h$	$V = \frac{4}{3}\pi r^3$	$V = \frac{1}{3}Bh$

By studying each formula and by comparing formulas between different solids, you can determine general relationships.

EXAMPLE 1: How would doubling the radius of a sphere affect the volume?

The volume of a sphere is $V = \frac{4}{3}\pi r^3$. Just by looking at the formula, can you see that by doubling the radius, the volume would increase by 2^3 or 8? So, a sphere with a radius of 2 would have a volume 8 times greater than a sphere with a radius of 1.

EXAMPLE 2: A cylinder and a cone have the same radius and the same height. What is the difference between their volumes?

Compare the formulas for the volume of a cone and the volume of a cylinder. They are identical except that the cone is multiplied by $\frac{1}{3}$. Therefore, the volume of a cone with the same height and radius as a cylinder would be one-third less. Or, the volume of a cylinder with the same height and radius as a cone would be three times greater.

EXAMPLE 3: If you double one dimensions of a rectangular prism, how will the volume be affected? How about doubling two dimensions? How about doubling all three dimensions?

Do you see that doubling just one of the dimensions of a rectangular prism will also double the volume? Doubling two of the dimensions will cause the volume to increase by 2^2 or 4. Doubling all three dimensions will cause the volume to increase by 2^3 or 8.

EXAMPLE 4: A cylinder holds 100 cubic centimeters of water. If you triple the radius of the cylinder but keep the height the same, how much water would you need to fill the new cylinder?

Tripling the radius of a cylinder causes the volume to increase by 3^2 or 9. The volume of the new cylinder would hold 9×100 or 900 cubic centimeters of water.

Answer the following questions by comparing the volumes of two solids that share some of the same dimensions.

1. If you have a cylinder with a height of 8 inches and a radius of 4 inches and you have a cone with the same height and radius, how many times greater is the volume of the cylinder than the volume of the cone?

2. 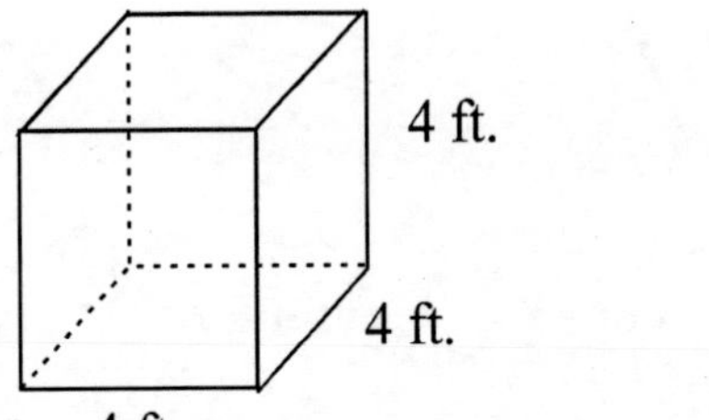

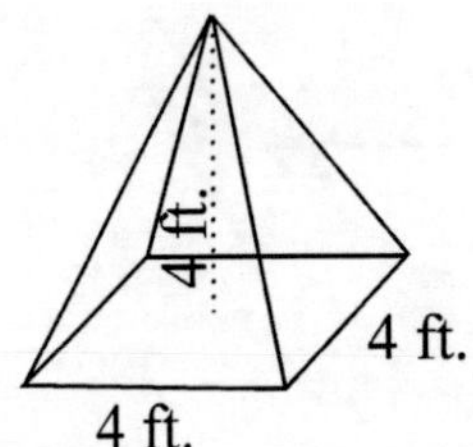

   In the two figures above, how many times larger is the volume of the cube than the volume of the pyramid?

3. How many times greater is the volume of a cylinder if you double the radius?

4. How many times greater is the volume of a cylinder if you double the height?

5. In a rectangular solid, how many times greater is the volume if you double the length?

6. In a rectangular solid, how many times greater is the volume if you double the length and the width?

7. In a rectangular solid, how many times greater is the volume if you double the length and the width and the height?

8. In the following two figures, how many cubes like Figure 1 will fit inside Figure 2?

   Figure 1 6 in. 6 in. 6 in.

   Figure 2 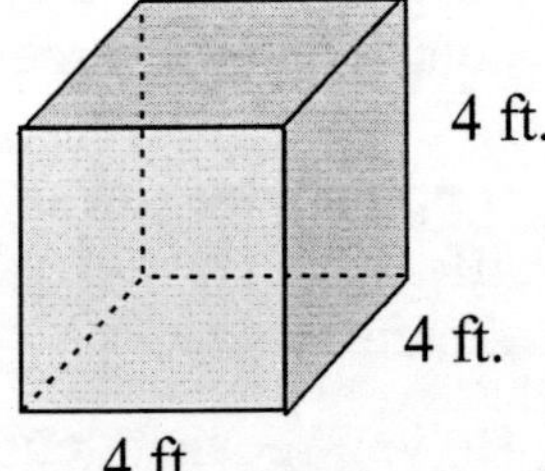

9. A sphere has a radius of 1. If the radius is increased to 3, how many times greater will the volume be?

10. It takes 2 liters of water to fill cone A below. If the cone is stretched so the radius is doubled but the height stays the same, how much water is needed to fill the new cone, B?

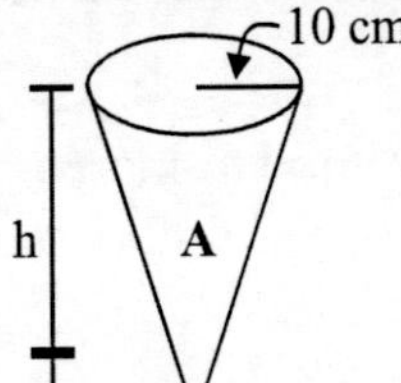

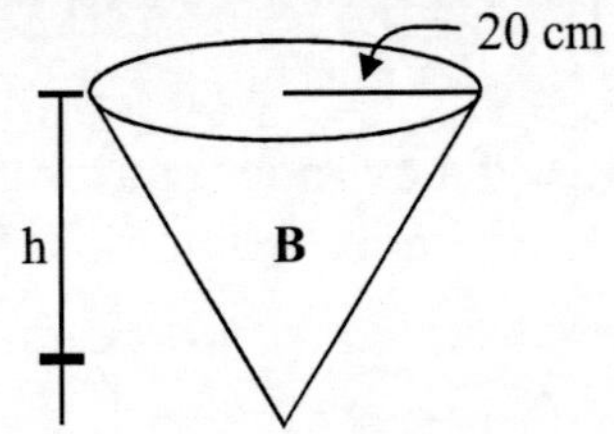

SURFACE AREA

The **surface area of a solid** is the total area of all the sides of a solid.

CUBE

There are six sides on a cube. To find the surface area of a cube, find the area of one side and multiply by 6.

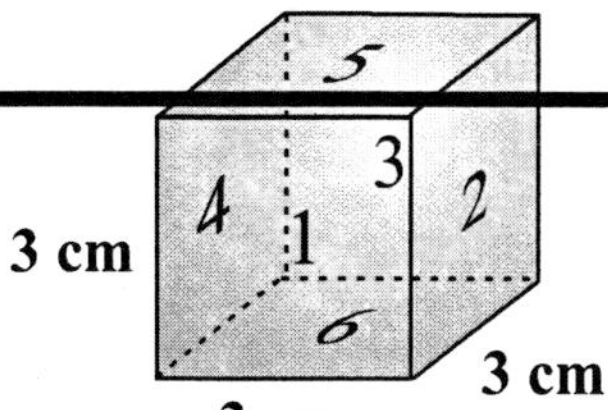

Area of each side of the cube:
$3 \times 3 = 9 \text{ cm}^2$

Total surface area: $9 \times 6 = 54 \text{ cm}^2$

RECTANGULAR PRISM

There are 6 sides on a rectangular prism. To find the surface area, add the areas of the six rectangular sides.

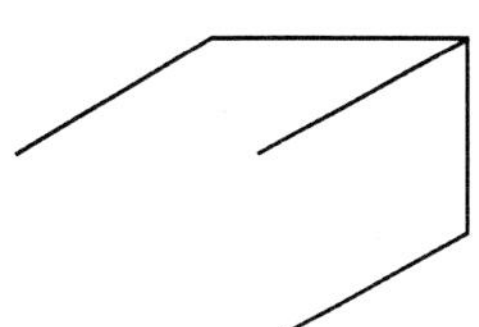

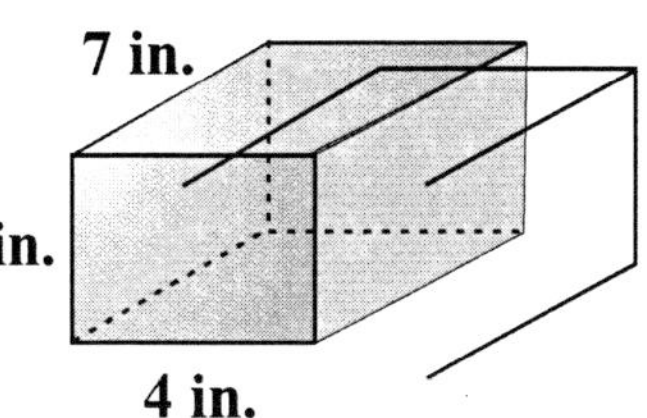

Top and Bottom

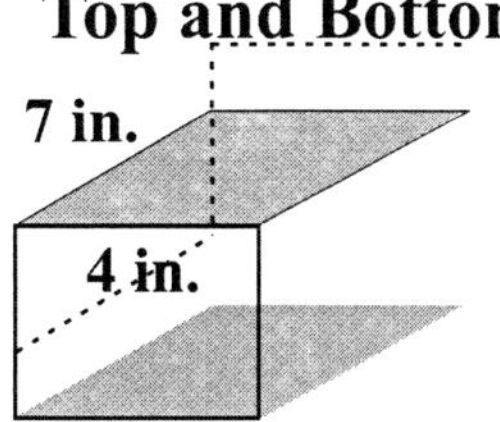

Front and Back

3 in.
4 in.

Left and Right

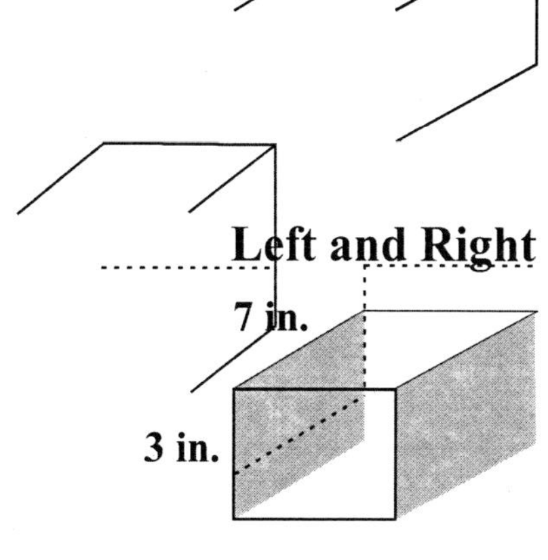

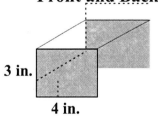

Area of top side:
$7 \text{ in.} \times 4 \text{ in.} = 28 \text{ in}^2$
Area of top and bottom:
$28 \text{ in.} \times 2 \text{ in.} = 56 \text{ in}^2$

Area of front:
$3 \text{ in.} \times 4 \text{ in.} = 12 \text{ in}^2$
Area of front and back:
$12 \text{ in.} \times 2 \text{ in.} = 24 \text{ in}^2$

Area of left side:
$3 \text{ in.} \times 7 \text{ in.} = 21 \text{ in}^2$
Area of left and right:
$21 \text{ in.} \times 2 \text{ in.} = 42 \text{ in}^2$

Total surface area: $56 \text{ in}^2 + 24 \text{ in}^2 + 42 \text{ in}^2 = 122 \text{ in}^2$

Find the surface area of the following cubes and prisms.

1. 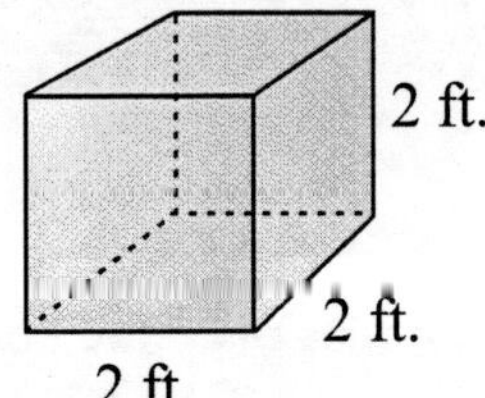

$SA =$ ____________

2. 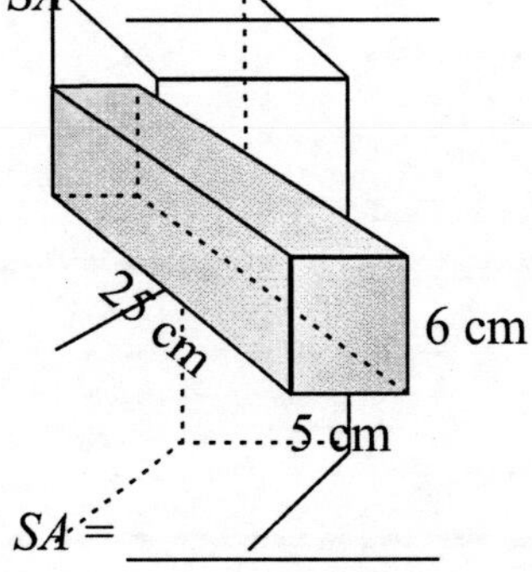

$SA =$ ____________

3. 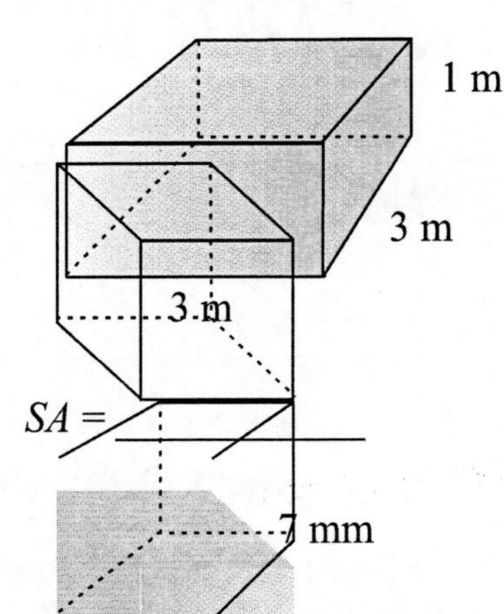

$SA =$ ____________

4. 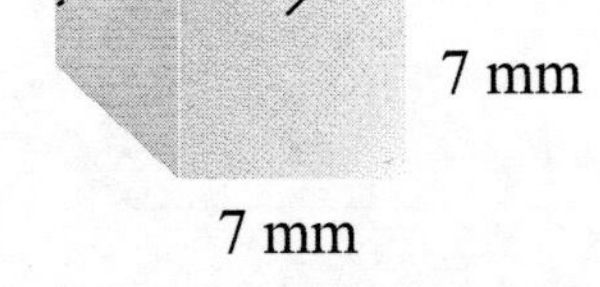

$SA =$ ____________

5. 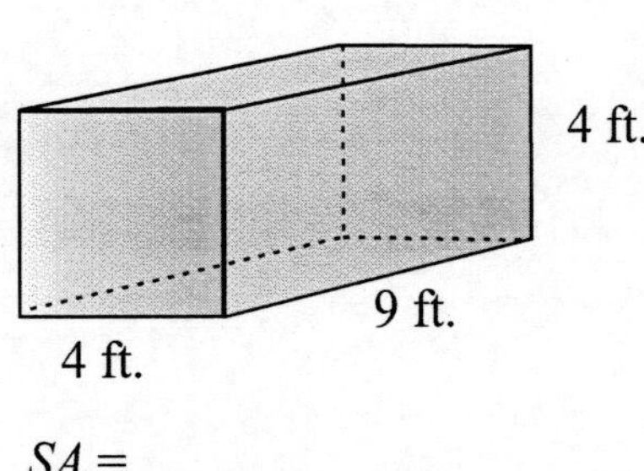

$SA =$ ____________

6. 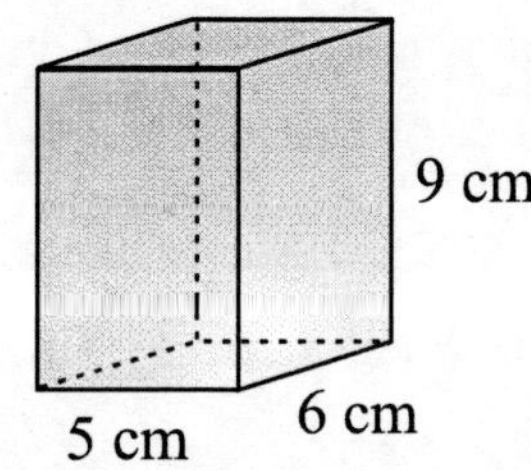

$SA =$ ____________

7. 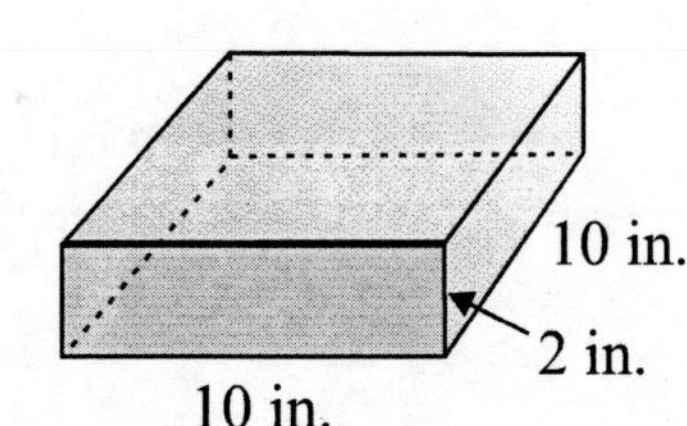

$SA =$ ____________

8. 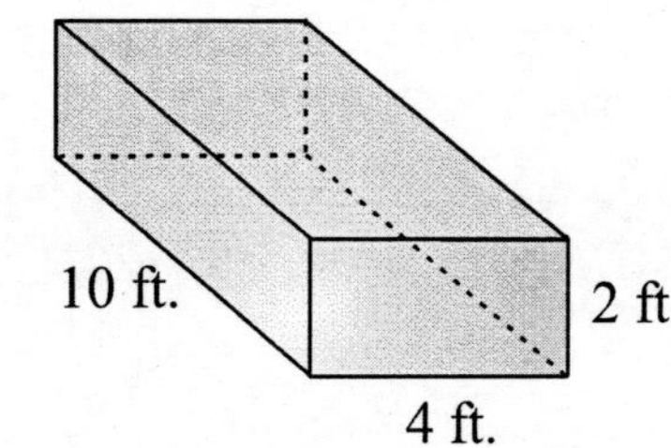

$SA =$ ____________

9. 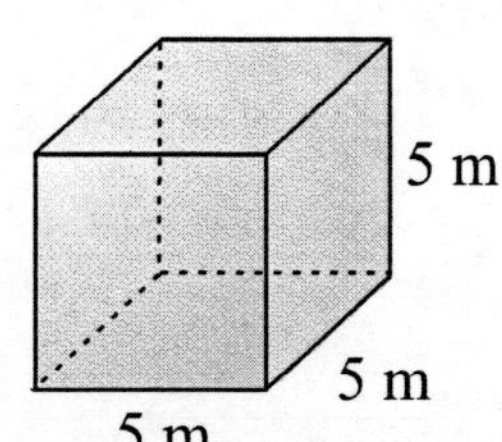

$SA =$ ____________

10. 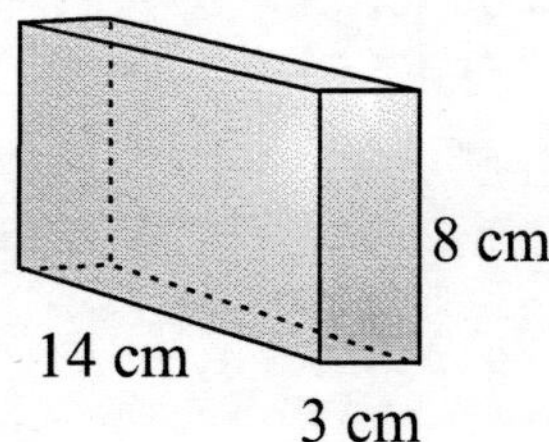

$SA =$ ____________

The pyramid below is made of a square base with 4 triangles on the sides.

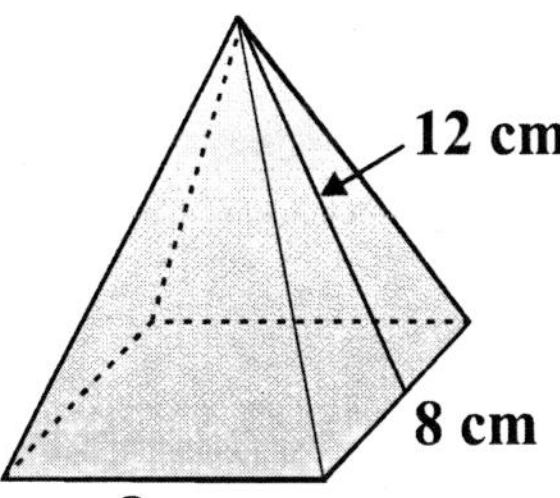

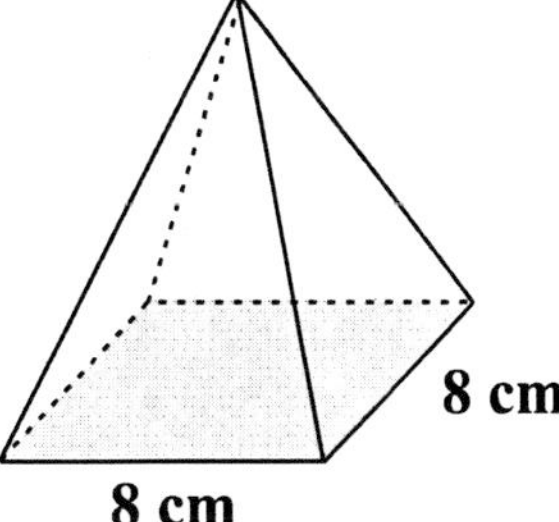

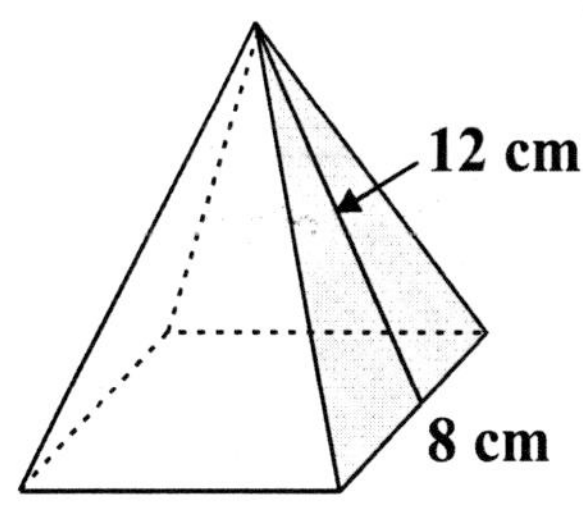

Area of square base:
$A = l \times w$
$A = 8 \times 8 = 64 \text{ cm}^2$

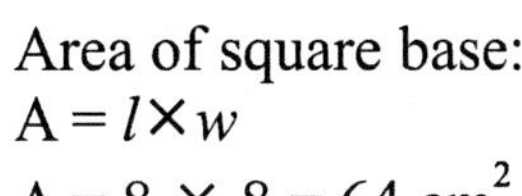

Area of sides:
Area of 1 side $= \frac{1}{2}Bh$
$A = \frac{1}{2} \times 8 \times 12 = 48 \text{ cm}^2$
Area of 4 sides $= 48 \times 4 = 192 \text{ cm}^2$

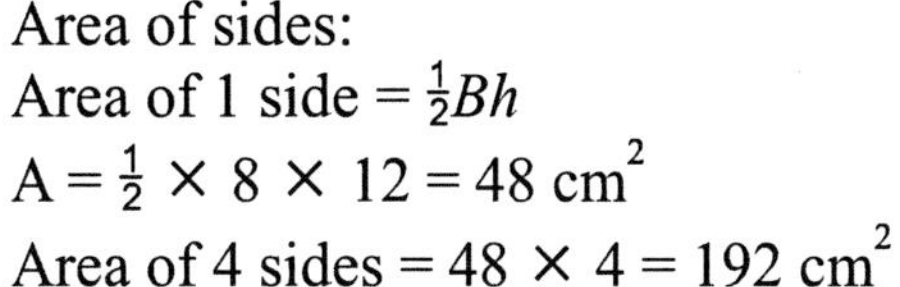

Total surface area: $64 + 192 = 256 \text{ cm}^2$

Find the surface area of the following pyramids.

1. 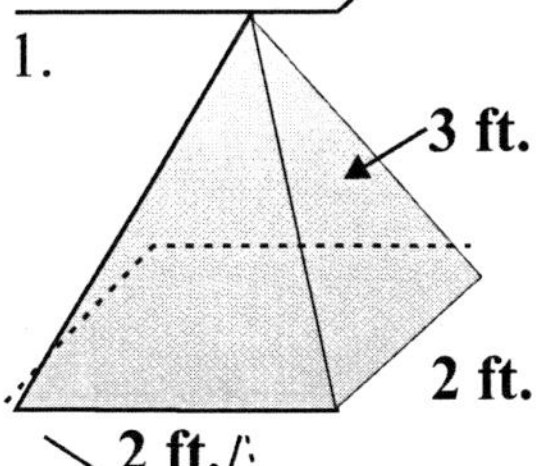

SA = ____________

4. 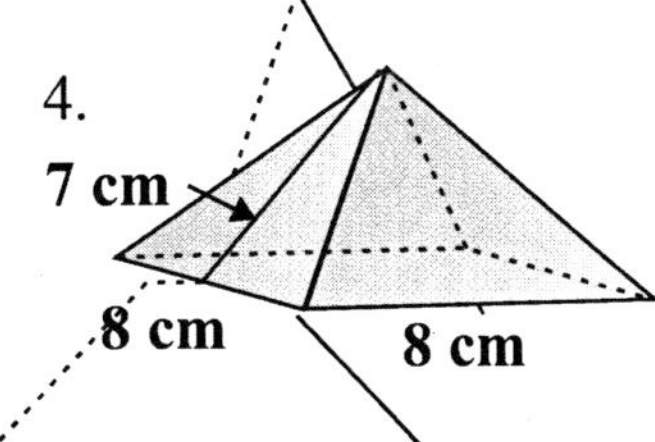

SA = ____________

7. 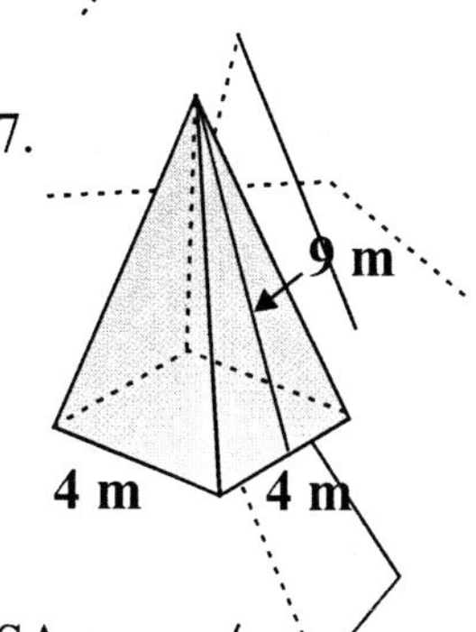

SA = ____________

2. 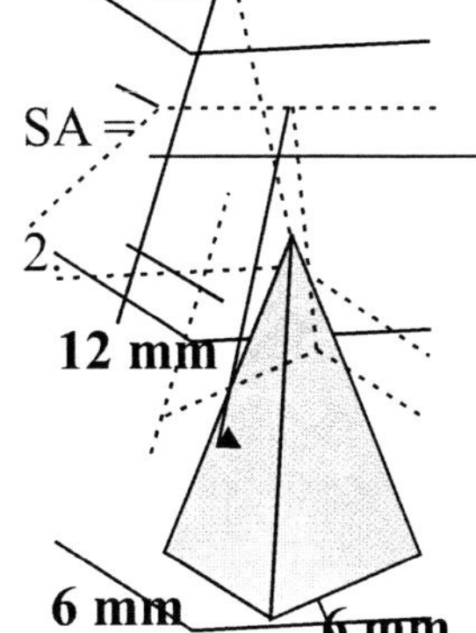

SA = ____________

5. 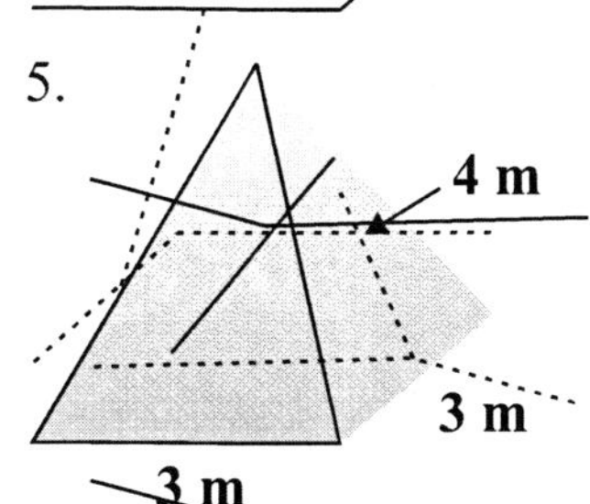

SA = ____________

8. 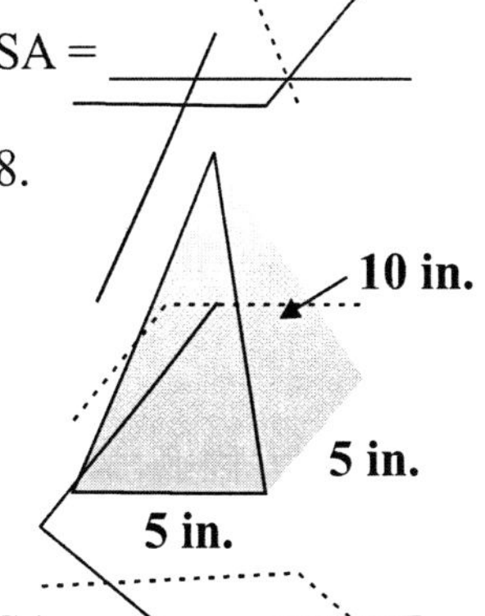

SA = ____________

3. 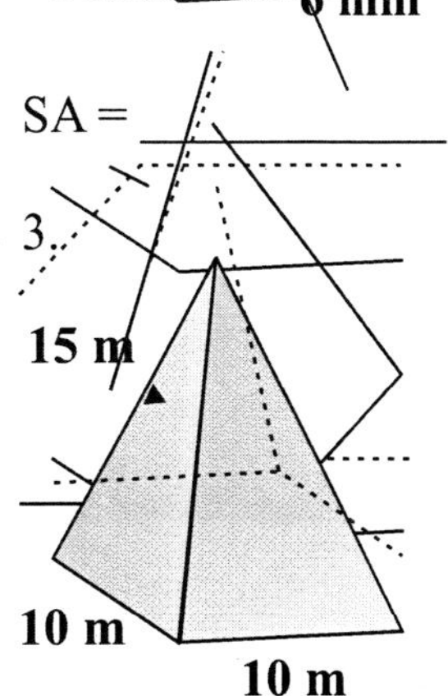

SA = ____________

6. 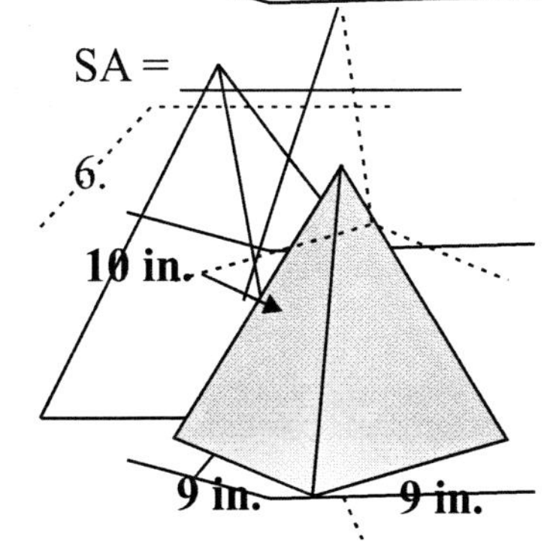

SA = ____________

9. 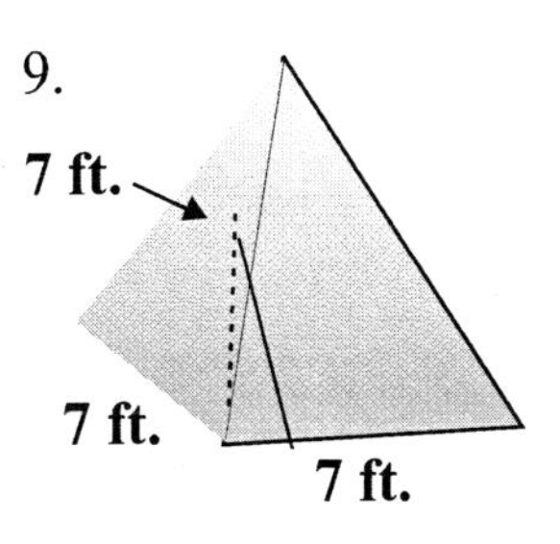

SA = ____________

CYLINDER

If the side of a cylinder were slit from top to bottom and laid flat, its shape would be a rectangle. The length of the rectangle is the same as the circumference of the circle that is the base of the cylinder. The width of the rectangle is the height of the cylinder.

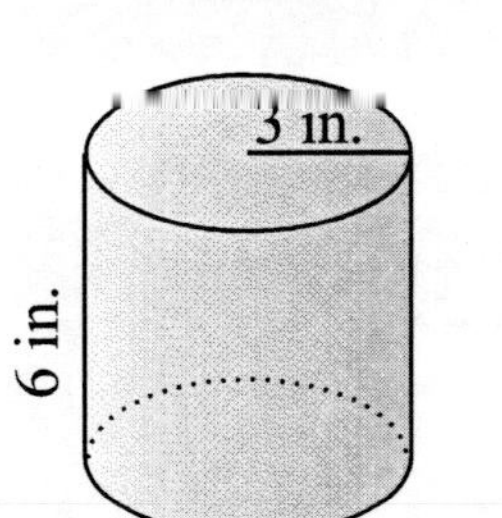

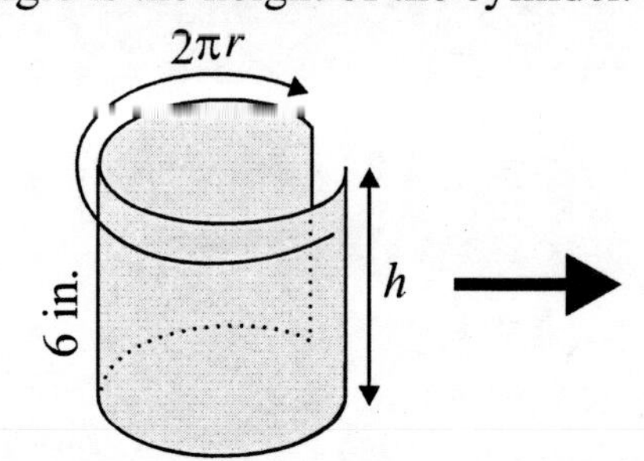

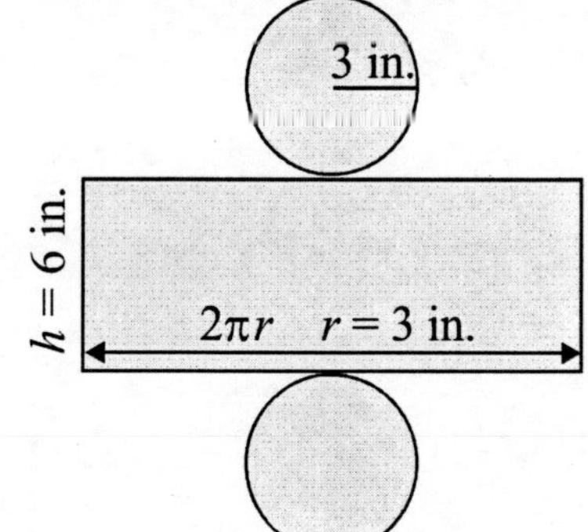

Total Surface Area of a Cylinder = $2\pi r^2 + 2\pi rh$

Area of top and bottom:
Area of a circle = πr^2
Area of top = $3.14 \times 3^2 = 28.26$ in.^{2}
Area of top and bottom = $2 \times 28.26 = 56.52$ in.^{2}

Area of side:
Area of rectangle = $l \times h$
$l = 2\pi r = 2 \times 3.14 \times 3 = 18.84$ in.
Area of rectangle = $18.84 \times 6 = 113.04$ in.^{2}

Total surface area = $56.52 + 113.04 = 169.56$ in^2

Find the surface area of the following cylinders. Use π = 3.14

1. 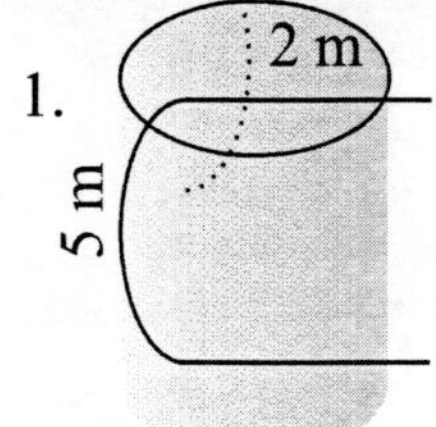

SA = ____________

2. 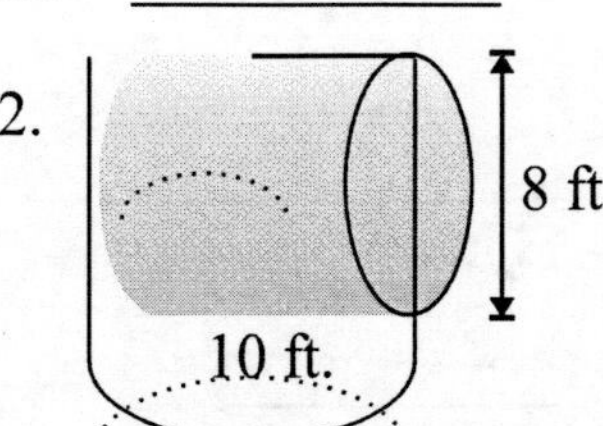

SA = ____________

3. 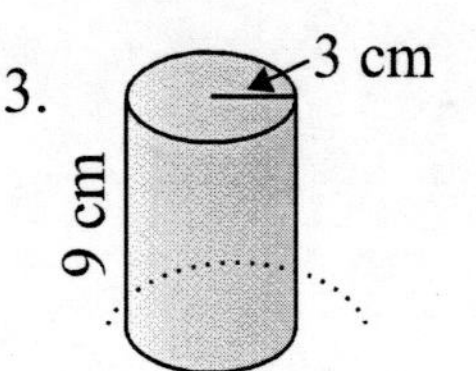

SA = ____________

4. 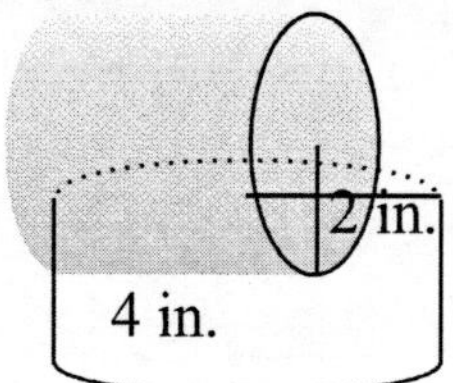

SA = ____________

5. 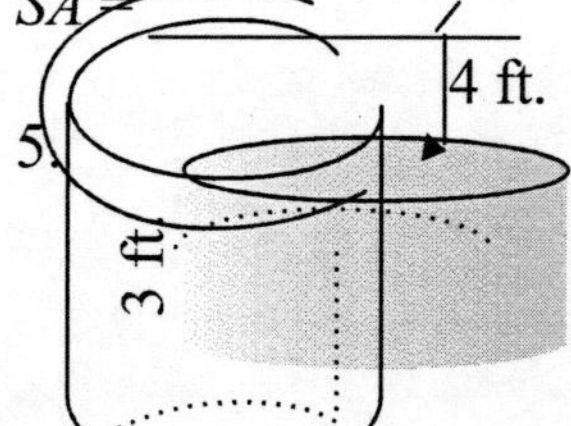

SA = ____________

6. 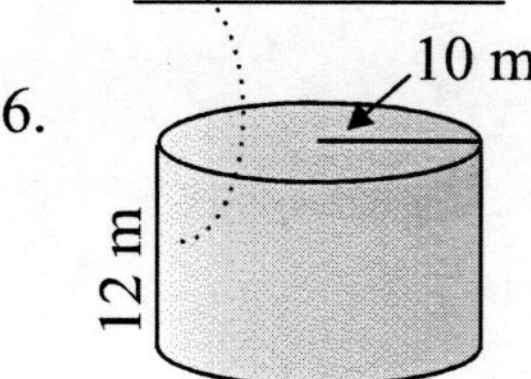

SA = ____________

7. 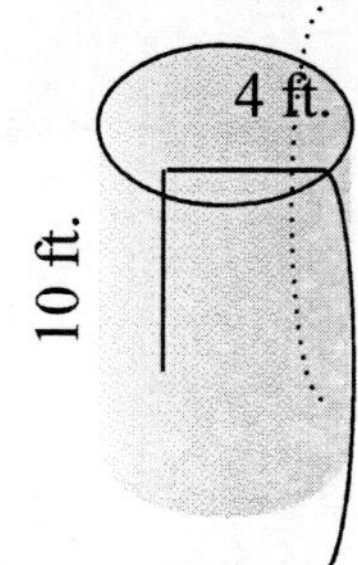

SA = ____________

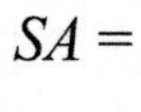

8. 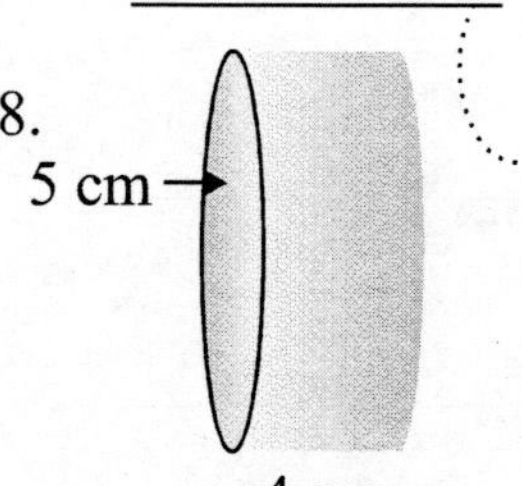

SA = ____________

9. 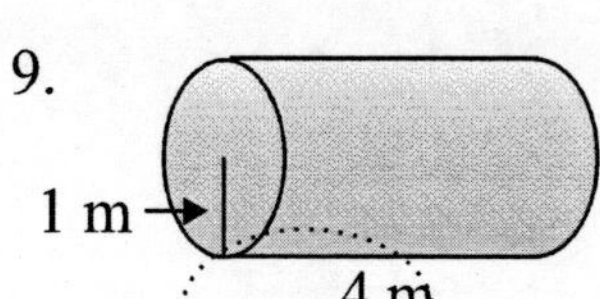

SA = ____________

SPHERE

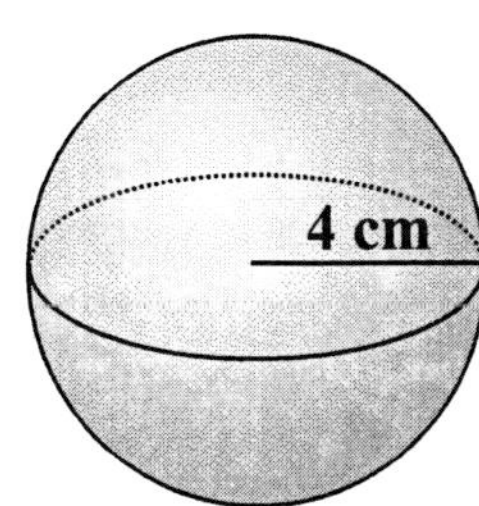

Surface area = $4\pi r^2$
Surface area = $4 \times 3.14 \times 4^2$
Surface area = 200.96 cm^2

Find the surface area of a sphere given the following measurements where *r* = radius and *d* = diameter. Use π = 3.14.

1. $r = 2$ in	SA = ______	7. $d = 14$ cm	SA = ______
2. $r = 6$ m	SA = ______	8. $r = \frac{1}{5}$ km	SA = ______
3. $r = \frac{3}{4}$ yd	SA = ______	9. $d = 3$ in	SA = ______
4. $d = 8$ cm	SA = ______	10. $d = \frac{2}{3}$ ft	SA = ______
5. $d = 50$ mm	SA = ______	11. $r = 10$ mm	SA = ______
6. $r = \frac{1}{4}$ ft	SA = ______	12. $d = 5$ yd	SA = ______

CONE

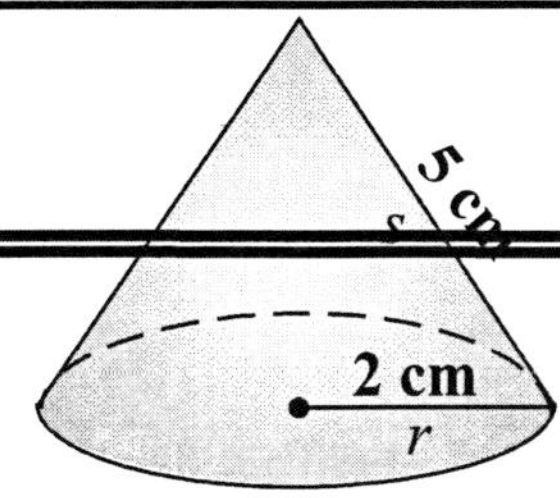

Total Surface Area : $T = \pi r(r + s)$

$\pi = 3.14$ r = radius of base s = slant height

$T = 3.14 \times 2(2 + 5)$
$T = 6.28 \times 7$
$T = 43.96$ cm^2

Find the surface area of the following cones. Use π = 3.14.

1. 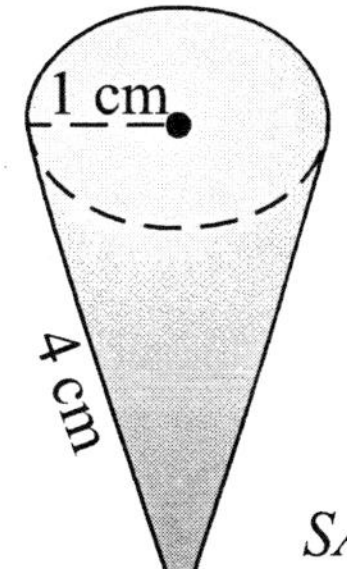

SA = ______

3. 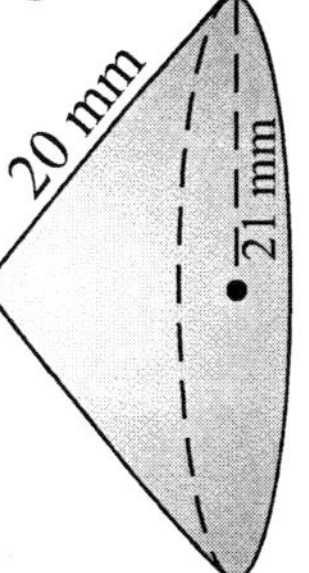

SA = ______

5. 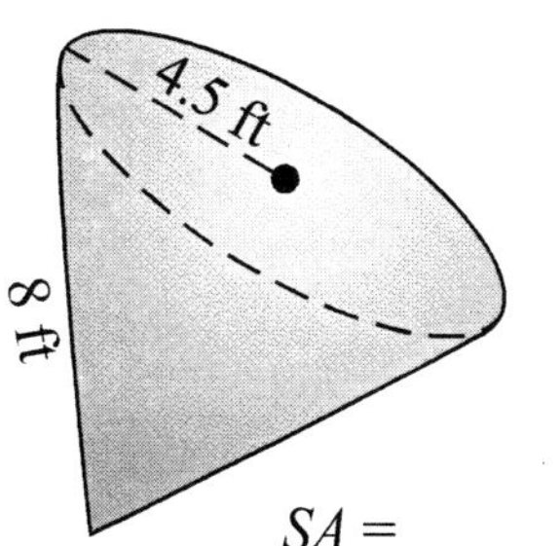

SA = ______

2. 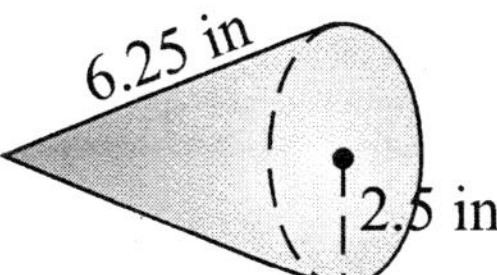

SA = ______

4. 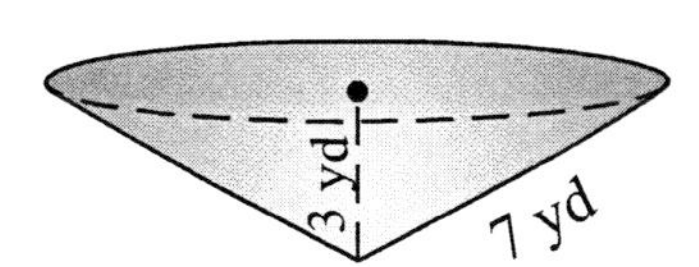

SA = ______

6. 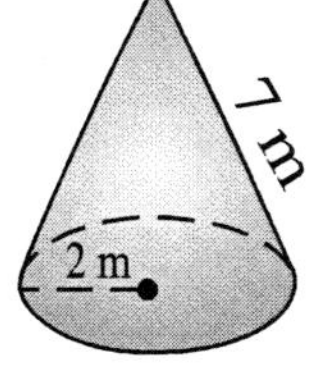

SA = ______

SOLID GEOMETRY WORD PROBLEMS

1. If an Egyptian pyramid has a square basc that mcasurcs 500 yards by 500 yards, and the pyramid stands 300 yards tall, what would be the volume of the pyramid? Use the formula for volume of a pyramid, $V = \frac{1}{3}Bh$ where B is the area of the base.

   V = ____________

2. Robert is using a cylindrical barrel filled with water to flatten the sod in his yard. The circular ends have a radius of 1 foot. The barrel is 3 feet wide. How much water will the barrel hold? The formula for volume of a cylinder is $V = \pi r^2 h$. Use $\pi = 3.14$.

   V = ____________

3. If a basketball measures 24 centimeters in diameter, what volume of air will it hold? The formula for volume of a sphere is $V = \frac{4}{3}\pi r^3$. Use $\pi = 3.14$.

   V = ____________

4. What is the volume of a cone that is 2 inches in diameter and 5 inches tall? The formula for volume of a cone is $V = \frac{1}{3}\pi r^2 h$. Use $\pi = 3.14$.

   V = ____________

5. Kelly has a rectangular fish aquarium that measures 24 inches wide, 12 inches deep, and 18 inches tall. What is the maximum amount of water that the aquarium will hold?

   V = ____________

6. Jenny has a rectangular box that she wants to cover in decorative contact paper. The box is 10 cm long, 5 cm wide, and 5 cm high. How much paper will she need to cover all 6 sides?

   SA = ____________

7. Gasco needs to construct a cylindrical, steel gas tank that measures 6 feet in diameter and is 8 feet long. How many square feet of steel will be needed to construct the tank? Use the following formulas as needed: $A = l \times w$, $A = \pi r^2$, $C = 2\pi r$. Use $\pi = 3.14$.

   SA = ____________

8. Craig wants to build a miniature replica of San Francisco's Transamerica Pyramid out of glass. His replica will have a square base that measures 6 cm by 6 cm. The 4 triangular sides will be 6 cm wide and 60 cm tall. How many square centimeters of glass will he need to build his replica? Use the following formulas as needed: $A = l \times w$ and $A = \frac{1}{2}bh$.

   SA = ____________

9. Jeff built a wooden, cubic toy box for his son. Each side of the box measures 2 feet. How many square feet of wood did he use to build the toy box? How many cubic feet of toys will the box hold?

   SA = ____________

   V = ____________

CHAPTER 16 REVIEW

Find the volume and/or the surface area of the following solids.

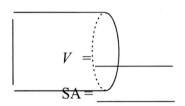

1.

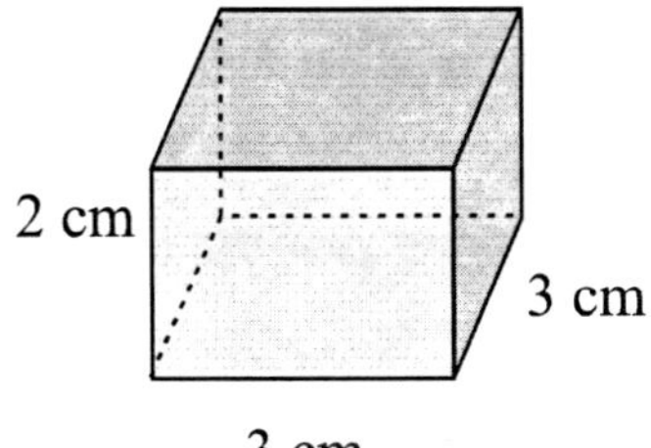

V = ________

SA = ________

2.

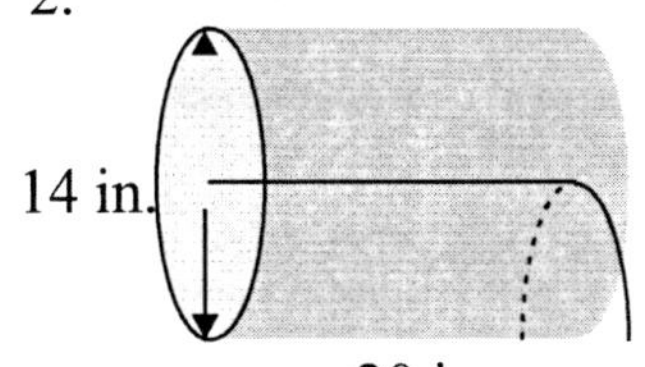

$V = \pi r^2 h$
SA $= 2\pi r^2 + 2\pi rh$

Use $\pi = 3.14$

V = ________

SA = ________

3.

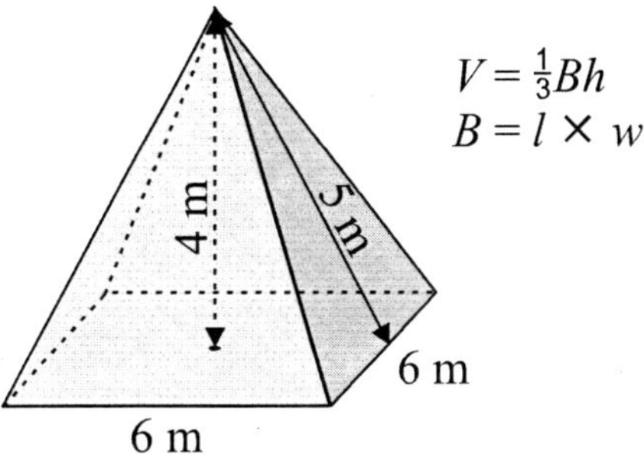

$V = \frac{1}{3}Bh$
$B = l \times w$

V = ________

SA = ________

4.

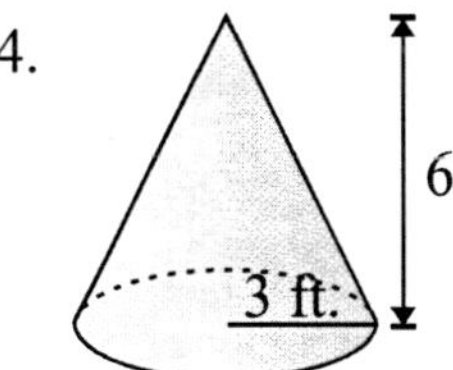

$V = \frac{1}{3}\pi r^2 h$

Use $\pi = 3.14$

V = ________

5.

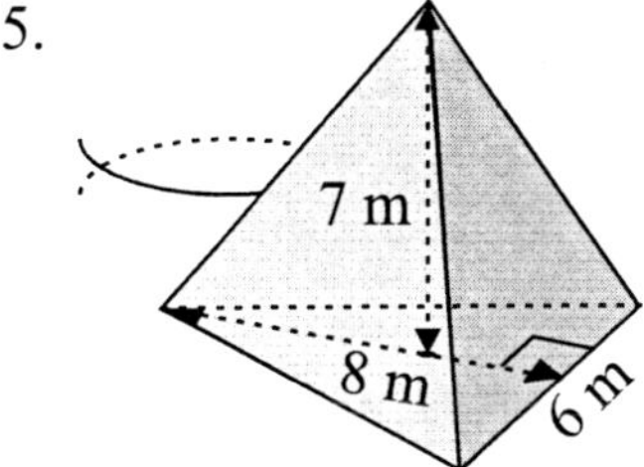

$V = \frac{1}{3}Bh$
B = area of the triangular base

V = ________

6.

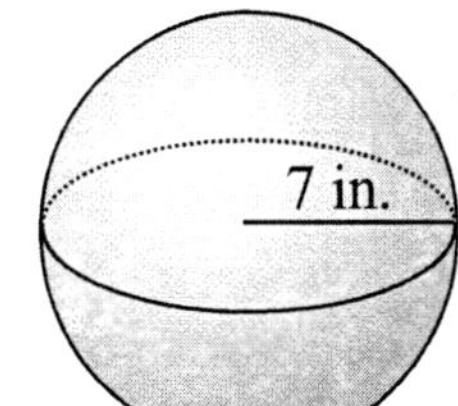

$V = \frac{4}{3}\pi r^3$

Use $\pi = 3.14$

V = ________

7. The sandbox at the local elementary school is 60 inches wide and 100 inches long. The sand in the box is 6 inches deep. How many cubic inches of sand are in the sandbox?

____

8. If you have cubes that are two inches on each edge, how many would fit in a cube that was 16 inches on each edge?

____

9. If you double each edge of a cube, how many times larger is the volume?

____

10. It takes 8 cubic inches of water to fill the cube below. If each side of the cube is doubled, how much water is needed to fill the new cube?

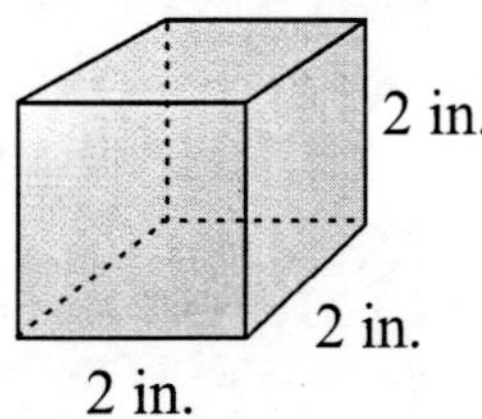

____

11. If a ball is 4 inches in diameter, what is its volume? Use $\pi = 3.14$

____

12. A grain silo is in the shape of a cylinder. If a silo has an inside diameter of 10 feet and a height of 35 feet, what is the maximum volume inside the silo?

Use $\pi = 3.14$

____

13. A closed cardboard box is 30 centimeters long, 10 centimeters wide, and 20 centimeters high. What is the total surface area of the box?

____

14. Siena wants to build a wooden toy box with a lid. The dimensions of the toy box are 3 feet long, 4 feet wide, and 2 feet tall. How many square feet of wood will she need to construct the box?

____

15. How many 1-inch cubes will fit inside a larger 1 foot cube? (Figures are not drawn to scale.)

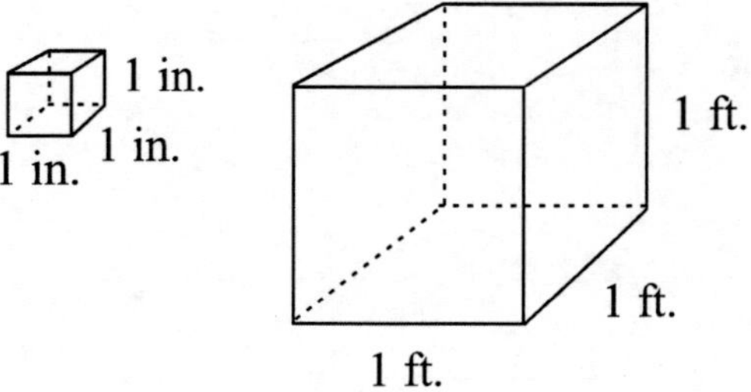

____

16. The cylinder below has a volume of 240 cubic inches. The cone below has the same radius and the same height as the cylinder. What is the volume of the cone?

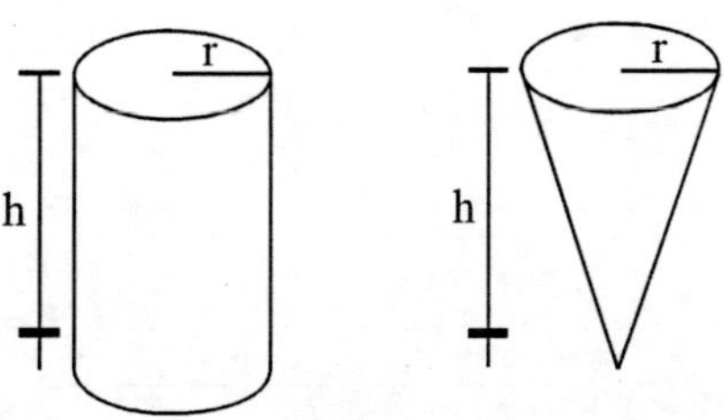

____

17. The height of a rectangular solid is four meters. The volume of the solid is 60 cubic meters. What is the area of the base of the rectangular solid?

____

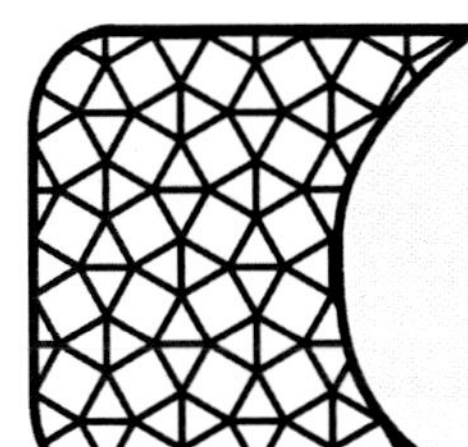

CHAPTER 17: SETS AND LOGIC

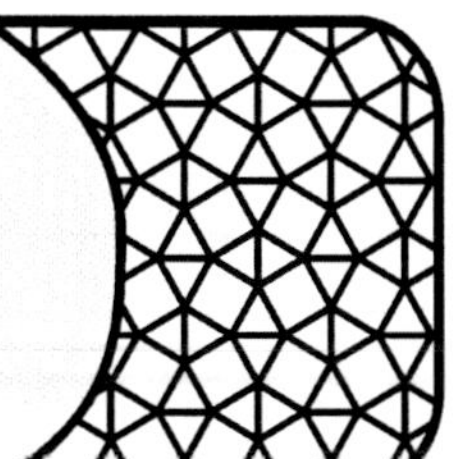

Standard 4.12.9

A set contains an object or objects that are clearly identified. The objects in a set are called **elements** or **members**. A set can also be **empty**. An **empty** set is also called the **null** set. The symbol ∅ or { } is used to denote the **null** set.

Members of a set can be described in two ways. One way is to give a **rule** of description that clearly defines the members. Another way is to make a **roster** of each member of the set, mentioning each member only <u>once</u> in any order. The **members** of a set are enclosed in braces { }and separated by commas.

Rule		**Roster**
{the letters in the word "school"}	=	{s,c,h,o,l}

The symbol ∈ is used to show that an object is the member of a set. For example, **3 ∈ {1,2,3,4}.**
The symbol ∉ means "is **not** a member of". For example, **8 ∉ {1,2,3,4}.**
The symbol ≠ means "is not equal to". For example,

{the cities in Texas} ≠ {New York, Philadelphia, Nashville}

Read each of the statements below and tell whether they are true or false.

		True/False
1.	{the days of the week} = {January, March, December}	
2.	{The first five letters of the alphabet} = {a, b, c, d, e}	
3.	5 ∉ {all odd numbers}	
4.	Friday ∈ {the days of the week}	
5.	{the last three letters of the alphabet} ≠ {x, y, z}	
6.	t ∉ {the letters in the word "yellow"}	
7.	the letters in the word "funny"∈ {the letters of the alphabet}	
8.	{The letters in the word "Alabama"} = {a, b, l, m}	
9.	{living unicorns} ≠ ∅	
10.	{the letters in the word "horse"} ≠ {the letters in the word "shore"}	

Identify each set by making a roster (see above). If a set has no members, use ∅.

11. {the letters in the word "hat" that are also in the word "thin"}
12. {the letters in the word "Mississippi"}
13. {the provinces of Canada that border the state of Texas}
14. {the letters in the word "kitchen" and also in the word "dinner"}
15. {the days of the week that have the letter "n"}
16. {the letters in the word "June" that are also in the word "April"}
17. {the letters in the word "instruments" that are not in the word "telescope"}
18. {the digits in the number "19,582" that are also in the number "56,871"}

SUBSETS

If every member of set *A* is also a member of set *B*, then set *A* is a **subset** of set *B*.

The symbol ⊂ means "is a subset of."

For example, every member of the set {1, 2, 3, 4} is a member of the set {1, 2, 3, 4, 5, 6, 7}. Therefore, {1, 2, 3, 4} ⊂ {1, 2, 3, 4, 5, 6, 7}.

The relationship can be pictured in the following diagram.

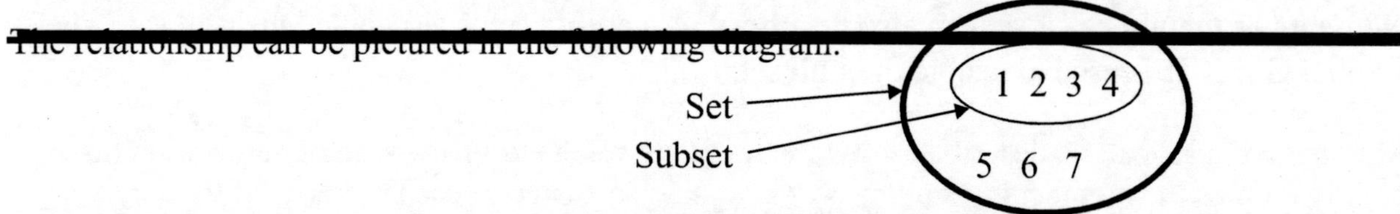

The symbol ⊄ means "is **not** a subset of."

For example, **not** every member of the set {Ann, John, Sue} is a member of the set {Ann, John, Cindy}. Therefore, {Ann, John, Sue} ⊄ {Ann, John, Cindy}.

Read each of the statements below and tell whether they are true or false.

		True/False
1.	{a, b, c} ⊄ {a, b, c, d}	
2.	{1, 3, 5, 7} ⊂ {1, 2, 4, 5, 6, 7}	
3.	{a, e, i, o, u,}⊂ {the vowels of the alphabet}	
4.	{dogs} ⊂ {poodles, bull dogs, collies}	
5.	{fruit} ⊄ {apples, grapes, bananas, oranges}	
6.	{10, 20, 30, 40} ⊂ {20, 30, 40, 50}	
7.	{English, Spanish, French, German} ⊂ {languages}	
8.	{duck, swan, penguin} ⊄ {mammals}	
9.	{1, 2, 5, 10} ⊂ {whole numbers}	
10.	{Atlanta} ⊄ {U.S. state capitals}	

The set {Pam} has two subsets: {Pam} and ∅ . The set {Emily, Brad} has 4 subsets: {Emily}, {Brad}, {Emily, Brad} and ∅. The ∅ is a subset of any set.

For each of the following sets, list all of the possible subsets.

1. {a}

2. {1, 2}

3. {r, s, t}

4. {1, 3, 5, 7}

5. {Joe, Ed}

INTERSECTION OF SETS

To find the intersection of two sets, you need to identify the members that the two sets share in common. The symbol for intersection is ∩. A **Venn diagram** shows how sets intersect.

Roster

{2, 4, 6, 8} ∩ {4, 6, 8, 10} = {4, 6, 8}

The shaded area is the intersection of the two sets. It shows which numbers both sets have in common.

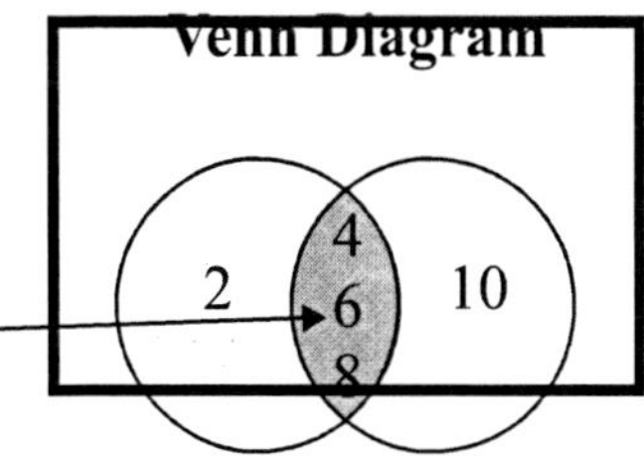

Find the intersection of the following sets:

1. {Ben, Jan, Dan, Tom} ∩ {Dan, Mike, Kate, Jan} =
2. {pink, purple, yellow} ∩ {purple, green, blue} =
3. {2, 4, 6, 8, 10} ∩ {1, 2, 3, 4, 5, 6, 7, 8, 9, 10} =
4. {a, e, i, o, u} ∩ {a, b, c, d, e, f, g, h, i, j, k} =
5. {pine, oak, walnut, maple} ∩ {maple, poplar} =
6. {100, 98, 95, 78, 62} ∩ {57, 82, 95, 98, 99} =
7. {orange, kiwi, coconut, pineapple} ∩ {pear, apple, orange} =

Look at the Venn diagram at the right to answer the questions below. Show your answers in roster form. Do the problem in parentheses first.

8. A ∩ B =
9. (A ∩ B) ∩ C =
10. A ∩ C =
11. B ∩ C =
12. (A ∩ C) ∩ B =

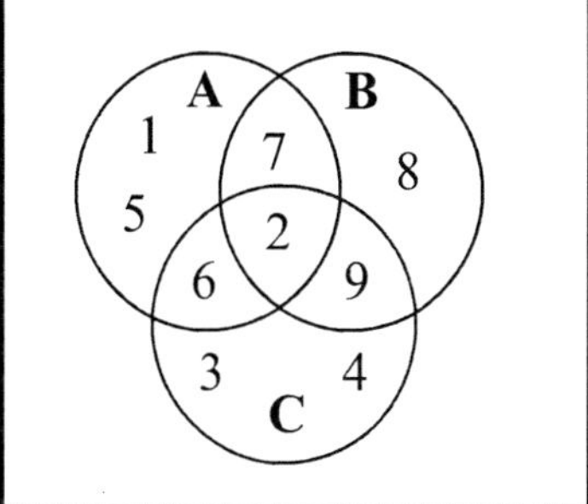

13. blue ∩ green =
14. (purple ∩ blue) ∩ green =
15. blue ∩ purple =
16. (blue ∩ green) ∩ purple =
17. green ∩ purple =

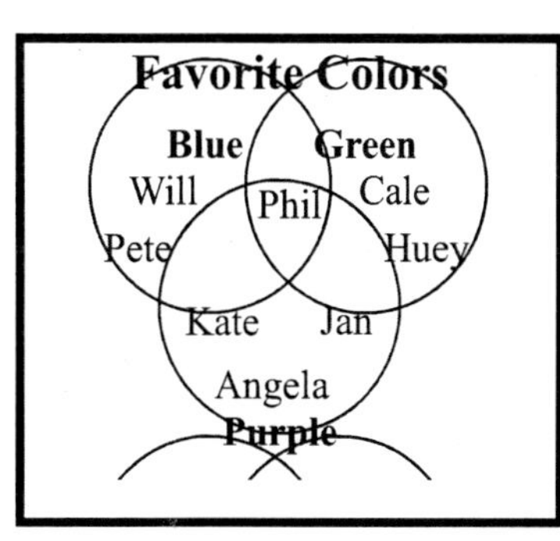

UNION OF SETS

The union of two sets means to put the members of two sets together into one set without repeating any members. The symbol for union is ∪.

{1, 2, 3, 4} ∪ {3, 4, 5, 6} = {1, 2, 3, 4, 5, 6}

The union of these two sets is the shaded area in the Venn diagram below.

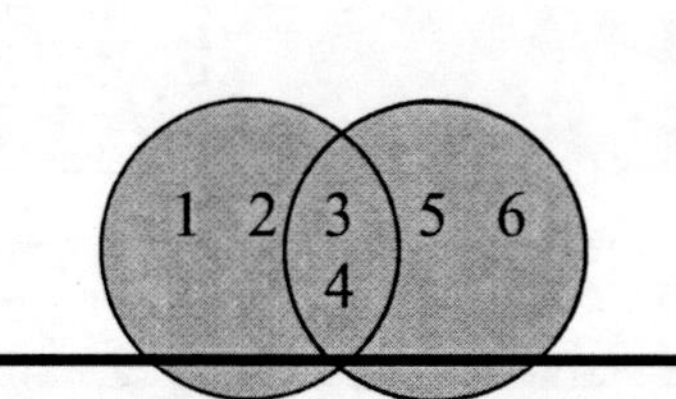

Find the union of the sets below.

1. {apples, pears, oranges} ∪ {pears, bananas, apples} =
2. {5, 10, 15, 20, 25} ∪ {10, 20, 30, 40} =
3. {Ted, Steve, Kevin, Michael} ∪ { Kevin, George, Kenny} =
4. {raisins, prunes, apricots} ∪ {peanuts, almonds, coconut} =
5. {sales, marketing, accounting} ∪ {receiving, shipping, sales} =
6. {beef, pork, chicken} ∪ {chicken, tuna, shark} =

Refer to the following Venn diagrams to answer the questions below. Identify each of the following sets by roster.

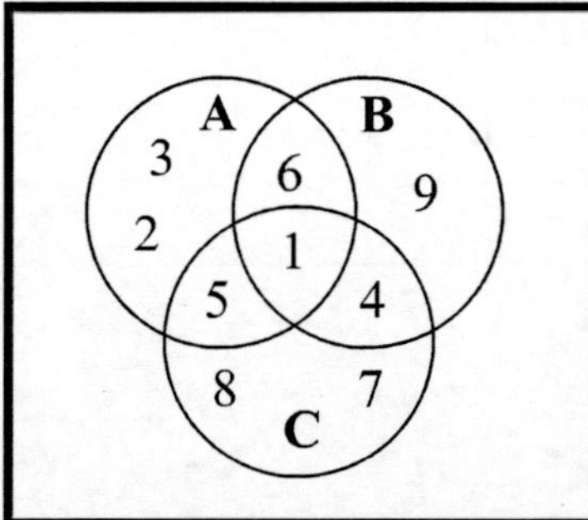

7. A ∪ C =
8. C ∪ B =
9. B ∪ A =
10. A ∪ B ∪ C =
11. Which salespeople worked in either the North and the East?
12. Which salespeople worked in either the North and the South?
13. Which salespeople worked in either the South and the East?

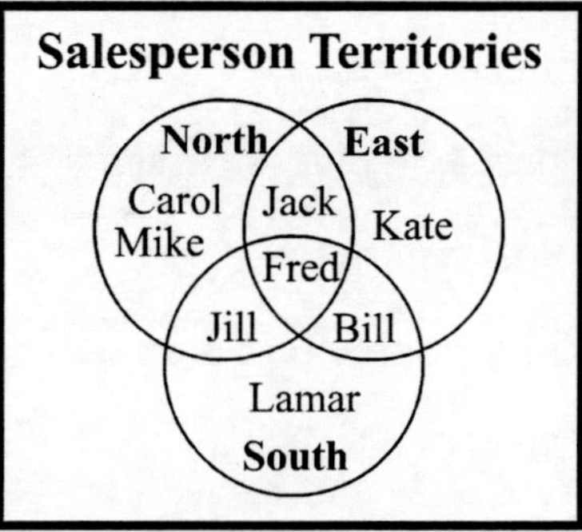

MATHEMATICAL REASONING/LOGIC

The ability to use reasoning and logic is an important skill for solving math problems, but it can also be helpful in real-life situations. For example, if you need to get to Park Street, and the Park Street bus always comes to the bus stop at 3 pm, then you know that you need to get to the bus stop at least by 3 pm. This is a real-life example of using logic, which many people would call "common sense."

There are many different types of statements which are commonly used to describe mathematical principles. However, using the rules of logic, the truth of any mathematical statement must be evaluated. Below are a list of tools used in logic to evaluate mathematical statements.

Logic is the discipline that studies valid reasoning. There are many forms of valid arguments, but we will just review a few here.

A **proposition** is usually a declarative sentence which may be true or false.

An **argument** is a set of two or more related propositions, called **premises**, that provide support for another proposition, called the **conclusion**.

Deductive reasoning is an argument which begins with general premises and proceeds to a more specific conclusion. Most elementary mathematical problems use deductive reasoning.

Inductive reasoning is an argument in which the truth of its premises make it likely or probable that its conclusion is true.

ARGUMENTS

Most of logic deals with the evaluation of the validity of arguments. An argument is a group of statements that includes a conclusion and at least one premise. A premise is a statement that you know is true or at least you assume to be true. Then, you draw a conclusion based on what you know or believe is true in the premise(s). Consider the following example:

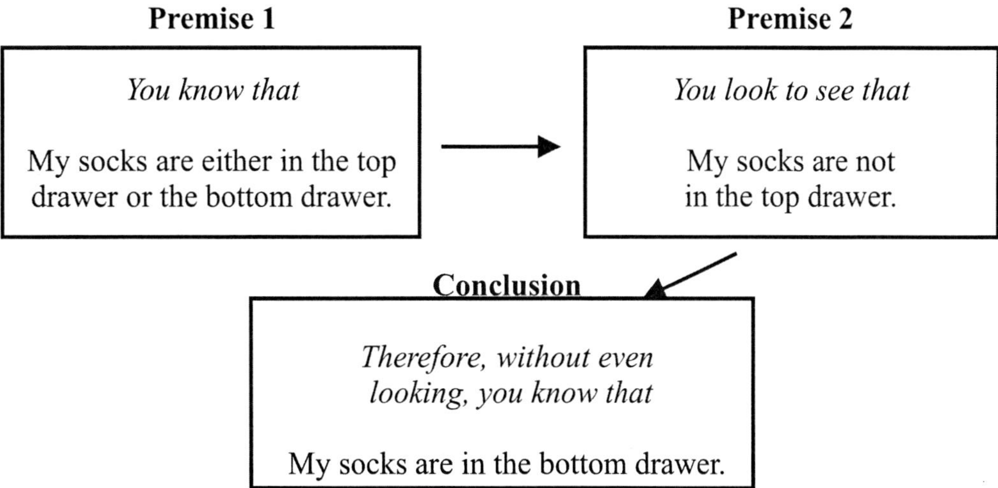

This argument is an example of deductive reasoning, where the conclusion is "deduced" from the premises and nothing else. In other words, if Premise 1 and Premise 2 are true, you don't even need to look in the bottom drawer to know that the conclusion is true.

DEDUCTIVE AND INDUCTIVE ARGUMENTS

In general, there are two types of logical arguments: **deductive** and **inductive.** Deductive arguments tend to move from general statements or theories to more specific conclusions. Inductive arguments tend to move from specific observations to general theories.

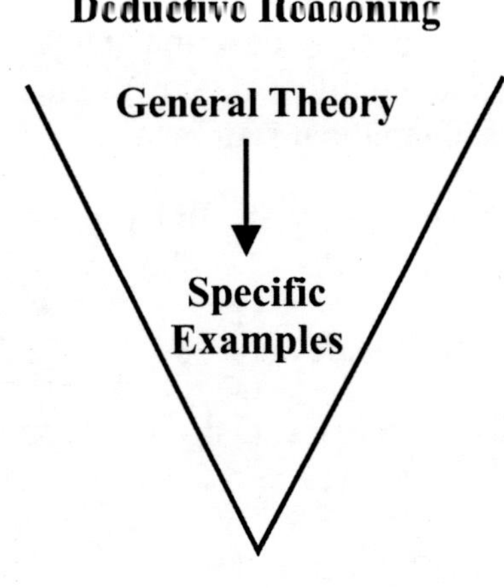

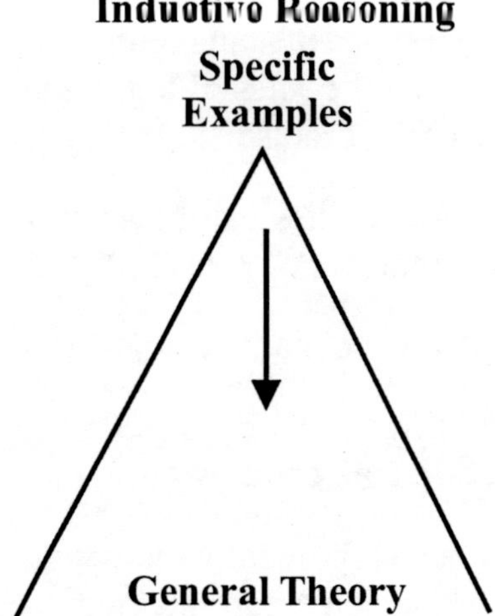

Compare the two examples below:

Deductive Argument

Premise 1	All men are mortal.
Premise 2	Socrates is a man.
Conclusion	Socrates is mortal.

Inductive Argument

Premise 1	The sun rose this morning.
Premise 2	The sun rose yesterday morning.
Premise 3	The sun rose two days ago.
Premise 4	The sun rose three days ago.
Conclusion	The sun will rise tomorrow.

An inductive argument cannot be proved beyond a shadow of a doubt. For example, it's a pretty good bet that the sun will come up tomorrow, but the sun not coming up presents no logical contradiction.

On the other hand, a deductive argument can have logical certainty, but it must be properly constructed. Consider the examples below.

True Conclusion for an Invalid Argument

All men are mortal.
Socrates is mortal.
Therefore, Socrates is a man.

Even though the above conclusion is true, the argument is based on invalid logic. Both men and women are mortal. Therefore, Socrates could be a woman.

False Conclusion from a Valid Argument

All astronauts are men.
Julia Roberts is an astronaut.
Therefore, Julia Roberts is a man.

In this case, the conclusion is false because the premises are false. However, the logic of the argument is valid because *if* the premises were true, then the conclusion would be true.

EXAMPLE 1: Which argument is valid?

If you speed on Hill Street, you will get a ticket.
If you get a ticket, you will pay a fine.

A. I paid a fine, so I was speeding on Hill Street.
B. I got a ticket, so I was speeding on Hill Street.
C. I exceeded the speed limit on Hill Street, so I paid a fine.
D. I did not speed on Hill Street, so I did not pay a fine.

Answer: C is valid.
A is incorrect. I could have paid a fine for another violation.
B is incorrect. I could have gotten a ticket for some other violation.
D is incorrect. I could have paid a fine for speeding somewhere else.

EXAMPLE 2: Assume the given proposition is true. Determine if each statement is true or false.

Given: If a dog is thirsty, he will drink.

A. If a dog drinks, then he is thirsty. T or F
B. If a dog is not thirsty, he will not drink. T or F
C. If a dog will not drink, he is not thirsty. T or F

Answer: A is false. He is not necessarily thirsty; he could just drink because other dogs are drinking or drink to show others his control of the water. This statement is the converse of the original.
B is false. The reasoning from A applies. This statement is the inverse of the original.
C is true. It is the **contrapositive** or the complete opposite of the original.

For numbers 1–5, what conclusion can be drawn from each proposition?

1. All squirrels are rodents. All rodents are mammals. Therefore,

2. All fractions are rational numbers. All rational numbers are real numbers. Therefore,

3. All squares are rectangles. All rectangles are parallelograms. All parallelograms are quadrilaterals. Therefore,

4. All Chevrolets are made by General Motors. All Luminas are Chevrolets. Therefore,

5. If a number is even and divisible by three, then it is divisible by six. Eighteen is divisible by six. Therefore,

For numbers 6–9, assume the given proposition is true. Then determine if the statements following it are true or false.

All squares are rectangles.

6. All rectangles are squares. T or F
7. All non-squares are non-rectangles. T or F
8. No squares are non-rectangles. T or F
9. All non-rectangles are non-squares. T or F

COUNTEREXAMPLE

A counterexample is an example in which the statement is true but the conclusion is false when we have assumed it to be true. If we said, "All cocker spaniels have blonde hair," then a counterexample would be a red-haired cocker spaniel. If we made the statement, "If a number is greater than 10, it is less than 20," we can easily think of a counterexample, like 35.

A **conditional statement** is a type of logical statement that has two parts, a **hypothesis** and a **conclusion**. The statement is written in "if-then" form, where the "if" part contains the hypothesis and the "then" part contains the conclusion. Shorthand: p→q. For example, let's start with the statement "Two lines intersect at exactly one point." We can rewrite this as a conditional statement in "if-then" form as follows:

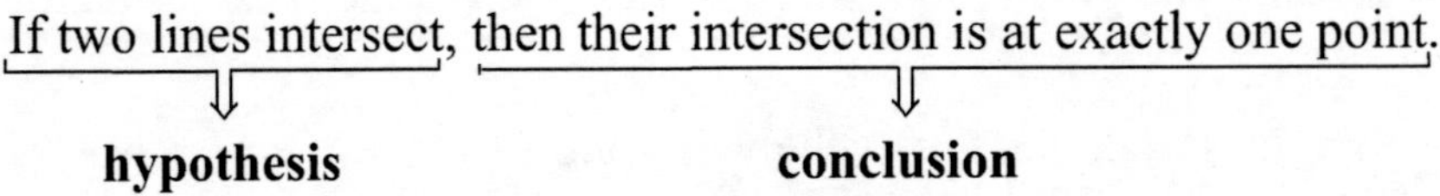

hypothesis **conclusion**

Conditional statements may be true or false. To show that a statement is false, you need only to provide a single **counterexample** which shows that the statement is not always true. To show that a statement is true, on the other hand, you must show that the conclusion is true for all occasions in which the hypothesis occurs. This is often much more difficult.

EXAMPLE: Provide a counterexample to show that the following conditional statement is false:

If $x^2 = 4$, then $x = 2$

To find a counterexample, think: Could x equal something else? Yes, x could be -2.

Counterexample: $(-2)^2 = 4$

Since we have a counterexample, we have shown that the conditional statement is false.

Answer the following problems about logic.

1. Rewrite the following as a conditional statement in "if–then" form: A number divisible by 8 is also divisible by 4.

2. Write the premise from the following statement: Two circles have equal radii; therefore, the circles are congruent.

3. Consider the conditional statement: If $x^4 = 81$, then $x = 3$. Is the statement true? Provide a counterexample if it is false.

CHAPTER 17 REVIEW

Read each of the statements below and tell whether they are true or false.

		True or False
1.	$\{\text{odd whole numbers}\} \subset \{\text{all whole numbers}\}$	
2.	$\{\text{yearly seasons}\} \neq \{\text{spring, summer, fall, winter}\}$	
3.	$\{\text{Monday, Tuesday, Wednesday}\} \subset \{\text{days of the week}\}$	
4.	United States of America $\notin$ $\{\text{countries in North America}\}$	
5.	$\{\text{green, purple}\} \not\subset \{\text{primary colors}\}$	
6.	$\{\text{plants with red flowers}\} = \varnothing$	
7.	{letters in "subsets"} = {b, u, s, e, t}	
8.	Milky Way $\notin$ $\{\text{galaxies in the universe}\}$	
9.	$\{\text{Houston, Dallas}\} \subset \{\text{cities in Texas}\}$	
10.	George Washington $\in$ $\{\text{former presidents of the United States}\}$	

Complete the following statements.

11. $\{3, 6, 9, 12, 15\} \cap \{0, 5, 10, 15, 20\} =$

12. {Felix, Mark, Kate} $\cup$ {Mark, Carol, Jack} =

13. {letters in "perfect"} $\cap$ {letters in "profit"} =

14. {Rome, London, Paris} $\cap$ {Italy, England, France} =

15. {black, white, gray} $\cup$ {red, white, blue} =

16. $\{1, 2, 3, 4, 5, 6\} \cup \{2, 4, 6, 8, 10, 12\} =$

Refer to the Venn diagram to complete the following statements. Answers should be in roster form.

17. basketball $\cup$ football $\cup$ baseball =

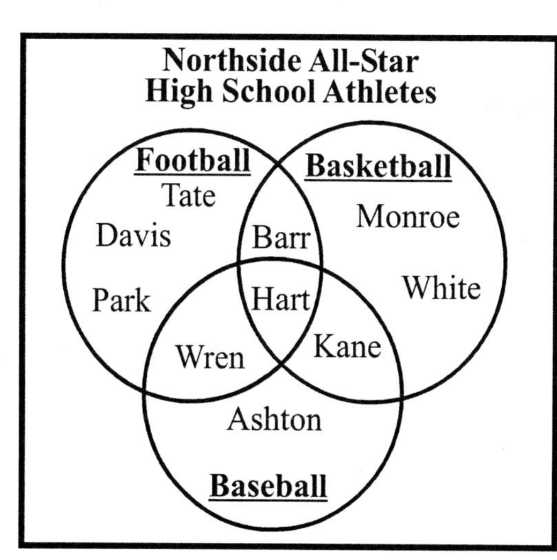

18. basketball $\cap$ football =

19. (football $\cap$ basketball) $\cap$ baseball =

20. baseball $\cup$ basketball =

21. football $\cup$ baseball =

22. baseball $\cap$ football =

23. basketball $\cap$ baseball =

24. football $\cup$ basketball =

Use the following Venn diagram to answer questions 25 and 26.

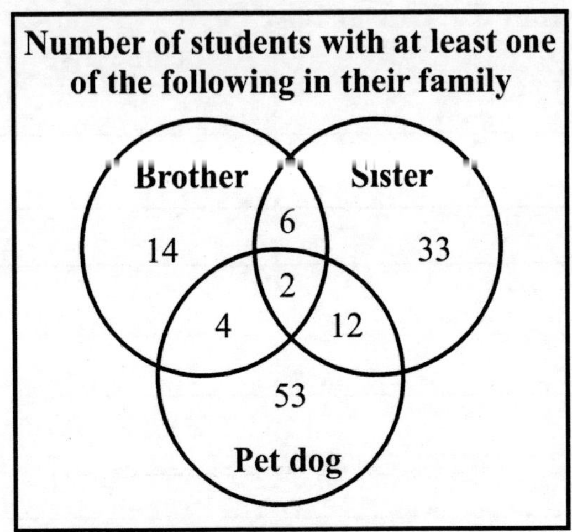

25. In the Venn diagram above, how many members are in the following set?
{sister ∪ pet dog}

A. 12
B. 14
C. 100
D. 110

26. In the Venn diagram above, how many members are in the following set?
{brother} ∩ {sister}

A. 6
B. 8
C. 53
D. 55

27. Which of the sets below does not contain {a, e, o} as a subset?

A. {letters of the alphabet}
B. {vowels in the alphabet}
C. {letters in the word "predator"}
D. {the first 5 letters of the alphabet}

28. Which of the following statements is true?

A. $\{1, 2, 4, 5\} \cup \varnothing = \varnothing$
B. $\{c\} \not\in \{c, a, t\}$
C. $\{f, g\} \subset \{f, g, h, i\}$
D. $\{k, l, m\} \cup \varnothing = \varnothing$

29. Which of the following statements is false?

A. {t, o} ⊂ {letters in the word "today"}
B. $8 \not\in \{0, 2, 4, 6, 8\}$
C. $\varnothing = \{ \}$
D. $\{a, b, c\} \cap \{b, c, d, e\} = \{b, c\}$

30.

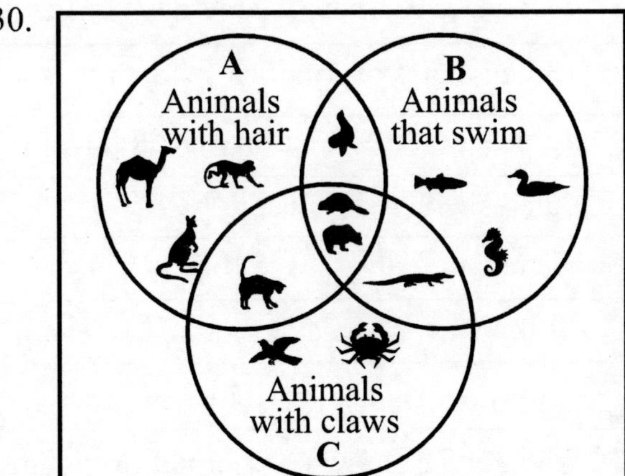

According to the above diagram, which one of the following statements is false?

A. $A \cap B = \{$, , $\}$
B. $\in B$
C. { , , } $\not\subset A$
D. $B \cup C \neq \varnothing$

31. Which of the following sets equals ∅?

A. {all whole numbers}
B. {all letters in the alphabet}
C. {all fish with feathers}
D. {all flowering plants}

32. {Karen, John, Sue} ∩ {John, Perry, Kay}

A. ∅
B. {John}
C. {Karen, Sue, Perry, Kay}
D. {Karen, John, Sue, Perry, Kay}

For numbers 33–36, assume the given proposition is true. Then determine if the statements following it are true or false.

All whales are mammals.

33.	All non-whales are non-mammals.	T or F
34.	If a mammal lives in the sea, it is a whale.	T or F
35.	All mammals are whales.	T or F
36.	All non-mammals are non-whales.	T or F

For 37–38, chose which argument is valid.

37. If I oversleep, I miss breakfast. If I miss breakfast, I cannot concentrate in class. If I do not concentrate in class, I make bad grades.

    A. I made bad grades today, so I missed breakfast.
    B. I made good grades today, so I got up on time.
    C. I could not concentrate in class today, so I overslept.
    D. I had no breakfast today, so I overslept.

38. Brad is asked to list five duties of the President. What type of logic is Brad using?

    A. mathematical reasoning
    B. inductive reasoning
    C. intuitive reasoning
    D. deductive reasoning

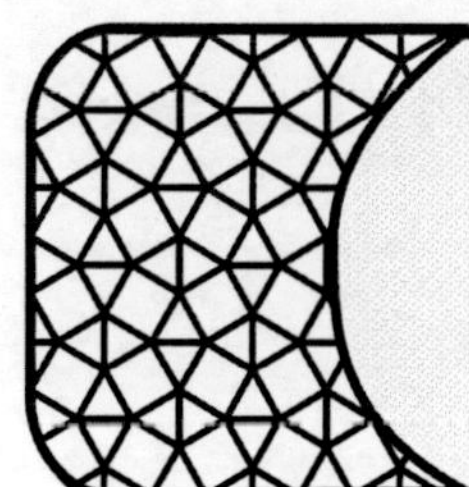

CHAPTER 18: CONSUMER MATH

Standard 3.12.4

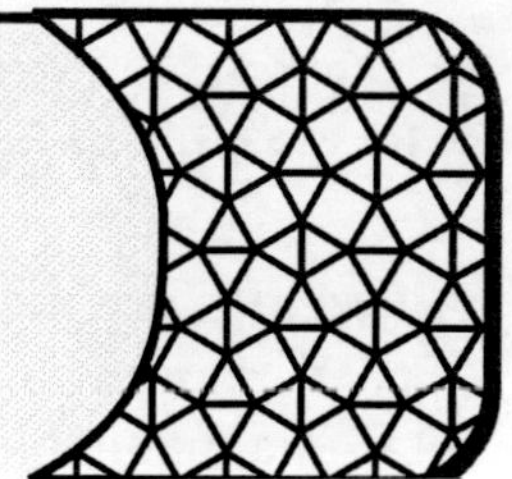

DEDUCTIONS - FRACTION OFF

Sometimes sale prices are advertised as $\frac{1}{4}$ off or $\frac{1}{3}$ off. To find out how much you will save, just multiply the original price by the fraction off.

EXAMPLE: CD players are on sale for $\frac{1}{3}$ off. How much can you save on a $240 CD player?

$$\frac{1}{\cancel{3}_1} \times \frac{\cancel{240}^{80}}{1} = 80$$ **You can save $80.00.**

Find the amount of savings in the problems below.

J.P. Nichols is having a liquidation sale on all furniture. Sale prices are $\frac{1}{2}$ off the regular price. How much can you save on the following furniture items?

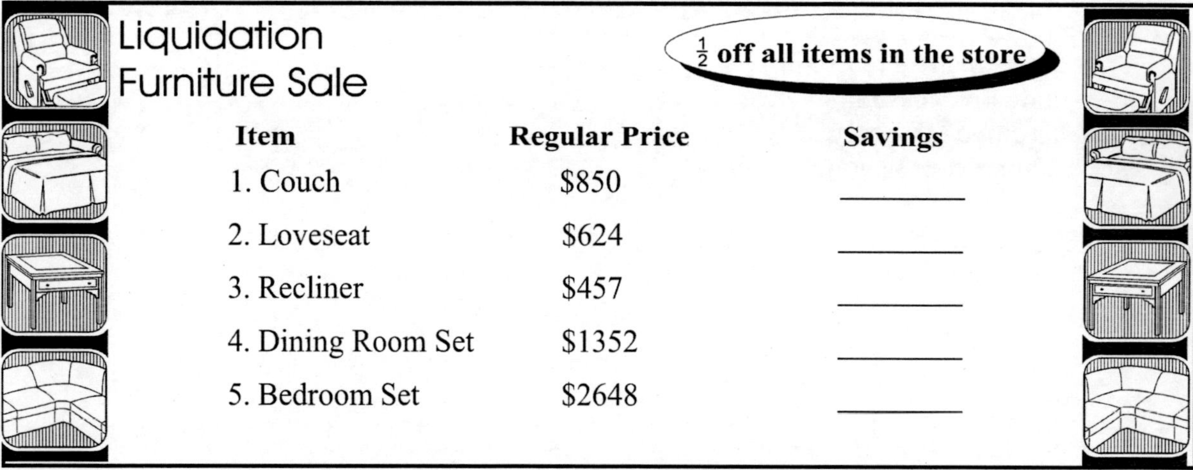

Liquidation Furniture Sale

$\frac{1}{2}$ **off all items in the store**

Item	Regular Price	Savings
1. Couch	$850	________
2. Loveseat	$624	________
3. Recliner	$457	________
4. Dining Room Set	$1352	________
5. Bedroom Set	$2648	________

Buy Rite Computer Store is having a $\frac{1}{3}$ off sale on selected computer items in the store. How much can you save on the following items?

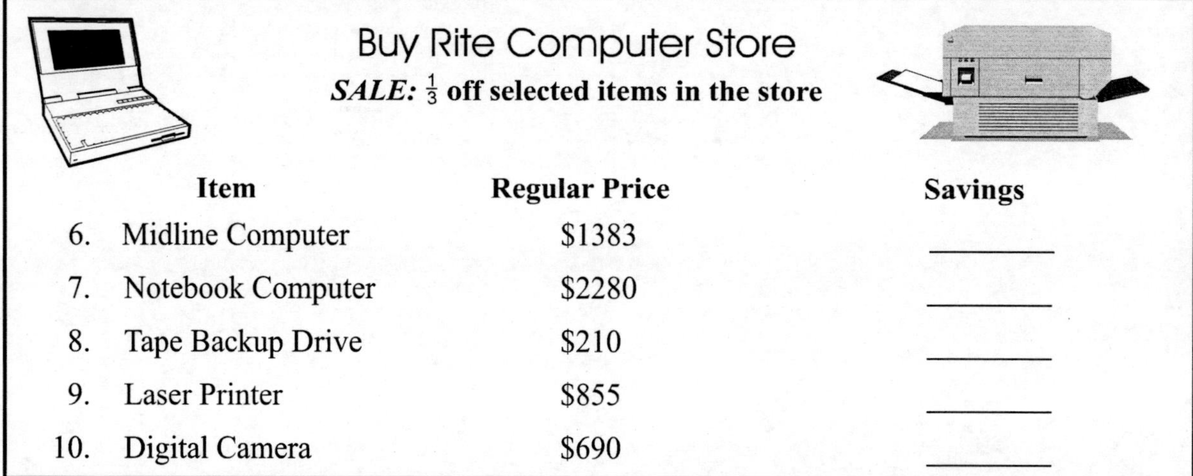

Buy Rite Computer Store

SALE: $\frac{1}{3}$ **off selected items in the store**

Item	Regular Price	Savings
6. Midline Computer	$1383	________
7. Notebook Computer	$2280	________
8. Tape Backup Drive	$210	________
9. Laser Printer	$855	________
10. Digital Camera	$690	________

Martin's Department Store is having its annual $\frac{1}{3}$ off all bedroom apparel sale. How much can you save on the following items?

Martin's Department Store

$\frac{1}{3}$ off all "Country Elegant" comforters and accessories

	Length	Regular Price	Savings
11.	Twin	$87	_______
12.	Full	$114	_______
13.	Queen	$141	_______
14.	King	$177	_______
15.	Accent Pillows	$24	_______
16.	Draperies	$78	_______

Durabelt Car Tires is placing its car tires on sale for $\frac{1}{4}$ off the regular price. Find the amount you can save on each tire.

Durabelt Car Tires

$\frac{1}{4}$ OFF ALL SIZES OF TIRES

	Size	Regular Price	Savings
17.	13"	$44	_______
18.	14"	$58	_______
19.	15"	$64	_______
20.	16"	$80	_______
21.	17"	$92	_______
22.	18"	$98	_______

FINDING A FRACTION OF A TOTAL

Mathematicians use the word "of" in word problems to indicate that you need to multiply to find the answer.

EXAMPLE 1: Two-thirds of male high school seniors will be taller than their fathers by the time they graduate. In a sample of 400 male seniors, about how many will be taller than their fathers on graduation day?

Solution: Multiply the fraction by the total. $\frac{2}{3} \times \frac{400}{1} = \frac{800}{3} = 267$

About 267 out of 400 male seniors will be taller than their fathers.

EXAMPLE 2:

Favorite Cake Flavors

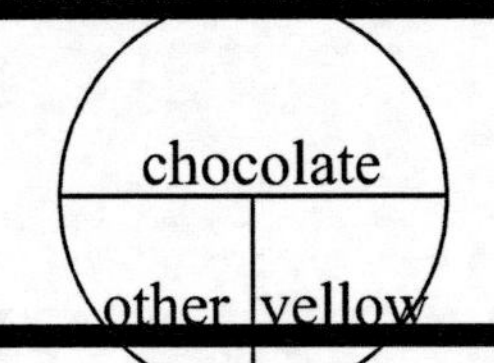

In a lunchroom survey of 360 students, about how many students preferred yellow cake?

Step 1: Estimate the fraction of the circle that shows the amount of yellow cake. It looks like about $\frac{1}{4}$ of the circle.

Step 2: Multiply $\frac{1}{4} \times 360$, the total number of students surveyed. About 90 students prefer yellow cake.

Solve the following problems.

1. This year $\frac{2}{3}$ of the seniors went to the prom. Out of a class of 438 seniors, how many went to the prom?

2. The North End Diner surveyed its customers one day to see which flavor of ice cream they preferred. Out of 512 customers, how many preferred chocolate chip?

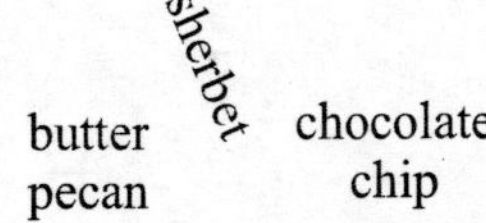

3. It has rained 5 out of 6 days for the last month! If there were 30 days in the last month, how many days did it rain?

4. Ryan worked 5 hours on his homework Tuesday. Two-thirds of the time was spent on algebra word problems. How much time did he spend on algebra?

5. Beth earned \$20 babysitting. She spent $\frac{2}{5}$ of it on a paperback book. How much did she spend on the book?

6. Three-fourths of the graduating class at Lakewood High School plan on going to college. There are 680 graduating seniors. How many are planning on going to college?

7. At West End High School, $\frac{1}{5}$ of the students in the band are also in the choir. There are 205 band members. How many are also in the choir?

8. Allen bought a $3\frac{1}{4}$ pound roast to cook for dinner. How much did the roast cost at \$2.00 per pound?

9. What is the cost of $4\frac{1}{3}$ yards of fabric if it sells for \$15 per yard?

10. Grandma made $4\frac{1}{2}$ dozen cookies. Her grandsons ate $\frac{2}{3}$ of them. How many cookies did they eat?

11. Randy needed to make a ditch $20\frac{1}{2}$ feet long. The first day he finished $\frac{2}{3}$ of it. How many feet did he finish digging?

12. The band members brought in 300 cupcakes to sell for a fund-raiser. Three-fifths of them were chocolate. How many chocolate cupcakes were there?

DETERMINING CHANGE

EXAMPLE: Jamie bought 2 T-shirts for $13.95 each and paid $1.68 sales tax. How much change should Jamie get from a $50.00 bill?

Step 1: Find the total cost of items and tax.

$$\begin{array}{r} \$13.95 \\ 13.95 \\ +\ 1.68 \\ \hline \$29.58 \end{array}$$

Step 2: Subtract the total cost from the amount of money given.

$$\begin{array}{r} \$50.00 \\ -\ 29.58 \\ \hline \$20.42 \end{array}$$

Change $20.42

Find the correct change for each of the following problems.

1. Kenya bought a leather belt for $22.89 and a pair of earrings for $4.69. She paid $1.38 sales tax. What was her change from $30.00?

2. Mark spent a total of $78.42 on party supplies. What was his change from a $100.00 bill?

3. The Daniels spent $42.98 at a steak restaurant. How much change did they receive from $50.00?

4. Myra bought a sweater for $49.95 and a dress for $85.89. She paid $9.51 in sales tax. What was her change from $150.00?

5. Roland bought a calculator for $22.78 and an extra battery for $5.69. He paid $1.56 sales tax. What was his change from $40.00?

6. For lunch, Daul purchased 2 hotdogs for $1.09 each, a bag of chips for $0.89, and a large drink for $1.39. He paid $0.18 sales tax. What was his change from $10.00?

7. Geri bought a dining room set for $2,265.99. She paid $135.96 sales tax. What was her change from $2500.00?

8. Juan purchased a bag of dog food for $5.89, a leash for $11.88, and a dog collar for $4.75. The sales tax on the purchase was $1.13. How much change did he get back from 25.00?

9. Celina bought a blouse for $15.46 and a shirt for $23.58. She paid $3.12 sales tax. What was her change from $50.00?

10. Jackie paid for four houseplants that cost $4.95 each. She paid $1.19 sales tax. How much change did she receive from 30.00?

11. Bo spent a total of $13.59 on school supplies. How much change did he receive from $14.00?

12. Fran bought 4 packs of candy on sale for 2 for $0.99 and 2 sodas for $0.65 each. She paid $0.13 sales tax. What was her change from a $5.00 bill?

GROSS PAY

Gross pay is the amount you earn before taxes, insurance, and other deductions are taken out.

EXAMPLE: Codie earns \$8.50 per hour. Last week he worked 38 hours. What was his gross pay?

Solution: Multiply the pay per hour by the number of hours worked.

$$\begin{array}{r} \$8.50 \\ \times\ 38 \\ \hline 6800 \\ +\ 2550 \\ \hline \$323.00 \end{array}$$

Find the gross pay (total earnings before deductions) in each of the following problems.

1. Ron earns \$8.25 per hour and works 35 hours each week. How much does he earn per week?
2. Casie earns \$13.00 per hour and worked 40 hours last week. How much did she earn last week?
3. Maria earns \$6.75 an hour at her part-time job. Last week she worked 15 hours. How much did she earn?
4. Roby worked 22.5 hours last week, and he earns \$8.60 per hour. How much was his gross pay?
5. Tikki worked 11 hours at her job that pays \$9.15 per hour. How much did she earn?
6. Murray worked 35 hours last week and makes \$6.45 per hour. What was his gross pay?
7. Paula's job pays \$12.00 per hour. Last week she worked 11.25 hours. How much did she earn last week?
8. Taylor earns \$6.50 per hour working in a fast food restaurant. If he works 23 hours per week, what is his gross pay per week?
9. Mark earns \$9.50 per hour painting houses. Last week he worked 36 hours. How much did he earn?
10. Kirby works in a greenhouse for \$8.75 per hour. He works 40 hours per week. What is his gross pay each week?
11. Julie earns \$25.00 per hour teaching tennis part time. If she works 8.5 hours in a week, how much will she earn?
12. Yvonne works in a boutique for 25 hours per week. The boutique pays her \$7.15 per hour. How much does she earn each week?

CALCULATING STARTING TIMES AND GROSS PAY

EXAMPLE: Doug started work at 8:45 Saturday morning. He finished work at 7:15 Saturday night. He earns $9.50 per hour. How much did he earn?

Step 1: First, figure the hours from 8:45 to noon, because at noon the clock starts counting from 0 again. To figure the hours from 8:45 to noon, subtract 8:45 from 12:00.

```
 11 60  ←— Remember when you borrow 1 hour,
 12:00      you borrow 60 minutes.
– 8:45
  3:15
```

Step 2: From noon to 7:15 is 7 hours and 15 minutes. Add the 3 hours and 15 minutes from morning hours worked.

```
   3:15
 + 7:15
  10:30   Doug worked a total of 10 hours and 30 minutes.
```

30 minutes $= \frac{30}{60} = \frac{1}{2} = 0.5$, so 10 hours 30 minutes = 10.5 hours

Step 3: Multiply the total number of hours worked by his pay rate of $9.50 per hour.

```
   10.5 hours
 × $9.50 per hour
   5250
 + 945
   99750 or $99.75     Doug earned $99.75 on Saturday.
```

Calculate how much each worker earned in each of the problems below.

1. Kyle started lawn mowing at 11:30 a.m. He finished at 4:15 p.m. He earned $6.70 per hour.
2. The electrician took 1 hour and 45 minutes to finish wiring a new air conditioner. He earned $40 per hour.
3. Jennifer works part-time in a shoe store. On Saturday she started work at 9:45 a.m. She finished working at 5:15 p.m. She gets paid $6.25 per hour.
4. Zach started working at 4:40 p.m. at the grocery store. He got off at 8:10 p.m. He earned $7.50 per hour.
5. Antonio began his accounting work at 10:30 a.m. He finished his work at 5:30 p.m. He earns $17.75 per hour.
6. Clarissa began her work as day manager at a local manicure shop at 7:30 a.m. She finished her work at 3:00 p.m. She earns $13.30 per hour.

UNIT COST

When products come in different sizes, you need to figure out the cost per unit to see which is the best buy. Often the box marked "economy size" is not really the best buy.

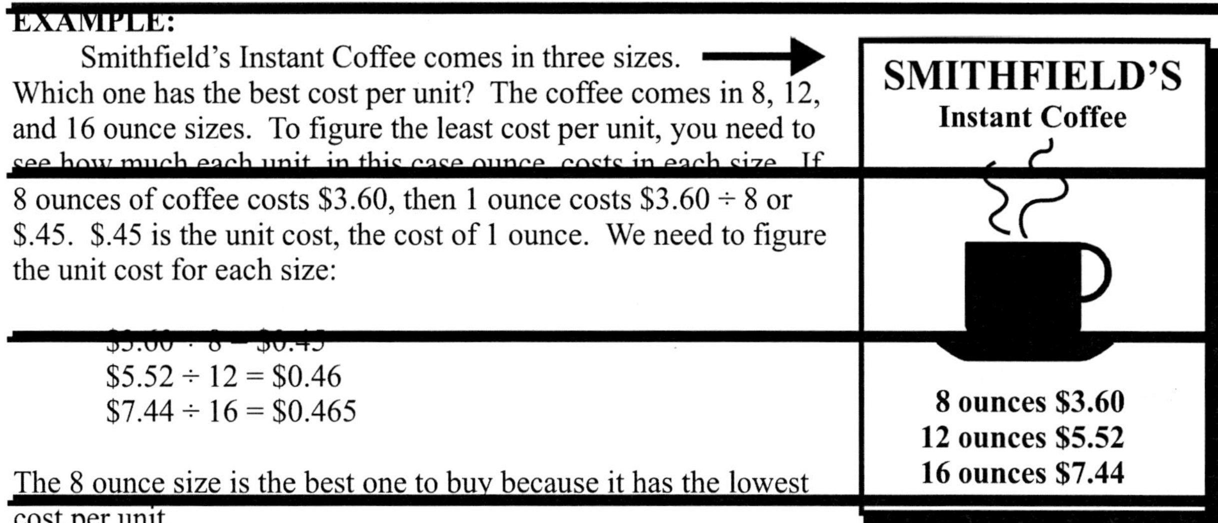

EXAMPLE:

Smithfield's Instant Coffee comes in three sizes. Which one has the best cost per unit? The coffee comes in 8, 12, and 16 ounce sizes. To figure the least cost per unit, you need to see how much each unit, in this case ounce, costs in each size. If 8 ounces of coffee costs \$3.60, then 1 ounce costs \$3.60 ÷ 8 or \$.45. \$.45 is the unit cost, the cost of 1 ounce. We need to figure the unit cost for each size:

$3.60 ÷ 8 = $0.45
$5.52 ÷ 12 = $0.46
$7.44 ÷ 16 = $0.465

The 8 ounce size is the best one to buy because it has the lowest cost per unit.

Figure the unit cost of each item in each question below to find the best buy. Underline the answer.

1. Which costs the most per ounce, 60 ounces of peanut butter for \$5.40, 28 ounces for \$2.24, or 16 ounces for \$1.76?
2. Which is the least per pound, 5 pounds of chicken for \$9.45, 3 pounds for \$5.97, or 1 pound for \$2.05?
3. Which costs the most per disk, a 10 pack of $3\frac{1}{2}$ inch floppy disks for \$5.99, a 25 pack for \$12.50, or a 50 pack for \$18.75?
4. Which is the best buy, 6 ballpoint pens for \$4.80 or 8 for \$6.48?
5. Which costs the least per ounce, a 20 ounce soda for \$0.60, 68 ounces for \$2.38, or 100 ounces for \$3.32?
6. Which costs more, oranges selling at 3 for \$1.00 or oranges selling 4 for \$1.36?
7. Which is the best buy, 1 roll of paper towels for \$2.13, 3 rolls for \$5.88, or 15 rolls for \$29.55?
8. Which costs the most per tablet, 50 individually wrapped pain reliever tablets for \$9.50, 100 tablets in a bottle for \$6.32, or 500 tablets in a bottle for \$13.42?
9. Which costs the least per can, a 24 pack of cola for \$5.52, a 12 pack of cola for \$2.64, or a 6 pack of cola for \$1.35?
10. Which costs less per bag, 18 tea bags for \$2.70 or 64 tea bags for \$9.28?
11. Which is the best buy, a 3 pack of correction fluid for \$2.97 or a 12 pack for \$11.76?
12. Which is the least per roll, a roll of masking tape for \$2.45, a 3-roll pack for \$7.38, or a 12-roll pack for \$29.16?

TIPS AND COMMISSIONS

Vocabulary

Tip: A tip is money given to someone doing a service for you such as a server, porter, cab driver, hair stylist, etc.

Commission: In many businesses, sales people are paid on commission - a percent of the total sales they make.

Problems requiring you to figure a tip, commission, or percent of a total are all done in the same way.

EXAMPLE: Ramon made a 4% commission on an $8,000 pickup truck he sold. How much was his commission?

TOTAL COST	**$8,000**
× RATE OF COMMISSION	**× 0.04**
COMMISSION	**$320.00**

Solve each of the following problems.

1. Whitney makes 12% commission on all her sales. This month she sold $9,000 worth of merchandise. What was her commission?

____

2. Marcus gives 25% of his income to his parents to help cover expenses. He earns $340 per week. How much money does he give his parents?

____

3. Jan pays $640 per month for rent. If the rate of inflation is 5%, how much can Jan expect to pay monthly next year?

____

4. The total bill at Jake's Catfish Place came to $35.80. Jim wanted to leave a 15% tip. How much money will he leave for the tip?

____

5. Rami makes $2,400 per month and puts 6% in a savings plan. How much does he save per month?

____

6. Cristina makes $2,550 per month. Her boss promised her a 7% raise. How much more will she make per month?

____

7. Out of 150 math students, 86% passed. How many students passed math class?

____

8. Marta sells Sue Ann Cosmetics and gets 20% commission on all her sales. Last month, she sold $560.00 worth of cosmetics. How much was her commission?

____

FINDING THE AMOUNT OF DISCOUNT

Sale prices are sometimes marked 30% off or, better yet, 50% off. A 30% DISCOUNT means you will pay 30% less than the original price. How much money you save is also known as the amount of the DISCOUNT. Read the **EXAMPLE** below to learn to figure the amount of a discount.

EXAMPLE: A $179.00 chair is on sale for 30% off. How much can I save if I buy it now?

Step 1: Change 30% to a decimal. 30% = 0.30

Step 2: Multiply the original price by the discount.

ORIGINAL PRICE	$179.00
× **% DISCOUNT**	× 0.30
SAVINGS	$ 53.70

Practice finding the amount of the discount. Round off answers to the nearest penny.

1. Tubby Tires is offering a 25% discount on tires purchased on Tuesday. How much can you save if you buy tires on Tuesday regularly priced at $225.00 any other day of the week? ________

2. The regular price for a garden rake is $10.97 at Sly's Super Store. This week, Sly is offering a 30% discount. How much is the discount on the rake? ________

3. Christine bought a sweater regularly priced at $26.80 with a coupon for 20% off any sweater. How much did she save? ________

4. The software that Marge needs for her computer is priced at $69.85. If shc waits until a store offers it at 20% off, how much will she save? ________

5. Ty purchased jeans that were priced $23.97. He received a 15% employee discount. How much did he save? ________

6. The Bakery Company offers a 60% discount on all bread made the day before. How much can you save on a $2.40 loaf made today if you wait until tomorrow to buy it? ________

7. A furniture store advertises a 40% off liquidation sale on all items. How much would the discount be on a $2530 dining room set? ________

8. Becky bought a $4.00 nail polish on sale for 30% off. What was the dollar amount of the discount? ________

9. How much is the discount on a $350 racing bike marked 15% off ? ________

10. Raymond receives a 2% discount from his credit card company on all purchases made with the credit card. What is his discount on $1575.50 worth of purchases? ________

FINDING THE DISCOUNTED SALE PRICE

To find the discounted sale price, you must go one step further than shown on the previous page. Read the **EXAMPLE** below to learn how to figure **discount** prices.

EXAMPLE: A $74.00 chair is on sale for 25% off. How much will it cost if I buy it now?

Step 1: Change 25% to a decimal. 25% = 0.25

Step 2: Multiply the original price by the discount.

ORIGINAL PRICE	$74.00
× **% DISCOUNT**	× 0.25
SAVINGS	$18.50

Step 3: Subtract the savings amount from the original price to find the sale price.

ORIGINAL PRICE	$74.00
− **SAVINGS**	− 18.50
SALE PRICE	$55.50

Figure the sale price of the items below. The first one is done for you.

ITEM	PRICE	% OFF	MULTIPLY	SUBTRACT	SALE PRICE
1. pen	$1.50	20%	1.50 ×0.2 = $0.30	1.50 – 0.30 = 1.20	$1.20
2. recliner	$325	25%			
3. juicer	$55	15%			
4. blanket	$14	10%			
5. earrings	$2.40	20%			
6. figurine	$8	15%			
7. boots	$159	35%			
8. calculator	$80	30%			
9. candle	$6.20	50%			
10. camera	$445	20%			
11. VCR	$235	25%			
12. video game	$25	10%			

SALES TAX

EXAMPLE: The total price of a sofa is $560.00 + 6% sales tax. How much is the sales tax? What is the total cost?

Step 1: You will need to change 6% to a decimal. 6% = 0.06

Step 2: Simply multiply the cost, $560, by the tax rate, 6%. 560 × 0.06 = 33.6
The answer will be $33.60. (You need to add a 0 to the answer. When dealing with money, there needs to be two places after the decimal point.)

COST	$560
× **6% TAX**	×0.06
SALES TAX	$33.60

Step 3: Add the sales tax amount, $33.60, to the cost of the item sold, $560. This is the total cost.

COST	$560.00
SALES TAX	+ 33.60
TOTAL COST	$593.60

NOTE: When the answer to the question involves money, you always need to round off the answer to the nearest hundredth (2 places after the decimal point). Sometimes you will need to add a zero.

Figure the total costs in the problems below. The first one is done for you.

ITEM	PRICE	% SALES TAX	MULTIPLY	ADD PRICE PLUS TAX	TOTAL
1. jeans	$42	7%	$42×0.07 = $2.94	42 + 2.94 = 44.94	$44.94
2. truck	$17,495	6%			
3. film	$5.89	8%			
4. T-shirt	$12	5%			
5. football	$36.40	4%			
6. soda	$1.78	5%			
7. 4 tires	$105.80	10%			
8. clock	$18	6%			
9. burger	$2.34	5%			
10. software	$89.95	8%			

UNDERSTANDING SIMPLE INTEREST *

I = PRT is a formula to figure out the **cost of borrowing money** or the **amount you earn** when you **put money in a savings account**. When you want to buy a used truck or car, you go to the bank to borrow the $7,000 you need. The bank will charge you interest on the $7,000. If the simple interest rate is 9% for four years, you can figure the cost of the interest with this formula.

First, you need to understand these terms:

I = Interest = The amount charged by the bank or other lender
P = Principal = The amount you borrow
R = Rate = The interest rate the bank is charging you
T = Time = How many years you will take to pay off the loan

EXAMPLE:

In the problem above: **I = PRT** This means the **interest** equals the **principal** times the **rate** times the **time** in **years.**

I = $7,000 × 9% × 4 years
I = $7,000 × 0.09 × 4
I = $2,520

Use the formula I = PRT to work the following problems.

1. Craig borrowed $1,800 from his parents to buy a stereo. His parents charged him 3% simple interest for 2 years. How much interest did he pay his parents? ________

2. Raul invested $5,000 in a savings account that earned 2% simple interest. If he kept the money in the account for 5 years, how much interest did he earn? ________

3. Bridgette borrowed $11,000 to buy a car. The bank charged 12% simple interest for 7 years. How much interest did she pay the bank? ________

4. A tax accountant invested $25,000 in a money market account for 3 years. The account earned 5% simple interest. How much interest did the accountant make on his investment? ________

5. Linda Kay started a savings account for her nephew with $2,000. The account earned 6% simple interest. How much interest did the account accumulate in 3 years? ________

6. Renada bought a living room set on credit. The set sold for $2,300, and the store charged her 9% simple interest for one year. How much interest did she pay? ________

7. Duane took out a $3,500 loan at 8% simple interest for 3 years. How much interest did he pay for borrowing the $3,500? ________

* Simple interest is not commonly used by banks and other lending institutions. Compound interest is more commonly used, but its calculations are more complicated .

CHAPTER 18 REVIEW

Solve the following word problems.

1. Tami wants to buy a sweater that is on sale for $\frac{1}{4}$ off the regular $56.00 price. How much will she save?

2. Whitlow's has a $\frac{3}{4}$ off the regular price clearance sale on dress shirts. How much can you save if you buy a shirt regularly priced at $32.00?

3. **Johnson Family Budget**

The diagram above shows where the money goes in the Johnson family. If the family brings home a total of $4,212 each month, about how much of the money goes into savings?

4. How many student lunches were sold if $\frac{2}{3}$ of the 1263 students bought their lunch?

5.

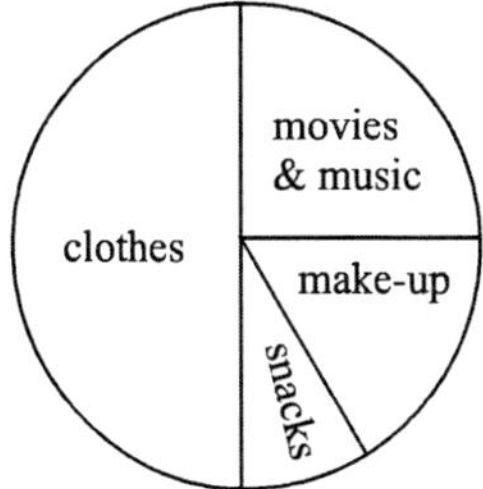

April gets $150 per month allowance from her parents. The graph above shows how she spends it. About how much does she spend on movies and music?

6. Super-X sells tires for $24.56 each. Save-Rite sells the identical tire for $21.97. How much can you save by purchasing a tire from Save-Rite?

7. White rice sells for $5.64 for a 20 pound bag. A three pound bag costs $1.59. Which is the better buy?

8. Xandra bought a mechanical pencil for $2.38, 3 pens for $0.89 each, and a pack of graph paper for $3.42. She paid $0.42 tax. What was her change from a ten dollar bill?

9. Charlie makes $13.45 per hour repairing lawn mowers part time. If he worked 15 hours, how much was his gross pay?

10. Tosha clocked in for work on Saturday at 8:30 a.m. and clocked out at 3:15. If she made $6.25 per hour, how much money did she earn on Saturday?

11. Linda took out a simple interest loan for $7,000 at 11% interest for 5 years. How much interest did she have to pay back?

12. Mark invested $1,340 in a simple interest account at 3.5% interest. How much interest did he earn after 4 years?

13. Jamie borrowed $800 from his parents as a simple interest loan at 5% interest. If he pays his parents back in two years, how much interest will he owe them?

14. Uncle Howard left his only niece 56% of his assets according to his will. If his assets totaled $564,000 when he died, how much did his niece inherit?

15. Celeste makes 6% commission on her sales. If her sales for a week total $4580, what is her commission?

16. Peeler's Jewelry is offering a 30% off sale on all bracelets. How much will you save if you buy a $45.00 bracelet during the sale?

17. Sales tax is 7% in Chester County. In Dobbs county the sales tax is 5.5%. How much could you save if you bought a $98 bicycle in Dobbs County?

18. How much would the $98 bicycle cost if it was on sale for 10% off in Dobbs County with the 5.5% sales tax?

19. Mindy buys a $36 sweater on a 15% off sale. How much will she pay if the sales tax is 6%?

20. Last week hundreds of people attended the opening of a movie. The theater which holds 900 was filled to 85% capacity. How many people attended the opening?

21. How much would an employee pay for a $724.00 stereo if the employee got a 15% discount?

22. Misha bought a CD for $14.95. If sales tax was 7%, how much did she pay total?

23. The Pep band made $640 during a fund-raiser. The band spent $400 of the money on new uniforms. What percent of the total did they spend on uniforms?

24. McMartin's is offering a deal on fitness club memberships. You can pay $999 up front for a 3 year membership, or pay $200 down and $30 per month for 36 months. How much would you save by paying up front?

25. Patton, Patton, and Clark, a law firm, won a malpractice law suit for $4,500,000. Sixty-eight percent went to the law firm. How much did the law firm make?

26. Jeneane earned $340.20 commission by selling $5670 worth of products. What percent commission did she earn?

27. Tara put $500 in a savings account that earned 3% simple interest. How much interest did she make after 5 years?

28. Clay bought a pair of basketball shoes for $79.99 plus 5% sales tax. What was his total cost for the shoes?

29. At Jefferson high school, the ratio of band members to non-band members is 2 to 5. What percent of the students are band members?

30.

$22,840
or
only **$2,000** down
$565 per month for 60 months

According to the advertisement, how much more would you pay to buy the car on credit?

31. A department store is selling all swimsuits for 40% off in August. How much would you pay for a swimsuit that is normally priced at $35.80?

32. High school students voted on where they would go on a field trip. For every 3 students who wanted to see Lincoln's boyhood home, 8 students wanted to go to the sand dunes. What percent of the students wanted to go to the sand dunes?

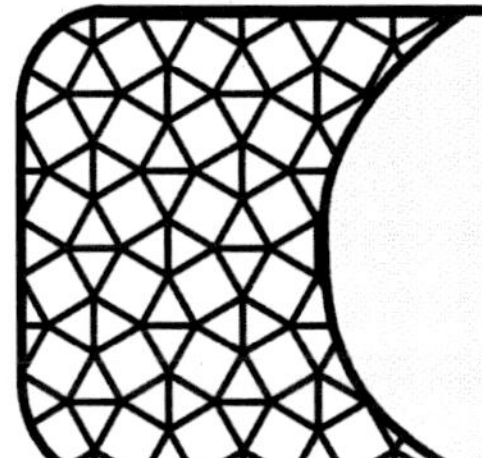

CHAPTER 19: DATA ANALYSIS

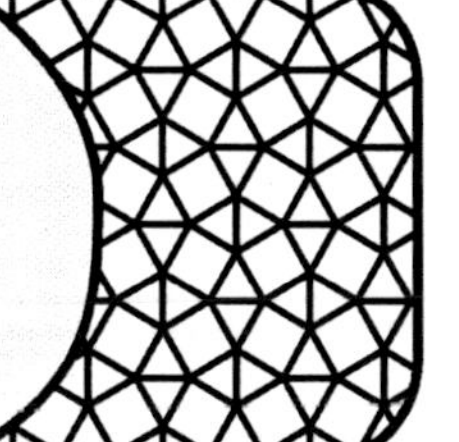

Standards 5.12.1 and 5.12.2

Statistics is a branch of mathematics. Using statistics, mathematicians organize data (numbers) into forms that are easily understood.

RANGE

In **statistics,** the difference between the largest number and the smallest number in a list is called the **range.**

EXAMPLE: Find the range of the following list of numbers: 16, 73, 26, 15, and 35.

The largest number is 73, and the smallest number is 15. 73 – 15 = 58
The range is 58.

Find the range for each list of numbers below.

1.	2.	3.	4.	5.	6.	7.
21	6	89	41	23	2	77
51	7	22	3	20	38	94
48	31	65	56	64	29	27
42	55	36	41	38	33	46
12	8	20	19	21	59	63
______	______	______	______	______	______	______

8.	9.	10.	11.	12.	13.	14.
51	65	84	84	21	45	62
62	54	59	65	78	57	39
32	56	48	32	6	57	96
16	5	21	50	97	14	45
59	63	80	71	45	61	14
______	______	______	______	______	______	______

15. 2, 15, 3, 25, and 17 range _____
16. 15, 48, 52, 41, and 8 range _____
17. 54, 74, 2, 86, and 75 range _____
18. 15, 61, 11, 22, and 65 range _____
19. 33, 18, 65, 12, and 74 range _____
20. 47, 12, 33, 25, and 19 range _____
21. 56, 10, 33, 7, 16, and 5 range _____
22. 46, 25, 78, 49, and 6 range _____
23. 45, 75, 63, and 21 range _____
24. 97, 23, 56, 12, and 66 range _____
25. 87, 44, 63, and 12 range _____
26. 84, 55, 66, 38, and 31 range _____
27. 35, 44, 81, 99, and 78 range _____
28. 95, 54, 62, 14, 8, and 3 range _____

MEAN

In statistics, the **mean** is the same as the **average**. To find the **mean** of a list of numbers, first, add together all the numbers in the list, and then divide by the number of items in the list.

EXAMPLE: Find the mean of 38, 72, 110, 548.

Step 1: First add 38 + 72 + 110 + 548 = **768**

Step 2: There are 4 numbers in the list, so divide the total by 4. $4\overline{)768}$ = 192

The mean is 192.

Practice finding the mean (average). Round to the nearest tenth if necessary.

1. Dinners served:
   489 561 522 450
   Mean = ________

4. Student absences:
   6 5 13 8 9 12 7
   Mean = ________

7. Long distance calls:
   33 14 24 21 19
   Mean = ________

2. Prices paid for shirts:
   \$4.89 \$9.97 \$5.90 \$8.64
   Mean = ________

5. Paychecks received:
   \$89.56 \$99.99 \$56.54
   Mean = ________

8. Train boxcars:
   56 55 48 61 51
   Mean = ________

3. Piglets born:
   23 19 15 21 22
   Mean = ________

6. Choir attendance:
   56 45 97 66 70
   Mean = ________

9. Cookies eaten:
   5 6 8 9 2 4 3
   Mean = ________

Solve the following word problems.

10. Val's science grades were 95, 87, 65, 94, 78, and 97. What was her average? _____

11. Ann runs a business from her home. The number of orders for the last 7 business days were 17, 24, 13, 8, 11, 15, and 9. What was the average number of orders per day? _____

12. Melissa tracked the number of phone calls she had per day: 8, 2, 5, 4, 7, 3, 6, 1. What was the average number of calls she received? _____

13. The Cheese Shop tracked the number of lunches they served this week: 42, 55, 36, 41, 38, 33, and 46. What was the average number of lunches served? _____

14. Leah drove 364 miles in 7 hours. What was her average miles per hour? _____

15. Tim saved \$680 in 8 months. How much did his savings average each month? _____

16. Ken made 117 passes in 13 games. How many passes did he average per game? _____

FINDING DATA MISSING FROM THE MEAN

EXAMPLE: Mara knew she had an 88 average in her biology class, but she lost one of her papers. The three papers she could find had scores of 98%, 84%, and 90%. What was the score on her fourth paper?

Step 1: Figure the total score on four papers with an 88% average. $0.88 \times 4 = 3.52$

Step 2: Add together the scores from the three papers you have. $0.98 + 0.84 + 0.90 = 2.72$

Step 3: Subtract the scores you know from the total score. $3.52 - 2.72 = 0.80$ She had 80% on her fourth paper.

Find the data missing from the following problems.

1. Gabriel earned 87% on his first geography test. He wants to keep a 92% average. What does he need to get on his next test to bring his average up?

2. Rian earned $68.00 on Monday. How much money must he earn on Tuesday to have an average of $80 earned for the two days?

3. Haley, Chuck, Dana, and Chris entered a contest to see who could bake the most chocolate chip cookies in an hour. They baked an average of 75 cookies. Haley baked 55, Chuck baked 70, and Dana baked 90. How many did Chris bake?

4. Four wrestlers made a pact to lose some weight before the competition. They lost an average of 7 pounds each, over the course of 3 weeks. Carlos lost 6 pounds, Steve lost 5 pounds, and Greg lost 9 pounds. How many pounds did Wes lose?

5. Three boxes are ready for shipment. The boxes average 26 pounds each. The first box weighs 30 pounds; the second weighs 25 pounds. How much does the third box weigh?

6. The five jockeys running in the next race average 92 pounds each. Nicole weighs 89 pounds. Jon weighs 95 pounds. Jenny and Kasey weigh 90 pounds each. How much does Jordan weigh?

7. Jessica made three loaves of bread that weighed a total of 45 ounces. What was the average weight of each loaf?

8. Celeste made scented candles to give away to friends. She had 2 pounds of candle wax which she melted, scented, and poured into 8 molds. What was the average weight of each candle?

9. Each basketball player has to average a minimum of 5 points a game for the next three games to stay on the team. Ben is feeling the pressure. He scored 3 points in the first game and 2 points in the second game. How many points does he need to score in the third game to stay on the team?

MEDIAN

In a list of numbers ordered from lowest to highest, the **median** is the middle number. To find the **median,** first arrange the numbers in numerical order. If there is an odd number of items in the list, the **median** is the middle number. If there is an even number of items in the list, the **median** is the **average of the two middle numbers.**

EXAMPLE 1: Find the median of 42, 35, 45, 37, and 41.

Step 1: Arrange the numbers in numerical order: 35 37 (41) 42 45.

Step 2: Find the middle number. **The median is 41.**

EXAMPLE 2: Find the median of 14, 53, 42, 6, 14, and 46.

Step 1: Arrange the numbers in numerical order: 6 14 (14 42) 46 53.

Step 2: Find the average of the 2 middle numbers.
(14 + 42) ÷ 2 = 28. **The median is 28.**

Circle the median in each list of numbers.

1. 35, 55, 40, 30, and 45
2. 7, 2, 3, 6, 5, 1, and 8
3. 65, 42, 60, 46, and 90
4. 15, 16, 19, 25, and 20
5. 75, 98, 87, 65, 82, 88, and 100
6. 33, 42, 50, 22, and 19
7. 401, 758, and 254
8. 41, 23, 14, 21, and 19
9. 5, 8, 3, 10, 13, 1, and 8

10.	11.	12.	13.	14.	15.	16.
19	9	45	52	20	8	15
14	3	32	54	21	17	40
12	10	66	19	25	13	42
15	17	55	63	18	14	32
18	6	61	20	16	22	28

Find the median in each list of numbers.

17. 10, 8, 21, 14, 9, and 12 ______
18. 43, 36, 20, and 40 ______
19. 5, 24, 9, 18, 12, and 3 ______
20. 48, 13, 54, 82, 90, and 7 ______
21. 23, 21, 36, and 27 ______
22. 9, 4, 3, 1, 6, 2, 10, and 12 ______

23.	24.	25.	26.	27.	28.	29.
2	11	13	75	48	22	17
10	22	15	62	45	19	30
6	25	9	60	52	15	31
18	28	35	52	30	43	18
20	10	29	80	35	34	14
23	23	33	50	58	28	25
______	______	______	______	______	______	______

MODE

In statistics, the **mode** is the number that occurs most frequently in a list of numbers.

EXAMPLE: Exam grades for a math class were as follows:
70 88 92 85 99 85 70 85 99 100 88 70 99 88 88 99 88 92 85 88.

Step 1: Count the number of times each number occurs in the list.

70 - 3 times
88 - 6 times
92 - 2 times
85 - 4 times
99 - 4 times
100 - 1 time

Step 2: Find the number that occurs most often.
The mode is 88 because it is listed 6 times. No other number is listed as often.

Find the mode in each of the following lists of numbers.

1.	2.	3.	4.	5.	6.	7.
88	54	21	56	64	5	12
15	42	16	67	22	4	41
88	44	15	67	22	9	45
17	56	78	19	15	8	32
18	44	21	56	14	4	16
88	44	16	67	14	7	12
17	56	21	20	22	4	12
mode ___	**mode** ___	**mode** ___	**mode** ___	**mode** ___	**mode** ___	**mode** ___

8. 48, 32, 56, 32, 56, 48, 56 **mode** _____
9. 12, 16, 54, 78, 16, 25, 20 **mode** _____
10. 5, 4, 8, 3, 4, 2, 7, 8, 4, 2 **mode** _____
11. 11, 9, 7, 11, 7, 5, 7, 7, 5 **mode** _____
12. 84, 22, 79, 22, 87, 22, 22 **mode** _____
13. 95, 87, 65, 94, 78, 95 **mode** _____
14. 8, 2, 5, 4, 7, 2, 3, 6, 1 **mode** _____
15. 89, 7, 11, 89, 17, 56 **mode** _____
16. 15, 48, 52, 41, 8, 48 **mode** _____
17. 22, 45, 48, 12, 22, 41, 22 **mode** _____
18. 62, 44, 78, 62, 54, 44, 62 **mode** _____
19. 54, 22, 54, 78, 22, 78, 22 **mode** _____
20. 14, 17, 33, 21, 33, 17, 33 **mode** _____
21. 65, 51, 8, 21, 8, 65, 70, 8 **mode** _____
22. 17, 24, 13, 8, 11, 8, 15, 9 **mode** _____
23. 51, 45, 84, 51, 65, 74, 51 **mode** _____
24. 8, 74, 65, 15, 9, 10, 74 **mode** _____
25. 62, 54, 2, 7, 89, 2, 7, 54, 2 **mode** _____

APPLYING MEASURES OF CENTRAL TENDENCY

On the Nevada test, you will be asked to solve real-world problems involving measures of central tendency.

EXAMPLE: Aida was shopping around for the best price on a 17" computer monitor. She traveled to seven stores and found the following prices: \$199, \$159, \$249, \$329, \$199, \$209, and \$189. When Aida went to the eighth and final store, she found the price for the 17" monitor was \$549. Which of the measures of central tendency, mean, median, or mode, will change the most as a result of the last price Aida found?

Step 1: Solve for all three measures of the seven values

Mean: $\frac{\$199 + \$159 + \$249 + \$329 + \$199 + \$209 + \$189}{7} = \$219$

Median: From least to greatest: \$159, \$189, \$199, \$199, \$209, \$249, \$329. The 4th value = \$199

Mode: The number repeated the most is \$199.

Step 2: Find the mean, median, and mode with the eighth value added

Mean: $\frac{\$199 + \$159 + \$249 + \$329 + \$199 + \$209 + \$189 + \$549}{8} = \$260.25$

Median: \$159, \$189, \$199, \$199, \$209, \$249, \$329, \$549. The avg. of 4th & 5th number = \$204

Mode: The number still repeated the most is \$199.

Answer: The measure which changed the most by adding the 8th value is the **mean**.

1. The Realty Company has the selling prices for 10 houses sold during the month of July. The following prices are given in thousands of dollars:
   176 89 525 125 107 100 525 61 75 114
   Find the mean, median, and mode of the selling prices. Which measure is most representative for the selling price of such homes? Explain.

2. A soap manufacturing company wants to know if the weight of its product is on target, meaning 4.75 oz. With that purpose in mind, a quality-control technician selects 15 bars of soap from production, 5 from each shift, and finds the following weights in oz.

   1st shift: 4.76, 4.75, 4.77, 4.77, 4.74
   2nd shift: 4.72, 4.72, 4.75, 4.76, 4.73
   3rd shift: 4.76, 4.76, 4.77, 4.76, 4.76

   a) What are the values for the measures of central tendency for the sample from each shift?
   b) Find the mean, median, and mode for the 24-hour production sample.
   c) Which is the most accurate measure of central tendency for the 24 hour production?
   d) Find the range of values for each shift. Is the range an effective tool for drawing a conclusion in this case? Why or why not?

STEM-AND-LEAF PLOTS

A **stem-and-leaf plot** is a way to organize and analyze statistical data. To make a stem-and-leaf plot, first draw a vertical line.

Final Math Averages

85	92	87	62	75	84	96	52
45	77	98	75	71	79	85	82
87	74	76	68	93	77	65	84
79	65	77	82	86	84	92	60
99	75	88	74	79	80	63	84
87	90	75	81	73	69	73	75
31	86	89	65	69	75	79	76

Stem	Leaves
3	1
4	5
5	2
6	0,2,3,5,5,5,8,9,9
7	1,3,3,3,4,4,5,5,5,5,5,5,6,6,7,7,7,9,9,9
8	0,1,2,2,4,4,4,4,5,5,6,6,7,7,7,8,9
9	0,2,2,3,6,8,9

On the left side of the line, list all the numbers that are in the tens place from the set of data. Next, list each number in the ones place on the right side of the line in ascending order. It is easy to see at a glance that most of the students scored in the 70's or 80's with a majority having averages in the 70's. It is also easy to see that the maximum average is 99, and the lowest average is 31. Stem-and-leaf plots are a way to organize data making it easy to read.

Make a stem-and leaf-plot from the data below, and then answer the questions that follow.

1. **Speeds on Turner Road**

CAR SPEED, mph

45	52	47	35	48	50	51	43
40	51	32	24	55	41	32	33
36	59	49	52	34	28	69	47
29	15	63	42	35	42	58	59
39	41	25	34	22	16	40	31
55	10	46	38	50	52	48	36
21	32	36	41	52	49	45	32
52	45	56	35	55	65	20	41

Stem	Leaves

2. What was the fastest speed recorded?

3. What was the slowest speed recorded?

4. Which speed was most often recorded?

5. If the speed limit is 45 miles per hour, how many were speeding?

6. If the speed limit is 45 miles per hour, how many were at least 20 mph over or under the speed limit?

MORE STEM-AND-LEAF PLOTS

Two sets of data can be displayed on the same stem-and-leaf plot.

EXAMPLE: The following is an example of a back-to-back stem-and-leaf plot.

Bryan's Math scores {60,65,72,78,85,90}
Bryan's English scores {78,88,89,89,92,95,100}

Math		English
5,0	6	
2,8	7	8
5	8	8,9,9
0	9	2,5
	10	0

2|7 means 72 8|9 means 89

Read the stem-and-leaf plot below and answer the questions that follow.

3rd grade Boys' Weights		3rd Grade Girls' Weights
8, 7, 5, 3, 2	4	0, 2, 4, 7
6, 4, 1, 0	5	1, 8, 8, 8, 9
5	6	0, 6, 6, 8, 8
0	9	8

4|5 means 54 6|8 means 68

1. What is the median for the girls' weights?
2. What is the median for the boys' weights?
3. What is the mode for the girls' weights?
4. What is the weight of the lightest boy?
5. What is the weight of the heaviest boy?
6. What is the weight of the heaviest girl?
7. Create a stem-and-leaf plot for the data given below.

Automobile Speeds on I-85

60	65	80	75	92	81	63
65	67	75	78	79	77	69
62	57	64	65	68	71	69
71	73	56	69	69	70	74

Automobile Speeds on I-75

72	56	62	65	63	60	58
55	57	70	69	59	53	61
58	61	63	67	57	63	67
56	58	59	62	64	63	69

8. What is the median speed for I-75?
9. What is the median speed for I-85?
10. What is the mode speed for I-75?
11. What is the mode speed for I-85?
12. What was the fastest speed on either interstate?

QUARTILES AND EXTREMES

In statistics, large sets of data are separated into four equal parts. These parts are called **quartiles**. The **median** separates the data into two halves. Then, the median of the upper half is the **upper quartile**, and the median of the lower half is the **lower quartile** variability of range. The differencc between the upper and lower quartile is the **interquartile range,** a type of variability.

The **extremes** are the highest and lowest values in a set of data. The lowest value is called the **lower extreme**, and the highest value is called the **upper extreme**.

EXAMPLE 1: The following set of data shows the high temperatures (in degrees Fahrenheit) in cities across the United States on a particular autumn day. Find the median, the upper quartile, the lower quartile, the upper extreme, and the lower extreme of the data.

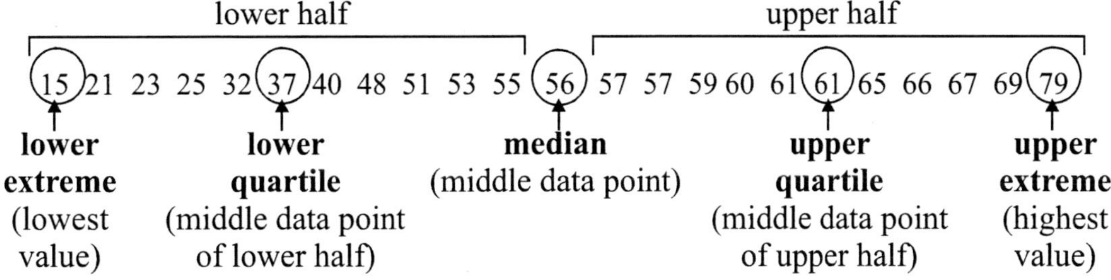

In Example 1, the interquartile range is 61-37= 24.

EXAMPLE 2: The following set of data shows the fastest race car qualifying speeds in miles per hour. Find the median, the upper quartile, the lower quartile, the upper extreme, and the lower extreme of the data.

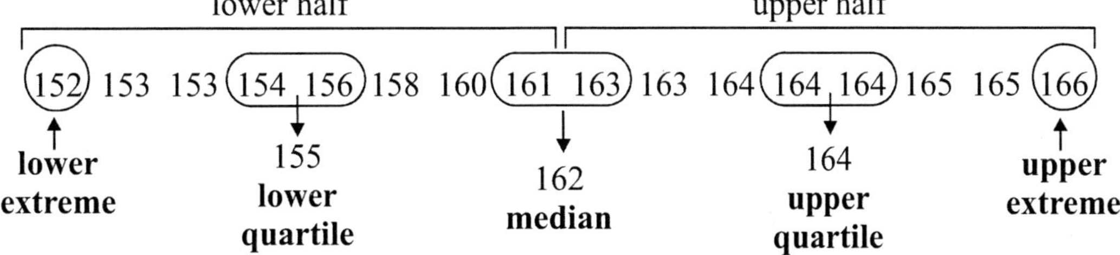

Note: When you have an even number of data points, the median is the average of the two middle points. The lower middle number is then included in the lower half of the data, and the upper middle number is included in the upper half.

Find the median, the upper quartile, the lower quartile, the upper extreme, and the lower extreme of each set of data given below.

1. 0 0 1 1 1 2 2 3 3 4 5
2. 15 16 18 20 22 22 23
3. 62 75 77 80 81 85 87 91 94
4. 74 74 76 76 77 78
5. 3 3 3 5 5 6 6 7 7 7 8 8
6. 190 191 192 192 194 195 196
7. 6 7 9 9 10 10 11 13 15
8. 21 22 24 25 27 28 32 35

BOX-AND-WHISKER PLOTS

Box-and-whisker plots are used to summarize data as well as to display data. A box-and-whisker plot summarizes data using the median, upper and lower quartiles, and the lower and upper extreme values. Consider the data below–a list of employees' ages at the Acme Lumber Company:

(21) 21 22 23 24 24 24 (25) 26 27 28 29 30 32 32 (33) 33 33 34 35 36 37 37 (38) 38 39 40 40 41 44 (48)

lower extreme (21) **lower quartile** (25) **median** (33) **upper quartile** (38) **upper extreme** (48)

Step 1: Find the median, upper quartile, lower quartile, upper extreme, and lower extreme just like you did on the previous page.

Step 2: Plot the 5 data points found in step 1 above on a number line as shown below.

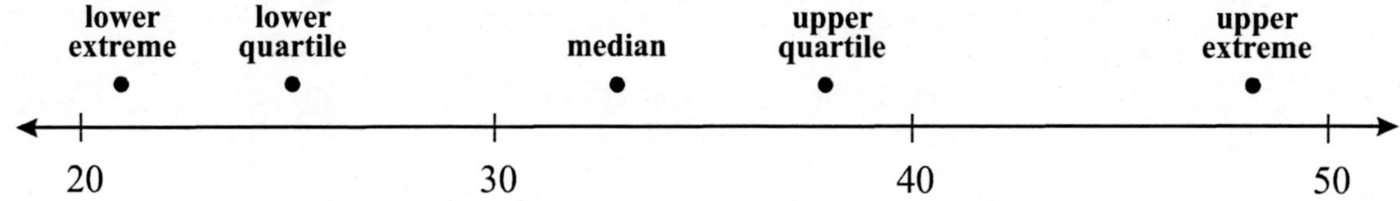

Step 3: Draw a box around the quartile values, and draw a vertical line through the median value. Draw whiskers from each quartile to the extreme value data points.

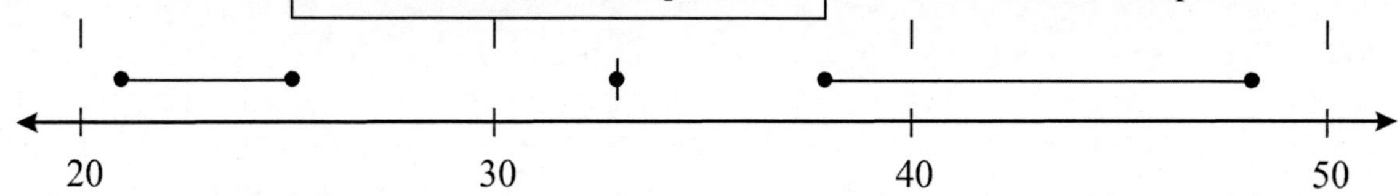

This box-and-whisker displays five types of information: lower extreme, lower quartile, median, upper quartile, and upper extreme.

Draw a box-and-whisker plot for the following sets of data.

1.
10 12 12 15 16 17 19 21 22 22 25 27 31 35 36 37 38 38 41 43 45 50 51 56 57 58 59

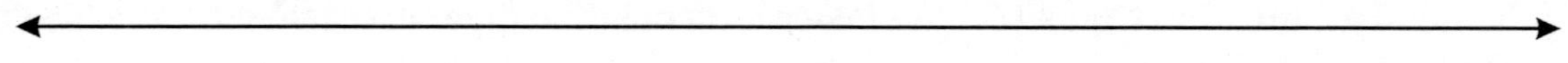

2.
5 5 6 7 9 9 10 11 12 15 15 16 17 18 19 19 20 22 24 26 27 27 30 31 31 35 37

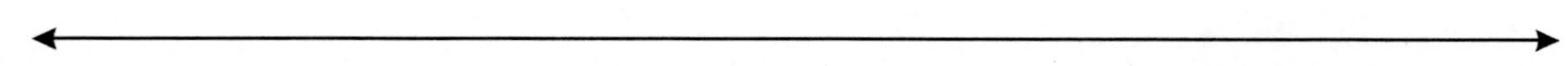

SCATTER PLOTS

A **scatter plot** is a graph of ordered pairs involving two sets of data. These plots are used to detect whether two sets of data, or variables, are truly related.

In the example to the right, two variables, income and education, are being compared to see if they are related or not. Twenty people were interviewed, ages 25 and older, and the results were recorded on the chart.

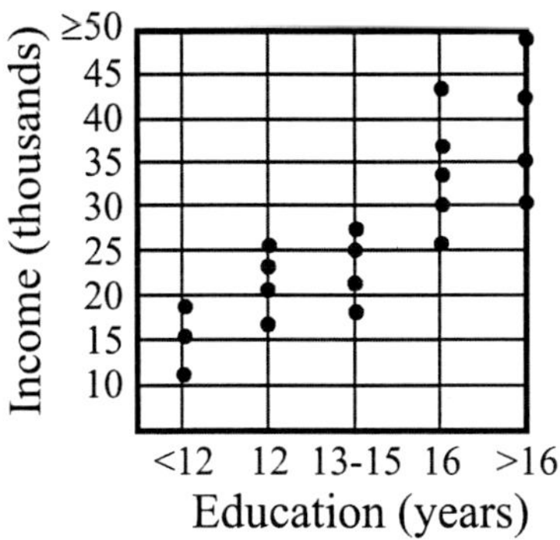

Imagine drawing a line on the scatter plot where half the points are above the line and half the points are below it. In the plot on the right, you will notice that this line slants upward and to the right. This line direction means there is a **positive** relationship between education and income. In general, for every increase in education, there is a corresponding increase in income.

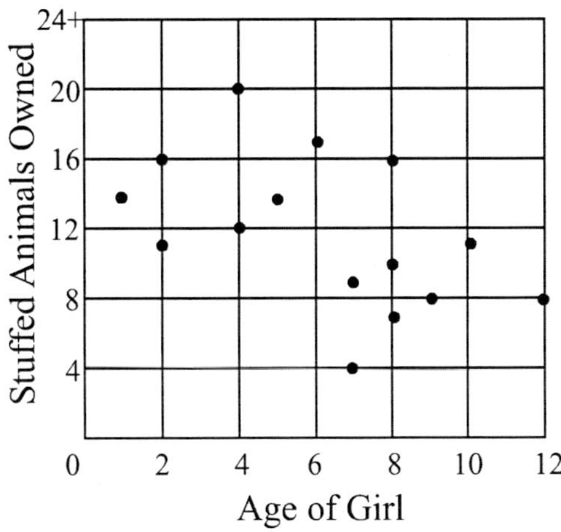

Now, examine the scatter plot on the left. In this case, 15 girls aged 2–12 were interviewed and asked, "How many stuffed animals do you currently have?" If you draw an imaginary line through the middle points, you will notice that the line slants downward and to the right. This plot demonstrates a **negative** relationship between the age of girls and their stuffed animal ownership. In general, as the girls' ages increase, the number of stuffed animals owned decreases.

Finally, look at the scatter plot shown on the right. In this plot, Rita wanted to see the relationship between the temperature in the classroom and the grades she received on tests she took at that temperature. As you look to your right, you will notice that the points are distributed all over the graph. Because this plot is not in a pattern, there is no way to draw a line through the middle of the points. This type of point pattern indicates there is **no** relationship between Rita's grades on tests and the classroom temperature.

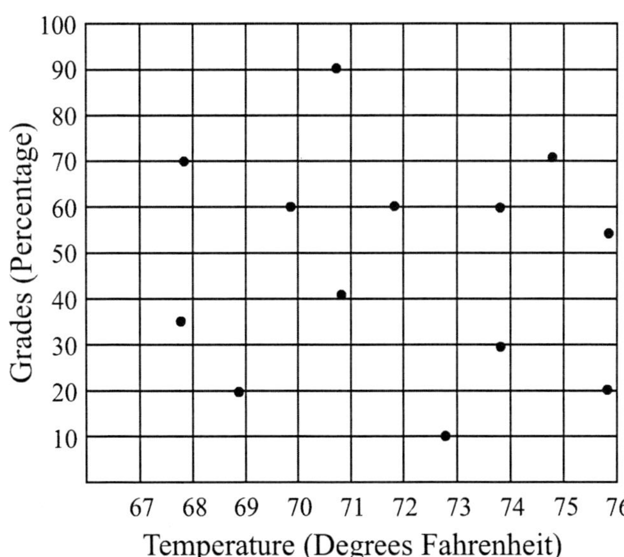

Examine each of the scatter plots below. On the line below each plot, write whether the relationship shown between the two variables is "positive", "negative", or "no relationship".

1.
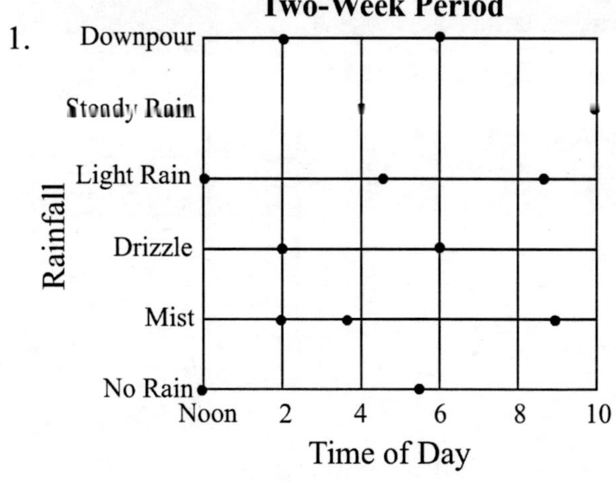

____

4.
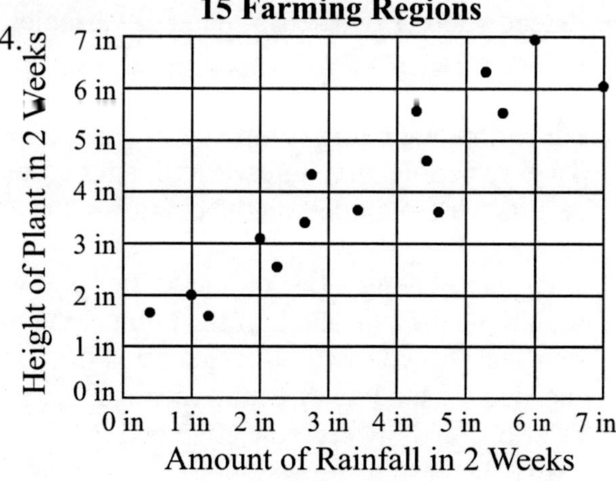

____

2.
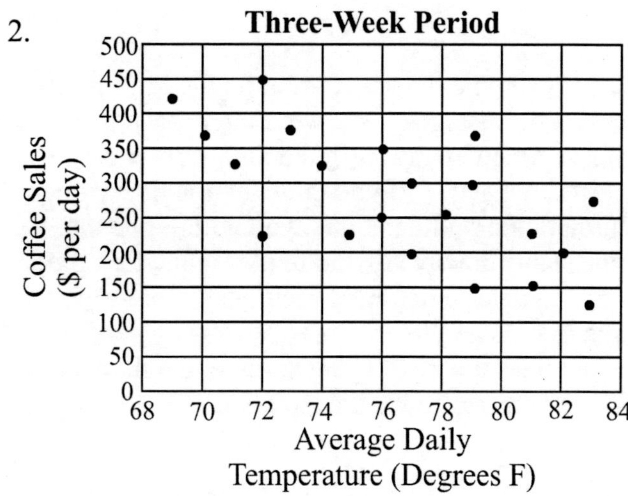

____

5.
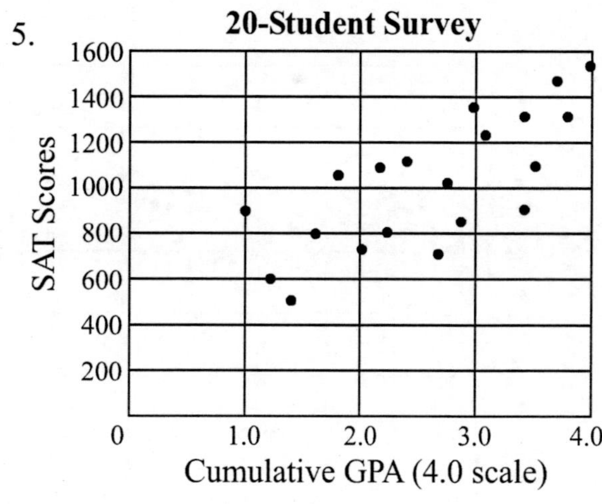

____

3.
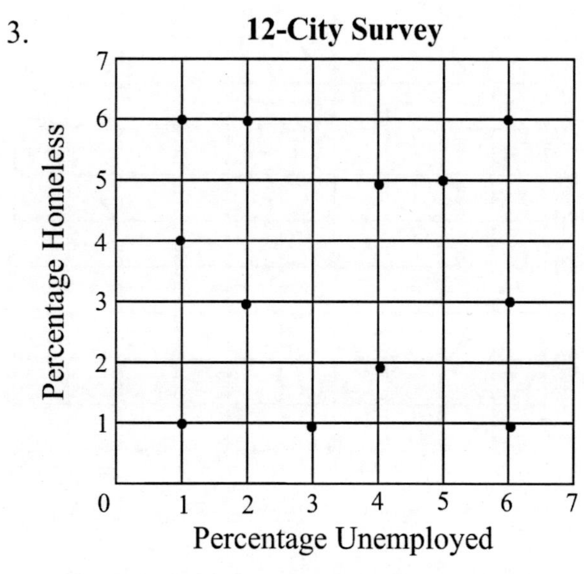

____

6.
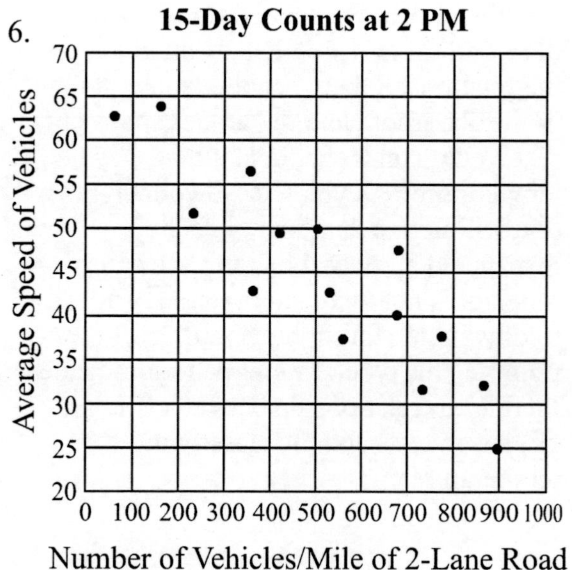

____

MISLEADING STATISTICS

As you read magazines and newspapers, you will see many charts and graphs which present statistical data. This data will illustrate how measurements change over time or how one measurement corresponds to another measurement. However, some charts and graphs are presented to make changes in data appear greater than they actually are. The people presenting the data create these distortions to make exaggerated claims.

There is one method to arrange the data in ways which can exaggerate statistical measurements. A statistician can create a graph in which the number line does not begin with zero.

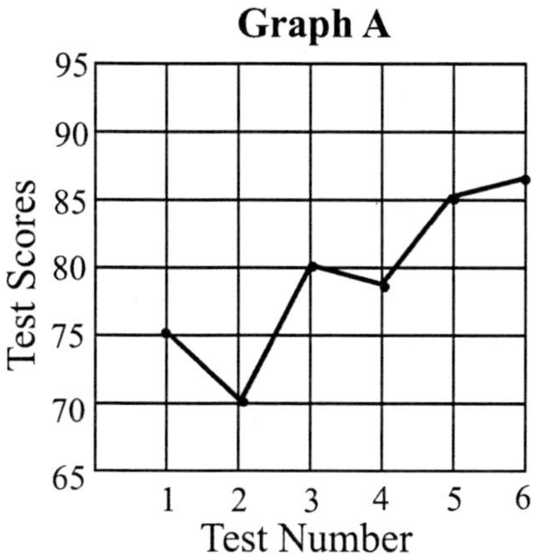

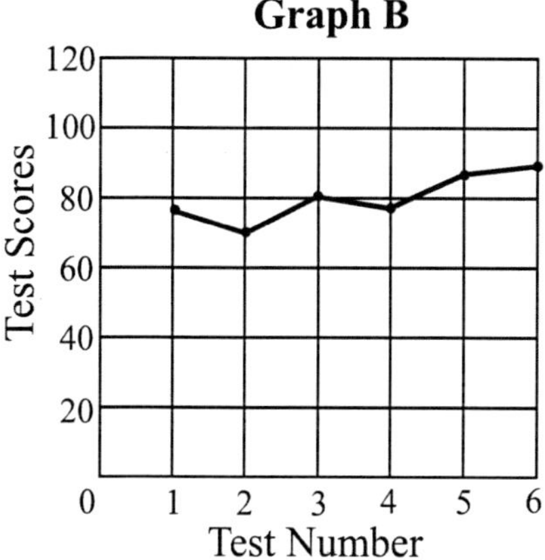

In the two graphs above, notice how each graph displays the same data. However, the way the data is displayed in graph A appears more striking than the data displayed in graph B. Graph A's data presentation is more striking because the test score numbers do not begin at zero.

Another form of misleading information is through the use of the wrong statistical measure to determine what is the median. For instance, the mean, or average, of many data measurements allows **outliers** (data measurements which lie well outside the normal range) to have a large effect. Examine the measurements in the chart below.

Address	**Household Income**	**Address**	**Household Income**
341 Spring Drive	$19,000	346 Spring Drive	$30,000
342 Spring Drive	$17,000	347 Spring Drive	$32,000
343 Spring Drive	$26,000	348 Spring Drive	$1,870,000
344 Spring Drive	$22,000	349 Spring Drive	$31,000
345 Spring Drive	$25,000	350 Spring Drive	$28,000

Average (Mean) Household Income: $210,000
Median Household Income: $27,000

In this example, the outlier, located at 348 Spring Drive, inflates the average household income on this street to the extent that it is over eight times the median income for the area.

Read the following charts and graphs, and then answer the questions below.

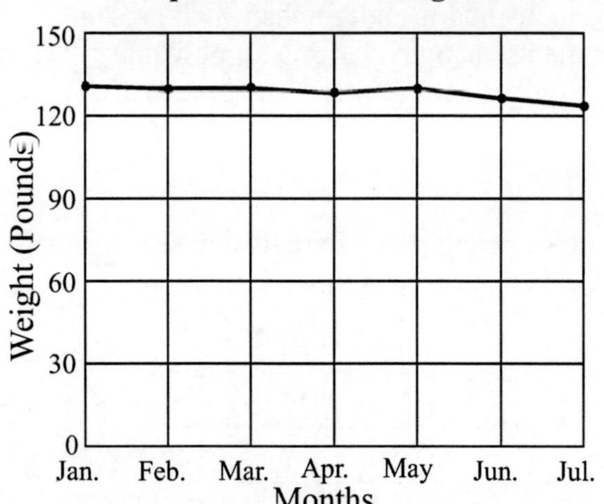

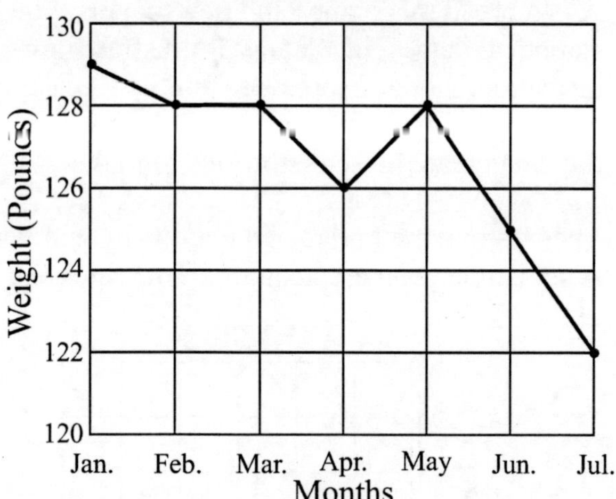

1. Which graph above presents misleading statistical information? Why is the graph misleading?

Twenty teenagers were asked how many electronic and computer games they purchased per year. The following table shows the results.

Number of Games	0	1	2	3	4	5	58
Number of Teenagers	4	2	5	3	4	1	1

2. Find the mean of the data.
3. Find the median of the data.
4. Find the mode of the data.
5. Which measurement is most misleading?
6. Which measurement would depict the data most accurately?
7. Is the *mean* of a set of data affected by outliers? Justify your answer with the example above.

Examine the two bar graphs below.

A.

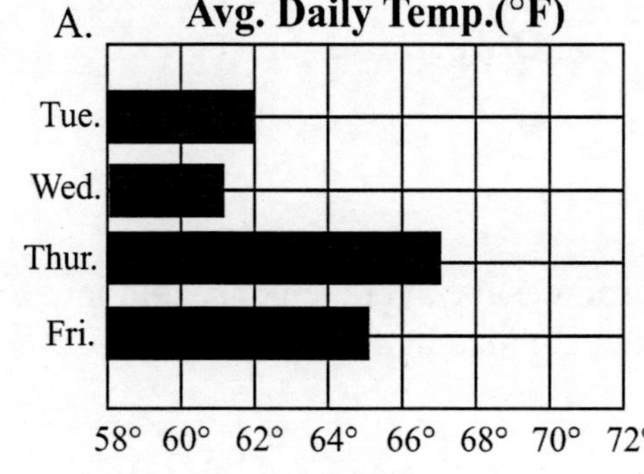

B.

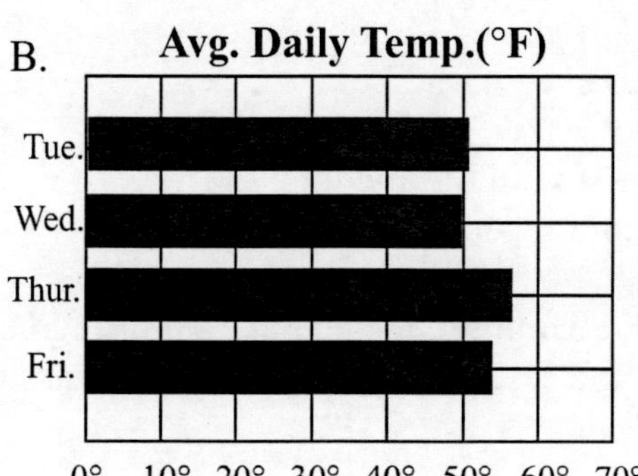

8. Which graph is misleading? Why?

CHAPTER 19 REVIEW

Find the mean, median, mode, and range for each of the following sets of data. Fill in the table below.

❶ Miles Run by Track Team Members	
Jeff	24
Eric	20
Craig	19
Simon	20
Elijah	25
Rich	19
Marcus	20

❷ 1992 SUMMER OLYMPIC GAMES Gold Medals Won			
Unified Team	45	Hungary	11
United States	37	South Korea	12
Germany	33	France	8
China	16	Australia	7
Cuba	14	Japan	3
Spain	13		

❸ Hardware Store Payroll June Week 2	
Erica	$280
Dane	$206
Sam	$240
Nancy	$404
Elsie	$210
Gail	$305
David	$280

Data Set Number	Mean	Median	Mode	Range
❶				
❷				
❸				

4. Jenica bowled three games and scored an average of 116 points per game. She scored 105 on her first game and 128 on her second game. What did she score on her third game?

5. Concession stand sales for each game in the season were $320, $540, $230, $450, $280, and $580. What was the mean sale per game?

6. Cedrick D'Amitrano works Friday and Saturday delivering pizza. He delivers 8 pizzas on Friday. How many pizzas must he deliver on Saturday to average 11 pizzas per day?

7. Long cooked three Vietnamese dinners that weighed a total of 40 ounces. What was the average weight for each dinner?

8. The Swamp Foxes scored an average of 7 points per soccer game. They scored 9 points in the first game, 4 points in the second game, and 5 points in the third game. What was their score for their fourth game?

9. Shondra is 66 inches tall, and DeWayne is 72 inches. How tall is Michael if the average height of these three students is 73 inches?

On the line below each plot, write whether the relationship shown between the two variables is "positive", "negative", or "no relationship".

10. 11.

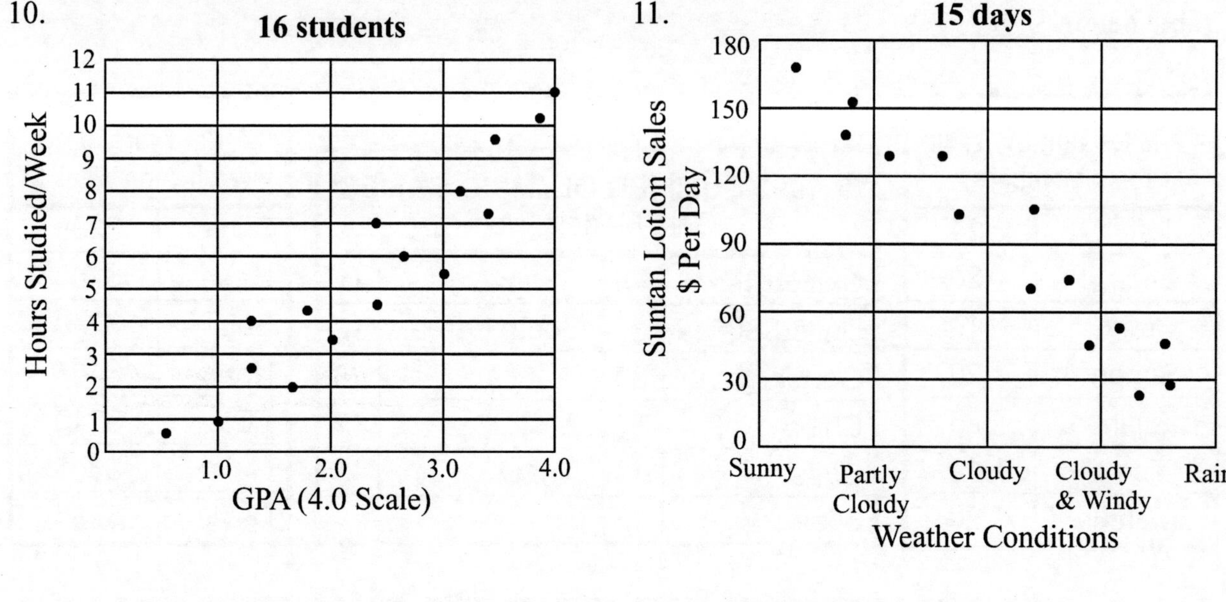

______________________ ______________________

Answer the following questions.

12. How do outliers affect the mean in a set of data? Why is this a bad thing?

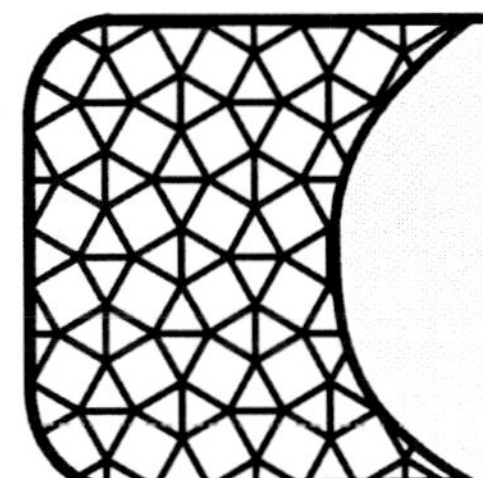

CHAPTER 20: PERMUTATIONS AND COMBINATIONS

Standard 5.12.4

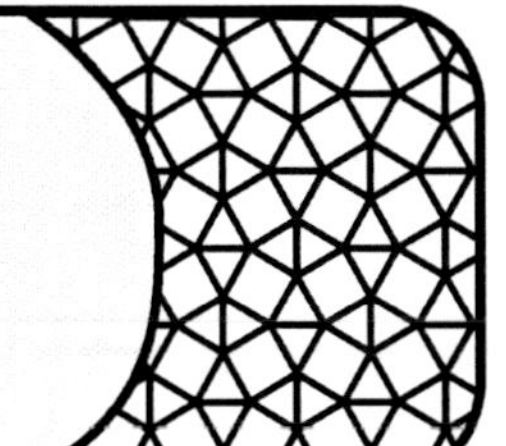

PERMUTATIONS

A **permutation** is an arrangement of items in a specific order. The formula ${}_nP_r = \frac{n!}{(n-r)!}$ is the formula for permutations. n is the number you have to choose from, and r is the number of objects you want to arrange. If a problem asks how many ways you can arrange 6 books on a bookshelf, it is asking you how many permutations there are for 6 items.

EXAMPLE 1: Ron has 4 items: a model airplane, a trophy, an autographed football, and a toy sports car. How many ways can he arrange the 4 items on a shelf?

Solution: The diagram below shows the permutations for arranging the 4 items on a shelf if he chooses to put the trophy first.

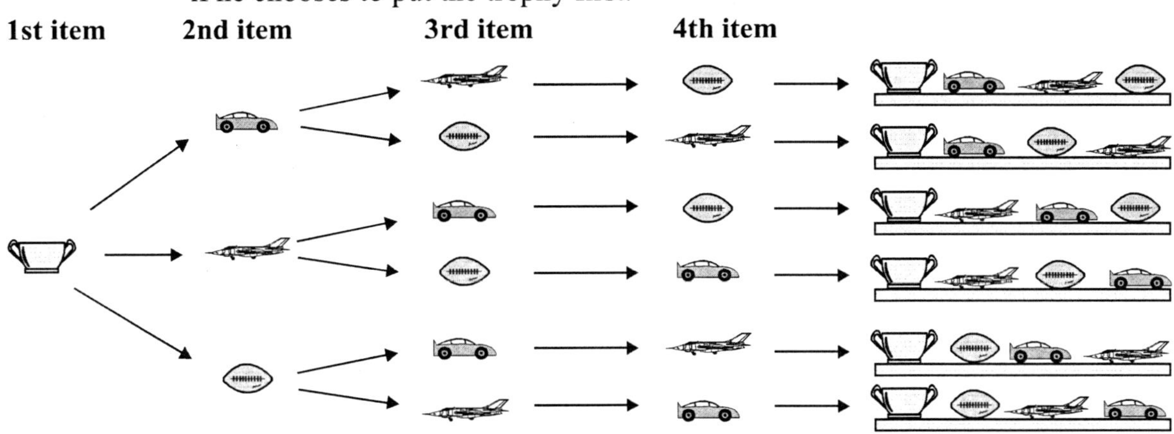

There are 6 permutations. Next, you could construct a tree diagram of permutations choosing the model car first. That tree diagram would also have 6 permutations. Then, you could construct a tree diagram choosing the airplane first. Finally, you could construct a pyramid choosing the football first. You would then have a total of 4 tree diagrams, each having 6 permutations. The total number of permutations is 6 × 4 = 24. There are 24 ways to arrange the 4 items on a bookshelf.

You probably don't want to draw tree diagrams for every permutation problem. For the problem above, Ron has 4 items to arrange. Therefore, multiply 4 × 3 × 2 × 1 = 24. Another way of expressing this calculation is 4!, stated as 4 factorial. 4! = 4 × 3 × 2 × 1.

Note: To find the permutation using the formula, ${}_nP_r = \frac{n!}{(n-r)!}$. n is the number you have to choose from, 4, and r is the number of objects you want to arrange, 4.

$${}_4P_4 = \frac{4!}{(4-4)!} = \frac{4!}{0!} = \frac{4!}{1} = 4! = 4 \times 3 \times 2 \times 1 = 24.$$

(Remember that 0! = 1.)

EXAMPLE 2: How many ways can you line up 6 students?

Solution: The number of permutations for 6 students = $6! = 6 \times 5 \times 4 \times 3 \times 2 \times 1 = 720$. There are 6 choices for the first position, 5 for the second, 4 for the third, 3 for the fourth, 2 for the fifth, and 1 for the sixth.

Note: The number of permutations for 6 students = ${}_6P_6 = \frac{6!}{(6-6)!} = \frac{6!}{0!} = \frac{6!}{1} = 6! = 6 \times 5 \times 4 \times 3 \times 2 \times 1 = 720$.

EXAMPLE 3: Shelley and her mom, dad, and brother are having cake for her birthday. Since it is Shelley's birthday she gets a piece first. How many ways are there to pass out the pieces of cake?

Solution: Since Shelley gets the first piece, the first spot is fixed. The second, third, and fourth spots are not fixed and anyone left can be in one of the three spots.

Spot	1	2	3	4
Choices of people	1	3	2	1

Now, multiply the choices together, $1 \times 1 \times 3 \times 2 \times 1 = 6$ ways to pass out cake.

Note: You can also use the permutation formula, but fix the first spot and just arrange the last three spots. $1 \times {}_3P_3 = 1 \quad \frac{3!}{(3-3)!} = \frac{3!}{0!} = \frac{3!}{1} = 3! = 3 \times 2 \times 1 = 6$ ways

---

1. How many ways can you arrange five books on a bookshelf? ______

2. Myra has six novels to arrange on a bookshelf. How many ways can she arrange the novels? ______

3. Seven sprinters signed up for the 100-meter dash. How many ways can the seven sprinters line up on the start line? ______

4. Keri wants an ice cream cone with one scoop of chocolate, one scoop of vanilla, and one scoop of strawberry. How many ways can the scoops be arranged on the cone if the top flavor is chocolate? ______

5. At Sam's party, the DJ has four song requests. In how many different orders can he play the four songs? ______

6. Yvette has five comic books. How many different ways can she stack the comic books? ______

7. Sandra's couch can hold three people. How many ways can she and her two friends sit on the couch? ______

8. How many ways can you arrange the numbers 1, 2, 3, 4, 5, 6, 7, 8, 9, 10 and always have 3 at position 1 and 10 at position 5? ______

MORE PERMUTATIONS

The formula $_nP_r = \frac{n!}{(n-r)!}$ can also be used if you are trying to arrange a specific number of objects, but have more than you want to arrange. n is the number you have to choose from, and r is the number of objects you want to arrange.

EXAMPLE: If there are 6 students, how many ways can you line up any 4 of them?

Step 1: Find all your variables. The formula is $_nP_r = \frac{n!}{(n=r)!}$.
n (the number you have to choose from) = 6
r (the number of objects you want to arrange) = 4

Step 2: Plug into the formula.

$$_6P_4 = \frac{6!}{(6-4)!} = \frac{6!}{2!} = \frac{6\times5\times4\times3\times\cancel{2\times1}}{\cancel{2\times1}} = 6\times5\times4\times3 = 360.$$

There are 360 ways to line up 4 of the 6 students.

Find the number of permutations for each of the problems below.

1. How many ways can you arrange four out of eight books on a shelf ?

2. How many 3 digit numbers can be made using the numbers 2, 3, 5, 8, and 9?

3. How many ways can you line up four students out of a class of twenty?

4. Kim worked in the linen department of a store. Eight new colors of towels came in. Her job was to line up the new towels on a long shelf. How many ways could she arrange the eight colors?

5. Terry's CD player holds 5 CDs. Terry owns 12 CDs. How many different ways can he arrange his CDs in the CD player?

6. Erik has eleven shirts he wears to school. How many ways can he choose a different shirt to wear on Monday, Tuesday, Wednesday, Thursday, and Friday ?

7. Deb has a box of twelve markers. The art teacher told her to choose three markers and line them up on her desk. How many ways can she line up three markers from the twelve?

8. Jeff went into an ice cream store serving 32 flavors of ice cream. He wanted a cone with two different flavors. How many ways could he order two scoops of ice cream, one on top of the other?

9. In how many ways can you arrange any three letters from the 26 letters in the alphabet?

COMBINATIONS

In a **combination**, the order does not matter. In a **permutation**, if someone picked two letters of the alphabet, **k, m** and **m, k,** they would be considered 2 different permutations. In a **combination, k, m** and **m, k** would be the same combination. A different order does not make a new combination. The formula for combinations is ${}_nC_r = \frac{n!}{(n-r)!\,r!}$ where n is the total number of objects you choose from and r is the number that you choose to arrange.

EXAMPLE: How many combinations of three letters from the set {a, b, c, d, e} are there?

Step 1: Find the **permutation** of 3 out of 5 objects.

Step 2: Divide by the permutation of the **number of objects** to be chosen from the total (3). This step eliminates the duplicates in finding the permutations.

Step 3: Cancel common factors and simplify.

$$\frac{5 \times \overset{2}{\cancel{4}} \times \cancel{3}}{\cancel{3} \times \cancel{2} \times 1} = 10$$

Note: Using the formula, find all your variables. The formula is ${}_nC_r = \frac{n!}{(n-r)!\,r!}$.

n (the number you choose from) = 5, r (the number of objects arranged) = 3

Now, plug into the formula.

$${}_5C_3 = \frac{5!}{(5-3)!\,3!} = \frac{5!}{2!\,3!} = \frac{5 \times 4 \times \cancel{3 \times 2 \times 1}}{(2 \times 1)\cancel{(3 \times 2 \times 1)}} = \frac{20}{2} = 10.$$

There can be 10 combinations of three letters from the set {a, b, c, d, e}.

Find the number of combinations for each problem below.

1. How many combinations of 4 numbers can be made from the set of numbers {2, 4, 6, 7, 8, 9}?

2. Johnston Middle School wants to choose 3 students at random from the 7th grade to take an opinion poll. There are 124 seventh graders in the school. How many different groups of 3 students could be chosen? (Use a calculator for this one.)

3. How many combinations of 3 students can be made from a class of 20?

4. Fashion Ware catalog has a sweater that comes in 8 colors. How many combinations of 2 different colors does a shopper have to choose from?

5. Angelo's Pizza offers 10 different pizza toppings. How many different combinations can be made of pizzas with four toppings?

6. How many different combinations of 5 flavors of jelly beans can you make from a store that sells 25 different flavors of jelly beans?

7. The track team is running the relay race in a competition this Saturday. There are 14 members of the track team. The relay race requires 4 runners. How many combinations of 4 runners can be formed from the track team?

8. Kerri got to pick 2 prizes from a grab bag containing 12 prizes. How many combinations of 2 prizes are possible?

MORE COMBINATIONS

Another kind of combination involves selection from several categories.

EXAMPLE: At Joe's Deli, you can choose from 4 kinds of bread, 5 meats, and 3 cheeses when you order a sandwich. How many different sandwiches can be made with Joe's choices for breads, meats, and cheeses if you choose 1 kind of bread, 1 meat, and 1 cheese for each sandwich?

JOE'S SANDWICHES

Breads	Meats	Cheeses
White	Roast Beef	Swiss
Pumpernickel	Corned Beef	American
Light rye	Pastrami	Mozzarella
Whole wheat	Roast Chicken	
	Roast Turkey	

Solution: Multiply the number of choices in each category. There are 4 breads, 5 meats, and 3 cheeses, so $4 \times 5 \times 3 = 60$. There are 60 combinations of sandwiches.

Find the number of combinations that can be made in each of the problems below.

1. Angie has 4 pairs of shorts, 6 shirts, and 2 pairs of tennis shoes. How many different outfit combinations can be made with Angie's clothes? ______

2. Raymond has 7 baseball caps, 2 jackets, 10 pairs of jeans, and 2 pairs of sneakers. How many combinations of the 4 items can he make? ______

3. Claire has 6 kinds of lipstick, 4 eye shadows, 2 kinds of lip liner, and 2 mascaras. How many combinations can she use to make up her face? ______

4. Clarence's dad is ordering a new truck. He has a choice of 5 exterior colors, 3 interior colors, 2 kinds of seats, and 3 sound systems. How many combinations does he have to pick from? ______

5. A fast-food restaurant has 8 kinds of sandwiches, 3 kinds of French fries, and 5 kinds of soft drinks. How many combinations of meals could you order if you ordered a sandwich, fries, and a drink? ______

6. In summer camp, Tyrone can choose from 4 outdoor activities, 3 indoor activities, and 3 water sports. He has to choose one of each. How many combinations of activities can he choose? ______

7. Jackie won a contest at school and gets to choose one pencil and one pen from the school store and an ice cream from the lunch room. There are 5 colors of pencils, 3 colors of pens, and 4 kinds of ice cream. How many combinations of prize packages can she choose? ______

CHAPTER 20 REVIEW

Answer the following permutation and combination problems.

1. Daniel has 7 trophies he has won playing soccer. How many different ways can he arrange them in a row on his bookshelf?

2. Missy has 12 colors of nail polish. She wears 1 color each day, 7 different colors a week. How many combinations of 7 colors can she make before she has to repeat the same 7 colors in a week?

3. Eileen has a collection of 12 antique hats. She plans to donate 5 of the hats to a museum. How many combinations of hats are possible for her donation?

4. Julia has 5 porcelain dolls. How many ways can she arrange 3 of the dolls on a display shelf?

5. Ms. Randal has 10 students. Every day she randomly draws the names of 2 students out of a bag to turn in their homework for a test grade. How many combinations of 2 students can she draw?

6. In the lunch line, students can choose 1 out of 3 meats, 1 out of 4 vegetables, 1 out of 3 desserts, and 1 out of 5 drinks. How many lunch combinations are there?

7. Andrea has 7 teddy bears in a row on a shelf in her room. How many ways can she arrange the bears in a row on her shelf ?

8. Adrianna has 4 hats, 8 shirts, and 9 pairs of pants. Choosing one of each, how many different clothes combinations can she make?

9. The buffet line offers 5 kinds of meat, 3 different salads, a choice of 4 desserts, and 5 different drinks. If you chose one food from each category, from how many combinations would you have to choose?

10. How many pairs of students can Mrs. Smith choose to go to the library if she has 20 students in her class?

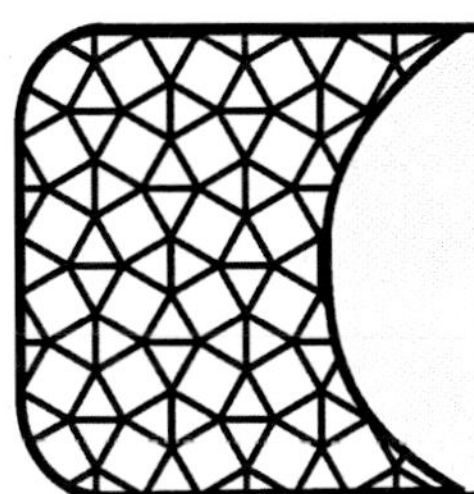

CHAPTER 21
PROBABILITY

Standards 5.12.3 and 5.12.5

PROBABILITY TERMS

Probability - the branch of mathematics that calculates the chance something will or will not happen.

Simple Event - one event. Tossing of one coin is an example.

Compound Event - multiple events. Tossing a coin more than once or rolling a die more than once are two examples.

Independent Events - the outcome of one event does not influence the outcome of the second event.

Dependent Events - the outcome of one event does influence the outcome of the second event.

Equally Likely Outcomes - all outcomes of the event have the same chance of occurring.

Law of Large Numbers - As the number of trials gets very large, the mean of experimental outcomes approaches the theoretical probability.

Population - an entire group or collection about which we wish to draw conclusions.

Sample - units selected to study from the population. The sample may be **biased**, which means that some knowledge of the sample is gained in advance. The sample can also be **unbiased**, which means that nothing is known about the sample before the study.

Rank - the number of a value in a list arranged in decreasing order. For example, a fifth number in a list.

Frequency - the number of times a value occurs in data that has been divided into classes.

***P*(*A*)** - notation used to mean the probability of outcome '*A*' occurring.

***P*′(*A*)** - notation used to mean the probability that outcome '*A*' does not occur.

PROBABILITY

Probability is the chance something will happen and is a theoretical result. An experimental result is what actually happens. Although the outcome (the experimental result) of a random event is uncertain, probability (the theoretical result) suggests a pattern that will emerge after many repetitions of the given event. Probability is most often expressed as a fraction, a decimal, a percent, or can also be written out in words.

EXAMPLE 1: Billy had 3 red marbles, 5 white marbles, and 4 blue marbles on the floor. His cat came along and batted one marble under the chair. What is the **probability** it was a red marble?

Step 1: The number of red marbles will be the top number of the fraction. → **3**

Step 2: The total number of marbles is the bottom number of the fraction. → **12**

The answer may be expressed in lowest terms. $P(A) = \frac{3}{12} = \frac{1}{4}$

EXAMPLE 2: Determine the probability that the pointer will stop on a shaded wedge or the number 1.

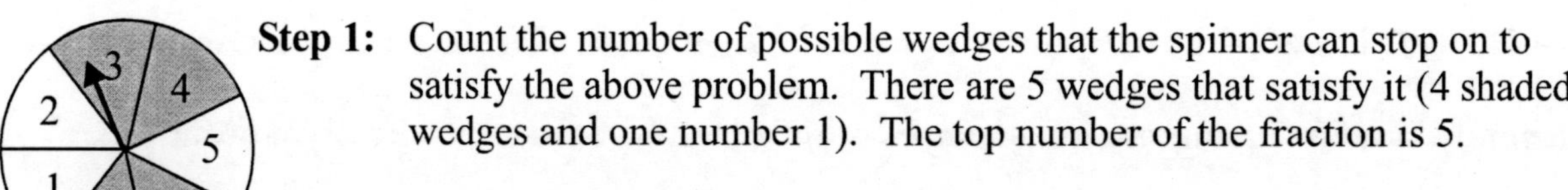

Step 1: Count the number of possible wedges that the spinner can stop on to satisfy the above problem. There are 5 wedges that satisfy it (4 shaded wedges and one number 1). The top number of the fraction is 5.

Step 2: Count the total number of wedges, 7. The bottom number of the fraction is 7.

$P(A) = \frac{5}{7}$ or **five out of seven**.

EXAMPLE 3: Refer to the spinner above. If the pointer stops on the number 7, what is the probability that it will **not** stop on 7 on the next spin?

Step 1: Ignore the information that the pointer stopped on the number 7 on the previous spin. The probability of the next spin does not depend on the outcome of the previous spin. Simply find the probability that the spinner will **not** stop on 7. Remember, if P is the probability of an event occurring, 1 – P is the probability of an event **not** occurring. In this example, the probability of the spinner landing on 7, is $\frac{1}{7}$.

Step 2: The probability that the spinner will not stop on 7 is $1 - \frac{1}{7}$ which equals $\frac{6}{7}$.

$P'(A) = \frac{6}{7}$ or **six out of seven**.

Find the probability of the following problems. Express the answer as a percent.

1. A computer chose a random number between 1 and 50. What is the probability of guessing the same number that the computer chose in 1 try?

   ____

2. There are 24 candy-coated chocolate pieces in a bag. Eight have defects in the coating that can be seen only with close inspection. What is the probability of pulling out a defective piece without looking?

   ____

3. Seven sisters have to choose which day each will wash the dishes. They put equal-sized pieces of paper each labeled with a day of the week in a hat. What is the probability that the first sister who draws will choose a weekend day?

   ____

4. For his garden, Clay has a mixture of 12 white corn seeds, 24 yellow corn seeds, and 16 bi-color corn seeds. If he reaches for a seed without looking, what is the probability that Clay will plant a bi-color corn seed first?

   ____

5. Mom just got a new department store credit card in the mail. What is the probability that the last digit is an odd number?

   ____

6. Alex has a paper bag of cookies that includes 8 chocolate chip, 4 peanut butter, 6 butterscotch chip, and 12 ginger. Without looking, his friend John reaches in the bag for a cookie. What is the probability that the cookie is peanut butter?

   ____

7. An umpire at a Little League baseball game has 14 balls in his pockets. Five of the balls are brand A, 6 are brand B, and 3 are brand C. What is the probability that the next ball he throws to the pitcher is a brand C ball?

   ____

8. What is the probability that the spinner arrow will land on an even number?

   ____

9. The spinner in the problem above stopped on a shaded wedge on the first spin and stopped on the number 2 on the second spin. What is the probability that it will not stop on a shaded wedge or on the 2 on the third spin?

   ____

10. A company is offering 1 grand prize, 3 second-place prizes, and 25 third-place prizes based on a random drawing of contest entries. If you entered one of the 500 total entries, what is the probability you will win a third-place prize?

    ____

11. In the contest problem above, what is the probability that you will win either the grand prize or a second-place prize?

    ____

12. A box of a dozen doughnuts has 3 lemon-cream-filled, 5 chocolate-cream-filled, and 4 vanilla-cream-filled. If the doughnuts look identical, what is the probability of picking a lemon-cream-filled?

    ____

INDEPENDENT AND DEPENDENT EVENTS

In mathematics, the outcome of an event may or may not influence the outcome of a second event. If the outcome of one event does not influence the outcome of the second event, these events are **independent**. However, if one event has an influence on the second event, the events are **dependent**. When someone needs to determine the probability of two events occurring, he or she will need to use an equation. These equations will change depending on whether the events are independent or dependent in relation to each other.

When finding the probability of two **independent** events, multiply the probability of each favorable outcome together.

EXAMPLE 1: One bag of marbles contains 1 white, 1 yellow, 2 blue, and 3 orange marbles. A second bag of marbles contains 2 white, 3 yellow, 1 blue, and 2 orange marbles. What is the probability of drawing a blue marble from each bag?

Solution: Probability of Favorable Outcomes

Bag 1: $\frac{2}{7}$ Bag 2: $\frac{1}{8}$

Probability of blue marble from each bag: $\frac{2}{7} \times \frac{1}{8} = \frac{2}{56} = \frac{1}{28}$

In order to find the probability of two **dependent** events, you will need to use a different set of rules. For the first event, you must divide the number of favorable outcomes by the number of possible outcomes. For the second event, you must subtract one from the number of favorable outcomes <u>only if</u> the favorable outcome is the <u>same</u>. However, you must subtract one from the number of total possible outcomes. Finally, you must multiply the probability of event one by the probability of event two.

EXAMPLE 2: One bag of marbles contains 3 red, 4 green, 7 black, and 2 yellow marbles. What is the probability of drawing a green marble, removing it from the bag, and then drawing another green marble?

	Favorable Outcomes	Total Possible Outcomes
Draw 1	4	16
Draw 2	3	15
Draw 1 × Draw 2	12	240

Answer: $P(A) = \frac{12}{240}$ or $\frac{1}{20}$

EXAMPLE 3: Using the same bag of marbles, what is the probability of drawing a red marble and then drawing a black marble?

	Favorable Outcomes	Total Possible Outcomes
Draw 1	3	16
Draw 2	7	15
Draw 1 × Draw 2	21	240

Answer: $P(A) = \frac{21}{240}$ or $\frac{7}{80}$

Find the probability of the following problems. Express the answer as a fraction.

1. Prithi has two boxes. Box 1 contains 3 red, 2 silver, 4 gold, and 2 blue combs. She also has a second box containing 1 black and 1 clear brush. What is the probability that Prithi selected a red brush from box 1 and a black brush from box 2?

2. Terrell cast his line into a pond containing 7 catfish, 8 bream, 3 trout, and 6 northern pike. He immediately caught a bream. What are the chances that Terrell will catch a second bream when he casts his line?

3. Gloria Quintero entered a contest in which the person who draws his or her initials out of a box containing all 26 letters of the alphabet wins the grand prize. Gloria reaches in and draws a "G", keeps it, then draws another letter. What is the probability that Gloria will next draw a "Q"?

4. Steve Marduke had two spinners in front of him. The first one was numbered 1–6, and the second was numbered 1–3. If Steve spins each spinner once, what is the probability that the first spinner will show an odd number and the second spinner will show a "1"?

5. Carrie McCallister flipped a coin twice and got heads both times. What is the probability that Carrie will get tails the third time she flips the coin?

6. Vince Macaluso is pulling two socks out of a washing machine in the dark. The washing machine contains three tan, one white, and two black socks. If Vince reaches in and pulls the socks out one at a time, what is the probability that Vince will pull out two tan socks in his first two tries?

7. John Salome has a bag containing 2 yellow plums, 2 red plums, and 3 purple plums. What is the probability that he reaches in without looking and pulls out a yellow plum and eats it, and then reaches in again without looking and pulls out a red plum to eat?

8. Artie Drake turns a spinner which is evenly divided into 11 sections numbered 1–11. On the first spin, Artie's pointer lands on "8." What is the probability that the spinner lands on an even number the second time he turns the spinner?

9. Leanne Davis played a game with a street entertainer. In this game, a ball was placed under one of three coconut halves. The vendor shifted the coconut halves so quickly that Leanne could no longer tell which coconut half contained the ball. She selected one and missed. The entertainer then shifted them around once more and asked Leanne to pick again. What is the probability that Leanne will select the coconut half containing the ball?

10. What is the probability that Jane Robelot reaches into a bag containing 1 daffodil and 2 gladiola bulbs and pulls out a daffodil bulb, and then reaches into a second bag containing 6 tulip, 3 lily, and 2 gladiola bulbs and pulls out a lily bulb?

MORE PROBABILITY

EXAMPLE: You have a cube with one number: 1, 2, 3, 4, 5, or 6, painted on each face of the cube. What is the probability that if you throw the cube 3 times, you will get the number 2 each time?

If you roll the cube once, you have a 1-in-6 chance of getting the number 2. If you roll the cube a second time, you again have a 1-in-6 chance of getting the number 2. If you roll the cube a third time, you again have a 1-in-6 chance of getting the number 2. The probability of rolling the number 2 three times in a row is:

$$P(A) = \frac{1}{6} \times \frac{1}{6} \times \frac{1}{6} = \frac{1}{216}$$

Find the probability that each of the following events will occur.

There are 10 balls in a box, each with a different digit on it: 0, 1, 2, 3, 4, 5, 6, 7, 8, or 9. A ball is chosen at random and then put back in the box.

1. What is the probability that if you picked out a number ball 3 times, you would get the number 7 each time?

2. What is the probability you would pick a ball with 5, then 9, and then 3?

3. What is the probability that if you picked out a ball four times, you would always get an odd number?

4. A couple has 4 children, ages 9, 6, 4, and 1. What is the probability that they are all girls?

There are 26 letters in the alphabet, allowing a different letter to be on each of 26 cards. The cards are shuffled. After each card is chosen at random, it is put back in the stack of cards, and the cards are shuffled again.

5. What is the probability that when you pick 3 cards, one at a time, that you will draw first a "y," then an "e," and then an "s"?

6. What is the probability that you will draw 4 cards and get the letter "z" each time?

7. What is the probability that you will draw twice and get a letter that is in the word "random" both times?

8. If you flipped a coin 3 times, what is the probability you would get heads every time?

9. Marie is clueless about 4 of her multiple-choice answers. The possible answers are A, B, C, D, E, or F. What is the probability that she will guess all four answers correctly?

TREE DIAGRAMS

Drawing a **tree diagram** is another method of determining the probability of events occurring.

EXAMPLE: **If you toss two six-sided numbered cubes that have 1, 2, 3, 4, 5, or 6 on each side, what is the probability you will get two cubes that add up to 9? One way to determine the probability is to make a tree diagram.**

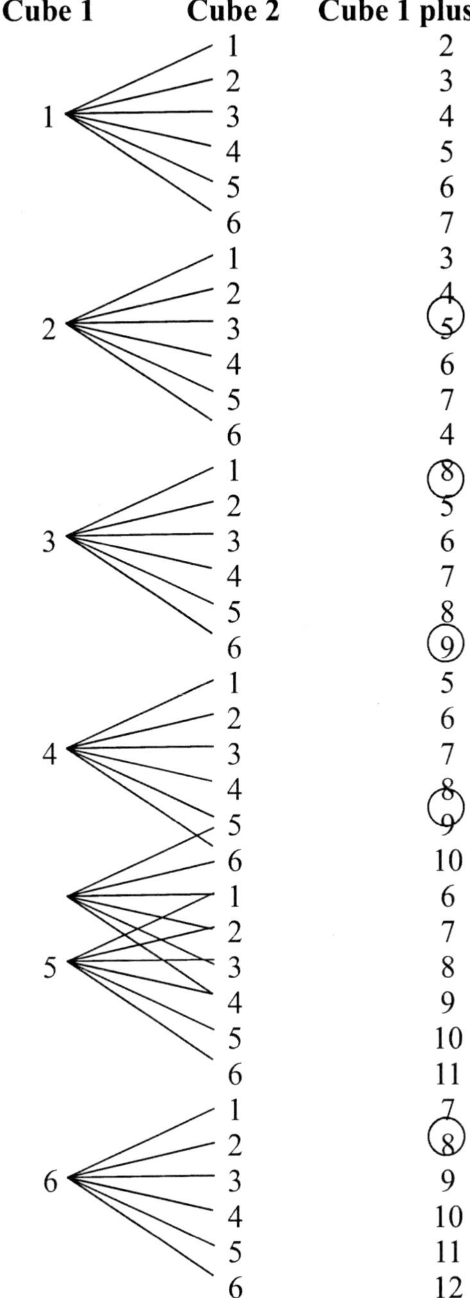

Alternative method

Write down all of the numbers on both cubes which would add up to 9.

Cube 1	Cube 2
4	5
5	4
6	3
3	6

Numerator = 4 combinations

For denominator: Multiply the number of sides on one cube times the number of sides on the other cube.

$6 \times 6 = 36$

$\dfrac{\text{Numerator:}}{\text{Denominator:}} \quad \dfrac{4}{36} = \dfrac{1}{9}$

There are 36 possible ways the cubes could land. Out of those 36 ways, the two cubes add up to 9 only 4 times. The probability you will get two cubes that add up to 9 is $\frac{4}{36}$ or $\frac{1}{9}$.

Read each of the problems below. Then answer the questions.

1. Jake has a spinner. The spinner is divided into eight equal regions numbered 1–8. In two spins, what is the probability that the numbers added together will equal 12?

2. Charlie and Libby each spin one spinner one time. The spinner is divided into 5 equal regions numbered 1–5. What is the probability that these two spins added together would equal 7?

3. Gail spins a spinner twice. The spinner is divided into 9 equal regions numbered 1–9. In two spins, what is the probability that the difference between the two numbers will equal 4?

4. Diedra throws two 10-sided numbered cubes. What is the probability that the difference between the two numbers will equal 7?

5. Cameron throws two six-sided numbered cubes. What is the probability that the difference between the two numbers will equal 3?

6. Tesla spins one spinner twice. The spinner is divided into 11 equal regions numbered 1–11. What is the probability that the two numbers added together will equal 11?

7. Samantha decides to roll two five-sided numbered cubes. What is the probability that the two numbers added together will equal 4?

8. Mary Ellen spins a spinner twice. The spinner is divided into 7 equal regions numbered 1–7. What is the probability that the product of the two numbers equals 10?

9. Conner decides to roll two six-sided numbered cubes. What is the probability that the product of the two numbers equals 4?

10. Tabitha spins one spinner twice. The spinner is divided into 9 equal regions numbered 1–9. What is the probability that the sum of the two numbers equals 10?

11. Darnell decides to roll two 15-sided numbered cubes. What is the probability that the difference between the two numbers is 13?

12. Inez spins one spinner twice. The spinner is divided into 12 equal regions numbered 1–12. What is the probability that the sum of two numbers equals 10?

13. Gina spins one spinner twice. The spinner is divided into 8 equal regions numbered 1–8. What is the probability that the two numbers added together equals 9?

14. Celia rolls two six-sided numbered cubes. What is the probability that the difference between the two numbers is 2?

15. Brett spins one spinner twice. The spinner is divided into 4 equal regions numbered 1–4. What is the probability that the difference between the two numbers will be 3?

SIMULATIONS

A **simulation** is usually generated by a computer program. It automatically produces the results of an experiment. To find probabilities, the simulation generates the results from a series of trials. The probabilities that are found from simulations are experimental and are not always accurate.

EXAMPLE: The chart below represents a computer simulation. It shows the frequencies of the results of flipping two coins. The two coins were flipped at the same time 100 times.

Outcome	TT	TH	HT	HH
Frequency	23	35	23	19

Find the theoretical probability of flipping one tail and one head, and find the experimental probability of flipping one tail and one head based on the computer simulation, then compare the two values.

Step 1: Find the theoretical probability. The probability of flipping a tail with coin one is $\frac{1}{2}$, and the probability of flipping a head with coin two is $\frac{1}{2}$. To find the probability of flipping a tail with coin one and flipping a head with coin two, you must multiply the two probabilities together, $\frac{1}{2}+\frac{1}{2}=\frac{1}{4}$. The probability of flipping a head with coin one and a tail with coin two is $\frac{1}{2}+\frac{1}{2}=\frac{1}{4}$. Since it does not matter which coin is tails and which is heads, add the two probabilities together.

$$\frac{1}{4}+\frac{1}{4}=\frac{1}{2}$$

The theoretical probability is 50%.

Step 2: Find the experimental probability. The frequency of TH is 35, so out of 100 flips, the probability is $\frac{35}{100}$. The frequency of HT is 23, so out of 100 flips, the probability is $\frac{23}{100}$. To find the theoretical probability of flipping one head and one tail, you need to add the two probabilities together.

$$\frac{35}{100}+\frac{23}{100}=\frac{58}{100}=\frac{29}{50}$$

The experimental probability based on the simulation is 58%.

Step 3: The difference between the theoretical probability and the experimental probability is 8%. Eight percent is not a huge difference. Since the two values are not too far apart, this means that the computer accurately simulates tossing two coins.

Use the simulations to find your answers.

1. A computer program simulated tossing three coins 500 times. The results are shown below.

HHH	50	HTT	66
HTH	76	THT	57
HHT	62	TTH	69
THH	64	TTT	56

a) Based on the computer simulation, what is the experimental probability of tossing two heads and a tail?

b) What is the theoretical probability of tossing two heads and one tail?

c) Based on the computer simulation, what is the experimental probability of tossing three tails?

d) What is the theoretical probability of tossing three tails?

e) Compare your answers from part a with part b and compare your answer from part c with part d. Based on this comparison, is this an accurate simulation of tossing three coins?

2. Below is a computer simulation of rolling one six-sided cube 50 times.

Outcome	1	2	3	4	5	6
Frequency	8	6	12	11	5	7

a) What is the theoretical probability of rolling a 3 or a 4?

b) Calculate the theoretical probability of rolling a 6.

c) Determine what the experimental probability of rolling a six based on the simulation.

d) Compare the theoretical and experimental probabilities of rolling a six from parts b and c, what are your conclusions?

PROBABILITY DISTRIBUTIONS

Many times in business or science situations, when data is collected, decisions are made by assigning probabilities to all possible outcomes of events taking place and evaluating the results. There are two types of outcomes that can be described: **discrete** and **continuous** probabilities. **Discrete probabilities** are those that must be counted, such as the toss of a coin or the roll of a die. **Continuous probabilities** are those that must be measured, such as time spent waiting in line at the bank or the height of a team of basketball players.

A **discrete probability distribution** is simply a table or graph which represents all possible outcomes of an experiment with the associated probability of each outcome. Consider rolling one six-sided die; there are six possible outcomes, each equally likely.

Outcome	1	2	3	4	5	6
$P(A)$	$\frac{1}{6}$	$\frac{1}{6}$	$\frac{1}{6}$	$\frac{1}{6}$	$\frac{1}{6}$	$\frac{1}{6}$

Using this data, you can see the probability of rolling a four is $\frac{1}{6}$. What is the probability of rolling 3 or less? It is the probability of rolling a three plus the probability of rolling a two plus the probability of rolling a one, which equals $\frac{1}{6} + \frac{1}{6} + \frac{1}{6} = \frac{1}{2}$.

A **continuous probability distribution** is a function of measured values rather than a table of all outcomes. The most frequently used of these is the **normal curve distribution**, which is commonly known as the bell curve.

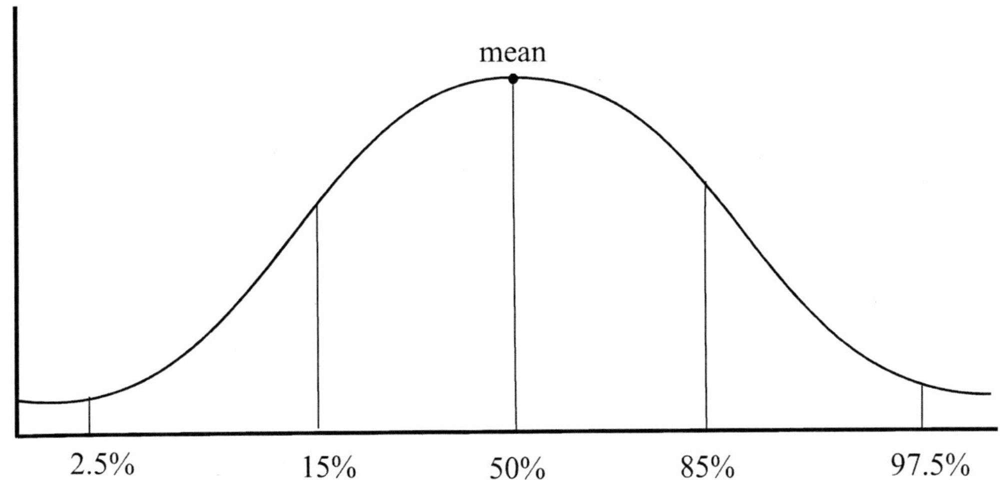

The **mean** of the data is the center of the curve, where the probability of the event is 50%. If this curve above measured the wait in minutes for a table at a restaurant instead of percent, then you can see that a few people waited 2.5 minutes or less, a few people waited 97.5 minutes or more, but most of the people waited about 50 minutes. These percentages are obtained using something known as the **empirical rule**, which determines how data is distributed in normal curves.

NOTE: You **CANNOT** find a probability of a specific value, like the percent of people who waited exactly 15 minutes to get a table.

The regions of the normal curve can be used to solve continuous probability questions such as the one posed on the previous page. One of interest to students is the distribution of grades in a particular class. Statisticians and educators alike are interested in the IQ distribution of a given population sample. These can be modeled by continuous probability distributions because these are outcomes that must be measured, not counted.

Problems.

1. What are the probabilities of the number of heads in three coin tosses? Construct the chart.

Outcomes	HHH							
$P(A)$	$\frac{1}{8}$							

2. What is the probability that only one head is obtained in three coin tosses? Use the chart constructed in Problem 1.

Use the curve below to answer the following questions. Apply the empirical rule percentages from the previous page.

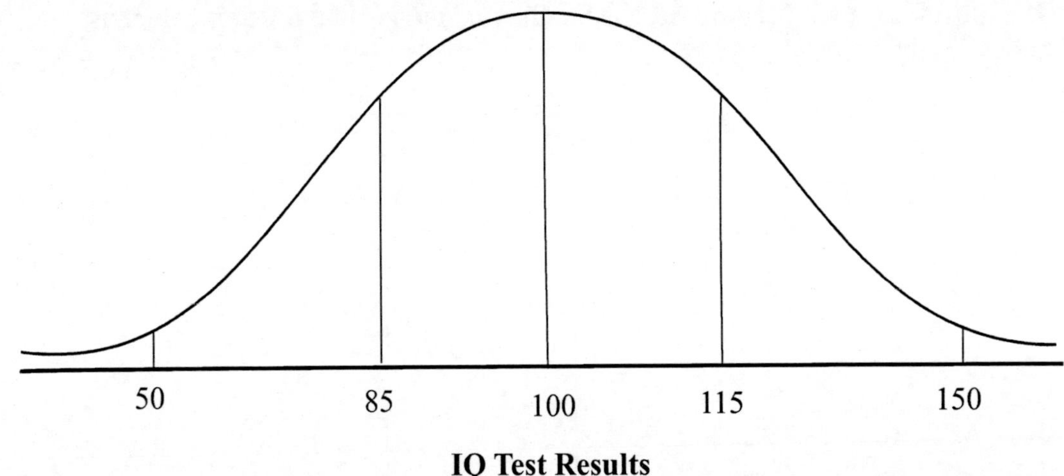

IQ Test Results

1. What percentage of adults taking the IQ test scored less than 115?
2. What percentage scored 85 or below?
3. What percentage scored 115 or above?
4. What percentage scored between 85 and 115?
5. How many scored exactly 95?

CHAPTER 21 REVIEW

1. There are 50 students in the school orchestra in the following sections:

   25 string section
   15 woodwind
   5 percussion
   5 brass

   One student will be chosen at random to present the orchestra director with an award. What is the probability the student will be from the woodwind section?

2. Fluffy's cat treat box contains 6 chicken-flavored treats, 5 beef-flavored treats, and 7 fish-flavored treats. If Fluffy's owner reaches in the box without looking and chooses one treat, what is the probability that Fluffy will get a chicken-flavored treat?

3. The spinner on the right stopped on the number 5 on the first spin. What is the probability that it will not stop on the number 5 on the second spin?

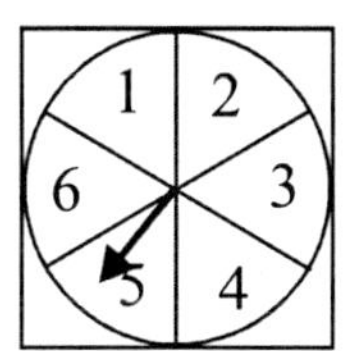

4. Three cakes are sliced into 20 pieces each. Each cake contains 1 gold ring. What is the probability that one person who eats one piece of cake from each of the 3 cakes will find 3 gold rings?

5. Brianna tossed a coin 4 times. What is the probability she got all tails?

6. Sherri turned the spinner on the right 3 times. What is the probability that the pointer always landed on a shaded number?

7. A box of a dozen doughnuts has 3 lemon-cream-filled, 5 chocolate-cream-filled, and 4 vanilla-cream-filled. If the doughnuts look identical, what is the probability that if you pick a doughnut at random, it will be chocolate-cream-filled?

8. Erica got a new credit card in the mail. What is the probability that the last four digits are all 5's?

9. There are 26 letters in the alphabet. What is the probability that the first two letters of your new license plate will be your initials?

10. Mary has 4 green mints and 8 white mints of the same size in her pocket. If she pulls one out, what is the probability it will be green?

Read the following, and answer questions 11–15.

There are 9 slips of paper in a hat, each with a number from 1 to 9. The numbers correspond to a group of students who must answer a question when the number for their group is drawn. Each time a number is drawn, the number is put back in the hat.

11. What is the probability that the number 6 will be drawn twice in a row?

12. What is the probability that the first 5 numbers drawn will be odd numbers?

13. What is the probability that the second, third, and fourth numbers drawn will be even numbers?

14. What is the probability that the first five times a number is drawn it will be the number 5?

15. What is the probability that the first five numbers drawn will be 1, 2, 3, 4, 5 in that order?

Answer the following probability questions.

16. If you toss two six-sided numbered cubes, what is the probability they will add up to 7? (Make a tree diagram.)

17. Make a tree diagram to show the probability of a couple with 3 children having a boy and two girls.

Solve the following word problems. Then, write whether the problem is "dependent" or "independent."

18. Felix Perez reaches into a 10-piece puzzle and pulls out one piece at random. This piece has two places where it could connect to other pieces. What is the probability that he will select another piece which fits the first one if he selects the next piece at random?

19. Barbara Stein is desperate for a piece of chocolate candy. She reaches into a bag which contains 8 peppermint, 5 butterscotch, 7 toffee, 3 mint, and 6 chocolate pieces and pulls out a toffee piece. Disappointed, she throws it back into the bag and then reaches back in and pulls out one piece of candy. What is the probability that Barbara pulled out a chocolate piece on the second try?

Match each term with the correct definition.

20. compound event
21. law of large numbers
22. probability
23. $P(A)$
24. unbiased sample
25. frequency
26. $P'(A)$
27. independent events
28. population
29. rank
30. continuous probabilities
31. equally likely outcomes
32. simple event
33. dependent events
34. discrete probabilities

A. a theoretical result

B. one event

C. probabilities that must be counted

D. multiple events

E. the outcome of one event does not influence the outcome of the second event

F. probabilities that must be measured

G. the outcome of one event does influence the outcome of the second event

H. all outcomes of the event have the same chance of occurring

I. units selected to study that are unknown

J. the number of a value in a list arranged in decreasing order

K. notation used to mean the probability of outcome '*A*' occurring

L. the mean of experimental outcomes that approach the theoretical probability as the number of trials gets very large.

M. an entire group or collection about which we wish to draw conclusions.

N. the number of times a value occurs in data that has been divided into classes.

O. notation used to mean the probability that outcome '*A*' does not occur.

Nevada High School Proficiency Exam in Math Practice Exams

The following Practice Exams will help you practice taking the Nevada High School Proficiency Exam in Mathematics. The following directions are similar to the directions you will read on the actual Nevada High School Proficiency Exit Exam in Mathematics. The actual Math Exam is timed, and you will be given a bubble sheet on which to mark your answers. If you are using a timed format for this diagnostic test and/or you are using a bubble sheet for scoring, follow the directions below carefully.

DIRECTIONS

DO NOT WASTE TIME ON DIFFICULT QUESTIONS. If a question is particularly difficult and taking a lot of time, go on to the next one and come back to it later. ***Be sure to skip this answer on your answer sheet as well!*** Make sure that the question that you are filling in on the answer sheet is ***the same number as the problem you are working on!***

ANSWER AS MANY QUESTIONS AS YOU CAN IN THE TIME PROVIDED. You should have plenty of time to answer all of the questions on this test. If the test administrator says that time is running out, however, you may wish to make ***your best guess*** on the questions that you have not yet completed.

EACH PROBLEM HAS FOUR POSSIBLE ANSWERS, LABELED A, B, C, AND D. Be sure that the problem number on the answer sheet matches the number on the test, and then mark your answer by filling in the space that contains the letter of the correct answer–either A, B, C, or D. ***Be sure to fill in only one answer on the answer sheet for each question, or the question will be marked wrong!***

IF YOU NEED TO CHANGE AN ANSWER ON YOUR ANSWER SHEET, BE SURE TO ERASE YOUR FIRST MARK COMPLETELY! There can be ***only one mark that shows for each question,*** or the question will be counted as wrong. ***Do not make any stray marks*** on the answer sheet!

Nevada HSPE Mathematics Formula Sheet

Note to student: You may use these formulas throughout the entire test. Feel free to use this sheet as needed during your testing time.

Parallelogram

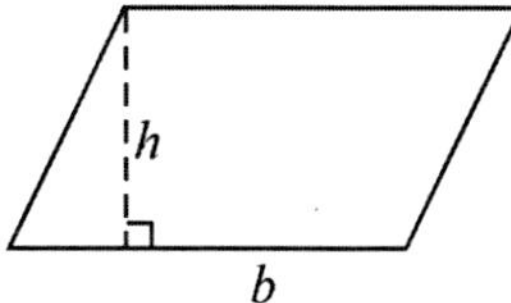

Area: $A = bh$

Trapezoid

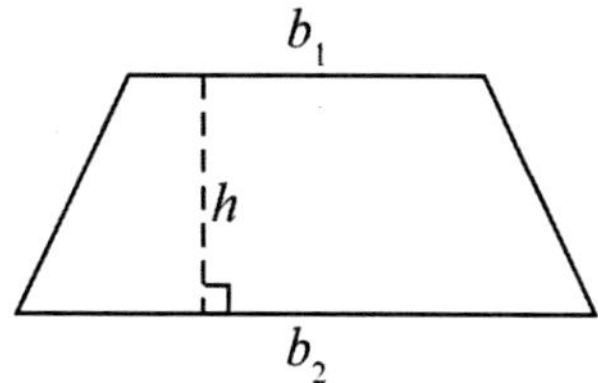

Area: $A = \frac{1}{2}(b_1 + b_2)h$

Circle

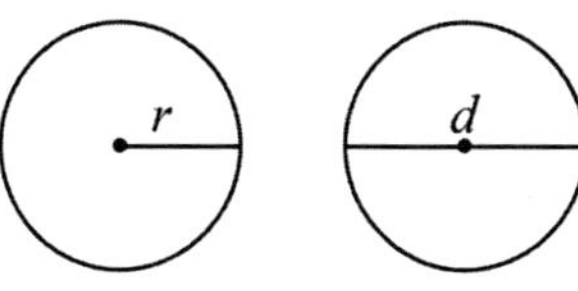

Circumference: $C = 2\pi r^2$

$C = \pi d$

Area: $A = \pi r^2$

Rectangular Solid

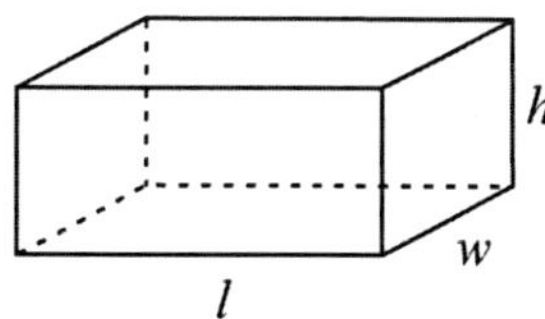

Volume: $V = lwh$

Surface Area: $SA = 2lw + 2lh + 2hw$

Pythogorean Theorem

$c^2 = a^2 + b^2$

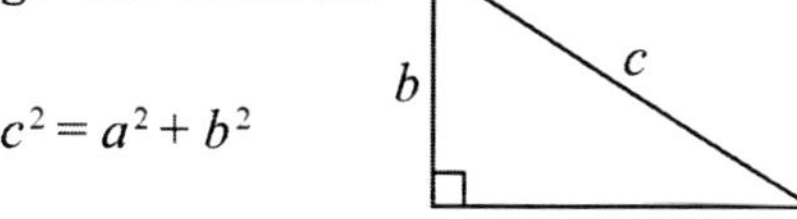

Cylinder

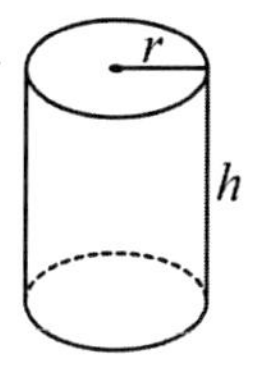

Volume: $V = \pi r^2 h$

Cone

h
r

Volume: $V = \frac{1}{3}\pi r^2 h$

Trigonomic Ratios

$\sin x = \frac{a}{c}$

$\cos x = \frac{b}{c}$

$\tan x = \frac{a}{b}$

a
c
x°
b

Permutations

$${}_nP_k = \frac{n!}{(n-k)!}$$

Combinations

$${}_nC_k = \frac{n!}{k!(n-k)!}$$

Special Right Triangles

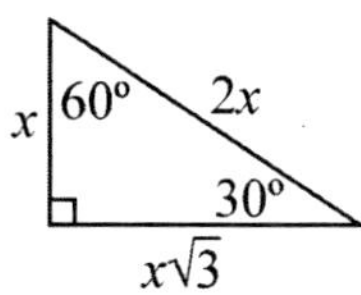

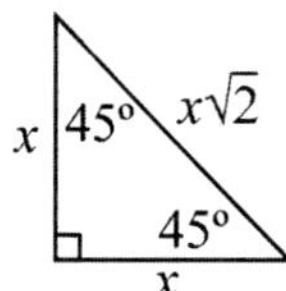

Temperature Formulas

$°F = \frac{9}{5}C + 32 \quad °C = \frac{5}{9}(F - 32)$

Math Practice Exam 1: Part 1

1. Which of the following is a number
1.12.7 which, when squared, results in a number less than itself?

A. -4
B. 4^{-2}
C. 4
D. $-\frac{1}{4}$

2. $\sqrt{6}$ is between
1.12.6

A. 5 and 6
B. 2 and 3
C. 4 and 5
D. 3 and 4

3. The figure below is a regular
4.12.1 decagon.

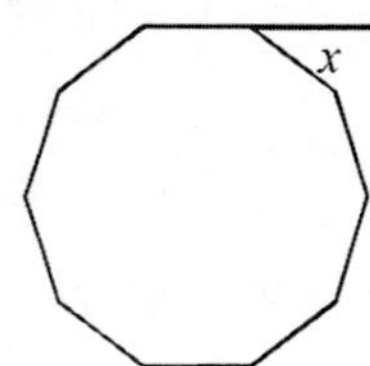

What is the measure of $\angle x$?

A. 10°
B. 18°
C. 25°
D. 36°

4. $4m(m - 5) + 3m(2m^2 - 6m + 4) =$
2.12.3

A. $6m^3 - 14m^2 - 8m$
B. $-8m^2 - 8m - 1$
C. $7m - 14m^2 - 1$
D. $10m^2 - 26m - 20$

5. A scientist wants to determine the half
2.12.1 life of iodine-131 experimentally. He started with 1 gram of iodine-131 and recorded the day when half of the iodine decomposed. He recorded the following data which shows that the mass decayed by half every eight days.

Half-life (n)	Mass
0	1 gram
1	$\frac{1}{2}$ gram
2	$\frac{1}{4}$ gram
3	$\frac{1}{8}$ gram

Which of the following functions describes the decay of mass for iodine-131?

A. $f(n) = 2^n$

B. $f(n) = \frac{1}{2}n$

C. $f(n) = \frac{1}{n}$

D. $f(n) = \frac{1}{2^n}$

6. Tim tossed three nickels on the ground.
5.12.5 What is the probability that all three will show "heads"?

A. $\frac{1}{8}$

B. $\frac{3}{8}$

C. $\frac{1}{6}$

D. $\frac{8}{9}$

7. The school cafeteria workers are
5.12.2 conducting a survey of student likes and dislikes for the foods they serve at lunch time. Once they tally the data and begin to order the food they need, which measure of central tendency will help them most to make their food purchases?

A. Mean
B. Mode
C. Median
D. Range

8. Susan rolled three six-sided dice at the
5.12.5 same time. What is the probability that all three dice came up 6?

A. $\frac{1}{648}$

B. $\frac{1}{216}$

C. $\frac{1}{72}$

D. $\frac{1}{36}$

9. The number of jelly beans in each of
5.12.2 10 bags is listed below.
61, 63, 59, 57, 57, 63, 64, 64, 63, 59
What is the mode of these numbers?

A. 61
B. 63
C. 57
D. 59

10. George has scores of 76, 78, 79, and
5.12.2 67 on four history tests. What is the lowest score George can have on the fifth test to have an average score of 80?

A. 85
B. 90
C. 95
D. 100

11. $(-7c^2 + 5c + 3) + (-c^2 - 7c + 2) =$
2.12.3

A. $-3c^3 - 3c^2 - 1$
B. $-8c^2 - 2c + 5$
C. $-6c^2 - 12c + 5$
D. $-8c^2 - 12c + 5$

12. For the following question, use the
3.12.3 formula $d = r \times t$. The formula for distance (d) is the rate traveled (r) times the time traveled (t). If Gracie drove 330 miles and averaged 50 miles per hour, how many hours was she driving?

A. 0.151515 hr
B. 6.6 hr
C. 280 hr
D. 16,500 hr

13. What is the perimeter of the rectangle
2.12.3 below?

$3x + 3$

$6x + 3$

A. $6x + 4$
B. $8(x^2 + 3x + 1)$
C. $16(2x^2 + 3x + 1)$
D. $2(9x + 6)$

14. Find $(4y^4 + 2y^2 + 7) + (2y^3 + 5y^2 - 4)$
2.12.3

A. $4y^4 + 2y^3 + 7y^2 + 3$

B. $4y^4 + 4y^3 + 5y^2 + 3$

C. $8y^7 + 10y^4 - 28$

D. $8y^{12} + 10y^4 + 3$

15. There are 21 students in Mr. Carlson's
5.12.4 home room. The office has asked him to send 2 students to the office to bring 2 boxes of candy bars back to the classroom. How many ways could he pick 2 students to go to the office?

A. 2

B. 42

C. 210

D. 420

16. Which of these sets of numbers is
1.12.6 ordered from least to greatest?

A. –1.6, –0.3, –0.09, 0.8, 1.2

B. –0.09, –0.03, –1.6, 0.8, 1.2

C. –0.09, –0.3, 0.8, 1.2, –1.6

D. 1.2, 0.8, –0.09, –0.3, –1.6

17. Use the formula below for this
3.12.3 question.

$I = Prt$
I = interest,
P = principal (the total amount),
r = rate of interest, and
t = time.

Tabitha put \$50 as the principal P into a savings account where it has earned an interest rate, r of 5% per year. She has earned \$25 of interest, I. How many years (t) has she had the money in the account?

A. 1 yr
B. 1.25 yrs
C. 4 yrs
D. 10 yrs

18. Simplify: $(3a^2)^3$
2.12.3

A. $27a^6$

B. $\frac{9a^3}{a^3}$

C. $9a^3$

D. $\frac{3a^6}{a^3}$

1.12.7

19. If 3 out of 4 people use a certain headache medicine, how many in a city of 150,400 will use this medicine?

A. 118,200
B. 37,600
C. 50,133
D. 112,800

20. Nicole works as an assistant
2.12.6 pharmacist. She is paid \$9.40 per hour for the first 40 hours per week with time-and-a-half for overtime. Which equation would be used to determine her salary (s) where r is her regular hours, and v is her overtime hours work?

A. $s = \$9.40(r + v) + .5v$
B. $s = \$9.40r + 1.5(\$9.40)v$
C. $s = 40r + 1.5(\$9.40)v$
D. $s = 40r + 1.5v$

21.
3.12.4
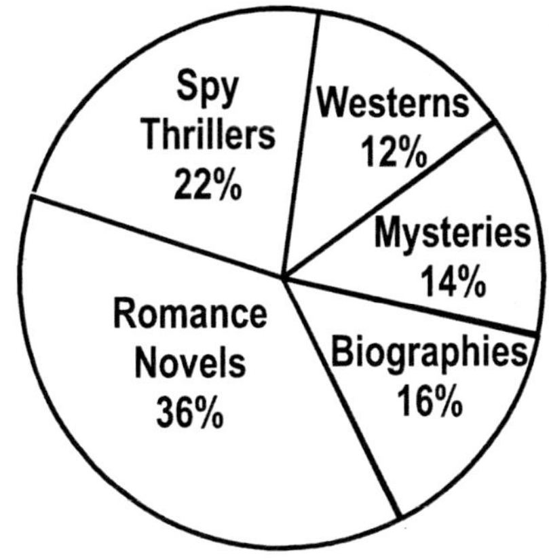

According to the pie graph, how much of the 1998 total sales came from romance novels?

A. \$120,000
B. \$160,000
C. \$220,000
D. \$360,000

22. Solve the equations
2.12.5 $-2x - 4y = -14$ and $5x + y = -1$ by substitution.

A. $(-1, 4)$
B. $(5, 1)$
C. $(10, -11)$
D. $(3, 2)$

23. Sally is seven years old and goes to the
4.12.9 store with her mother every Saturday morning. Every time they go, she notices that her mother gives check out clerk money before they leave the store. Sally concluded that everyone has to give money in exchange for things they want to take out of the store. This an example of:

A. deductive reasoning
B. inductive reasoning
C. analytical reasoning
D. none of the above

24. Carol is arranging rows of tables for
2.12.1 discussion groups.

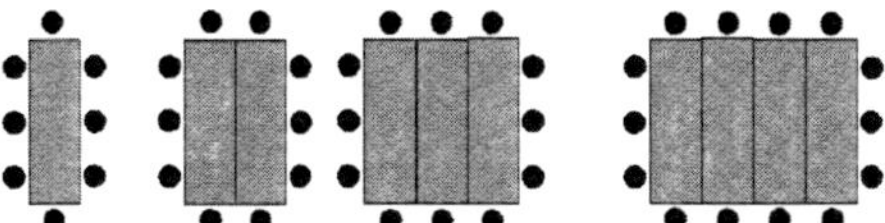

The table below shows the relationship between the number of tables (T) in a group and the maximum participants (P) in the group.

Number of Tables (T)	Maximum Participants (P)
1	8
2	10
3	12
4	14

Which of these functions generalizes the pattern of data in the table?

A. $P = 2T + 6$
B. $P = 3T + 2$
C. $P = 3T + 5$
D. $P = 4T + 2$

25. The sum of two numbers is fourteen.
2.12.6 The sum of six times the smaller number and two equals four less than the product of three and the larger number. Find the two numbers.

A. 6 and 8
B. 5 and 9
C. 3 and 11
D. 4 and 10

26. How many different ways can Mrs.
5.12.4 Smith choose 2 students to go to the library if she has 20 students in her class?

A. 40
B. 190
C. 380
D. 3,800

27. A school store buys packages of graph
3.12.4 paper to sell to students for a small profit. The school buys 50 packages of graph paper for \$8.00. They sell the packs of graph paper for 25¢ each. How many packages of graph paper do they have to sell to generate a profit of \$18.00?

A. 200
B. 225
C. 104
D. 72

28. Ryan and Trevor went to the school
2.12.5 store. The table below shows what they bought and the amount they paid.

	Pens	Pencils	Total Cost
Ryan	2	4	\$0.74
Trevor	1	1	\$0.29

What was the cost of one pencil?

A. \$0.05

B. \$0.06

C. \$0.07

D. \$0.08

29. What is the range of the function
2.12.4 $y = x^2 + 2$ as represented in the graph?

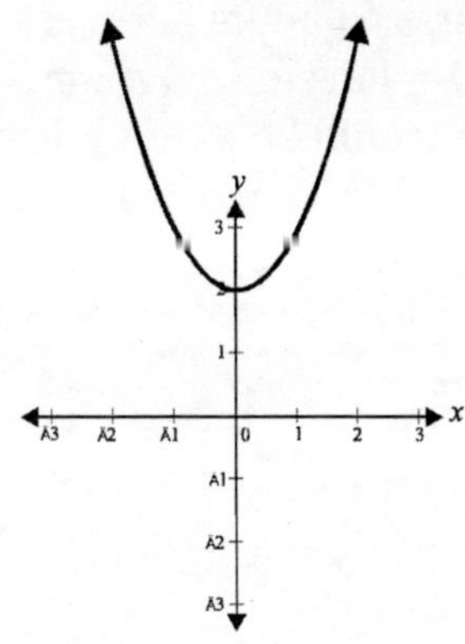

A. $-3 \le R \le 3$

B. $R \ge 0$

C. All real numbers ≥ 2

D. All real numbers

30. Find the point of intersection of the
2.12.4 two equations by adding and/or subtracting.

$$x + y = 4$$
$$2x - y = 5$$

A. (3, 1)

B. (−3, 1)

C. (1, 3)

D. (−1, −3)

31. Which of the following graphs shows a
4.12.5 line with slope $\frac{1}{2}$ that passes through the point $(-1, 1)$?

A.

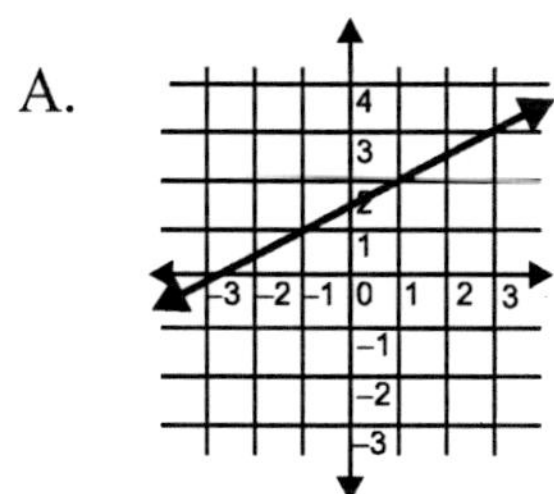

B.

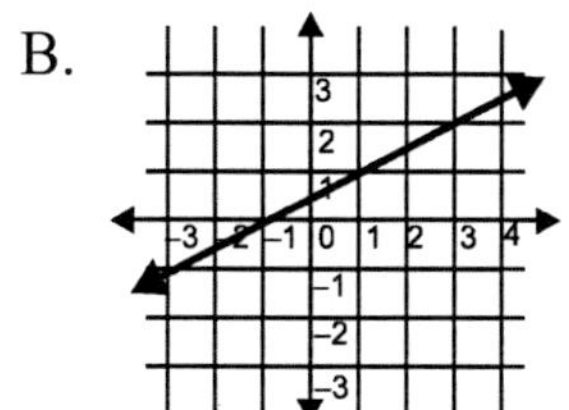

C.

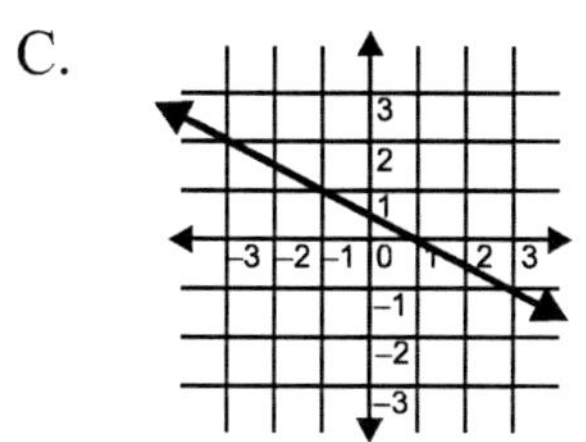

D.

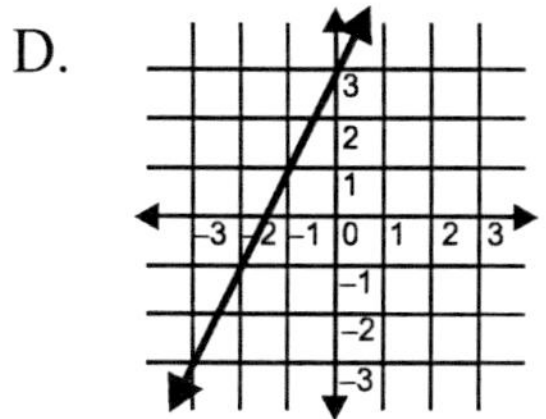

32. Marcus is concerned about the
5.12.3 possibility of his school losing funding for its band program so that the school can start a women's lacrosse team. He took a survey asking 40 students who sing in the school chorus if they would favor keeping the band program. Of the chorus students surveyed, 34 said "yes." Marcus concluded that an overwhelming 85% of students want to keep the funding for band. Which of these explains why his conclusion is invalid?

A. sample is biased in favor of music programs
B. computation is incorrect
C. sample size is too large
D. Sample is random

Math Practice Exam 1: Part 2

33.
3.12.4

Dobbins Family	
Months	**Electric**
JANUARY	\$ 89.15
FEBRUARY	\$ 99.58
MARCH	\$ 78.99
APRIL	\$ 72.47
MAY	\$ 99.23
JUNE	\$124.69

Look at the chart above. About how much did the Dobbins family spend for electricity during the first three months of the year?

A. \$240.00
B. \$250.00
C. \$260.00
D. \$270.00

34. Find the area of the triangle below.
4.12.7

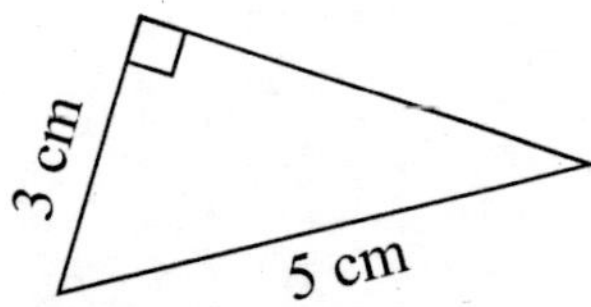

A. 6 cm^2
B. 7.5 cm^2
C. 12 cm^2
D. 15 cm^2

35. Consider the following equations:
2.12.5 $f(x) = 3x + 2$ and $f(x) = 3x - 7$. Which of the following statements is true concerning the graphs of these equations?

A. The graphs of the equations are lines that are perpendicular to each other.
B. The graph of the line represented by the equation $f(x) = 3x + 2$ always remains above the x-axis, while the graph of the line represented by the equation $f(x) = 3x - 7$ always remains below the x-axis.
C. The graphs of the equations are lines that are parallel to each other, but they have different y-intercepts.
D. The graphs of the lines intersect at the point $(2, -7)$.

36. Which of the following is the greatest
3.12.2 tolerance for error for the measurements 70 mL and 76 mL, measured in a container marked in milliliters?

A. ±3 mL
B. ±0.5 mL
C. ±0.01 mL
D. ±10 mL

37. The following trapezoids are similar.
4.12.6 What is the measure of $\angle A$ on the larger trapezoid?

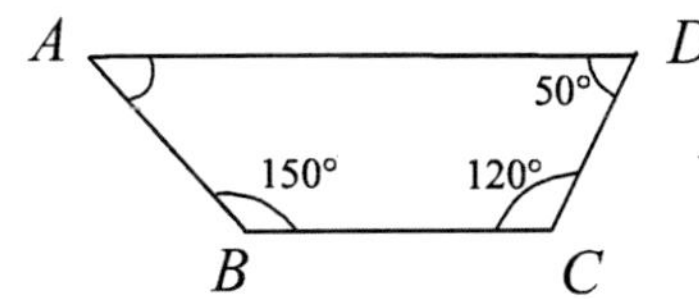

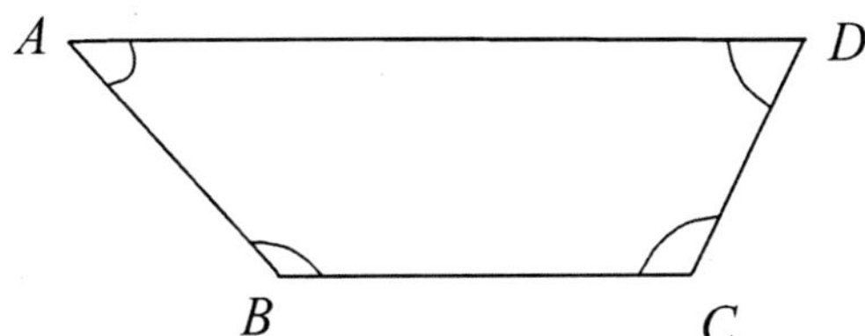

A. 150°
B. 120°
C. 40°
D. 50°

38. If $3x + 4y = 9$, then x equals
2.12.2

A. $3 - 4y$

B. $3 + 4y$

C. $\frac{9 + 4y}{3}$

D. $\frac{9 - 4y}{3}$

39. Factor $y^2 - 81$.
2.12.3

A. $(y + 9)(y + 9)$
B. $(y - 9)(y - 9)$
C. $(y + 9)(y - 9)$
D. $(y + 3)(y - 3)$

40. What is the perimeter of the next
2.12.1 rectangle in the pattern below?

$x + 1$

$2x + 1$

$2x + 2$

$4x + 2$

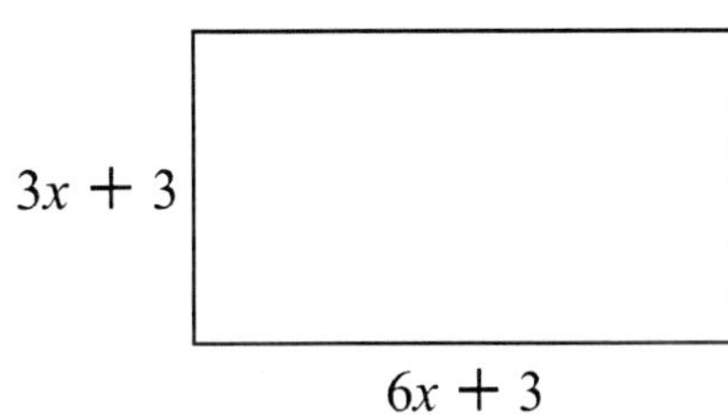

A. $6x + 4$
B. $8(x^2 + 3x + 1)$
C. $4(2x^2 + 6x + 1)$
D. $8(3x + 2)$

41. Rhonda is beating her favorite video
5.12.1 game, Monkey Madness. Each level is timed and you collect points until the time expires (or your character is defeated). The first level is 30 minutes. The next three levels are 15 minutes each. Which of these tables shows the relationship between the number of levels completed and the total amount of time played?

A.

levels completed	**1**	**2**	**3**	**4**
minutes played	**15**	**30**	**45**	**60**

B.

levels completed	**1**	**2**	**3**	**4**
minutes played	**30**	**15**	**15**	**30**

C.

levels completed	**1**	**2**	**3**	**4**
minutes played	**15**	**45**	**60**	**75**

D.

levels completed	**1**	**2**	**3**	**4**
minutes played	**30**	**45**	**60**	**75**

42. The band teacher took a survey of
5.12.2 reasons students took his class. He made the following bar graph from the data he collected.

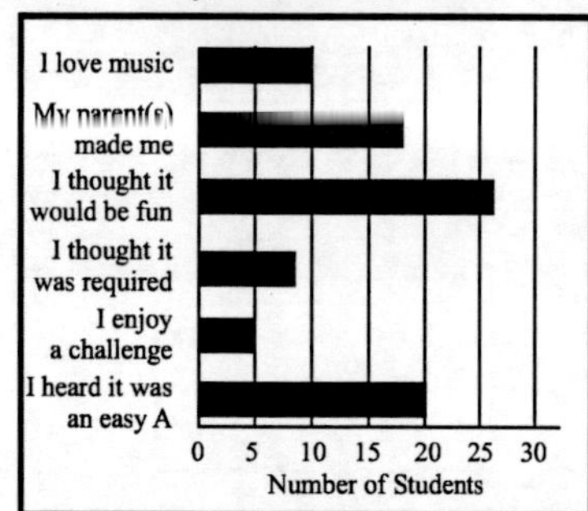

Which reason for taking band is the mode of the data?

A. My parents made me.
B. I love music.
C. I thought it would be fun.
D. I heard it was an easy A.

D.1

43. What is the distance from Point P to
4.12.7 the origin?

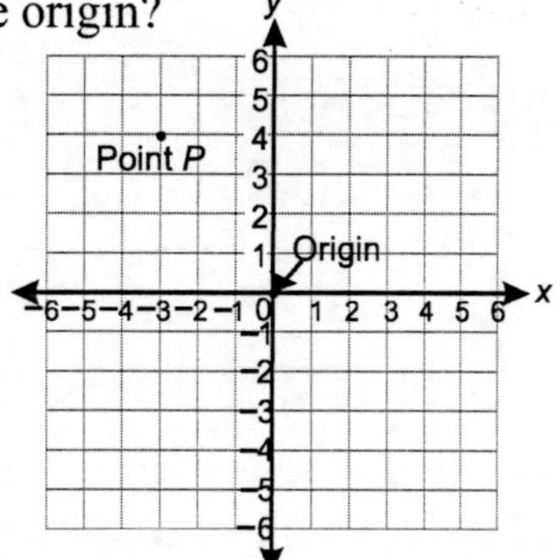

A. 5
B. $\sqrt{35}$
C. 7
D. 9

44. Which of the following measurements
3.12.2 is most precise?

A. 1 lb.
B. 30 lb.
C. 7 oz.
D. 3.2 oz.

45. If $x = -3$, find $3x^2 - 5x$
2.12.3

A. 12
B. −6
C. 42
D. 3

46. What is the sum of the measures of the
4.12.1 interior angles of a hexagon if each side measures 7 inches?

A. 360°
B. 540°
C. 720°
D. 960°

47. Hannah earns 12% commission on any
4.12.6 jewelry sale she makes. How much commission will she earn on a $45 sale?

A. $5.40
B. $33.00
C. $3.75
D. $12.00

48. Given the conditional statement, "If
4.12.9 $m\angle A = 110°$, then $\angle A$ is obtuse," which of the following is the contrapositive of the statement?

A. If $\angle A$ is not obtuse, then $\angle A$ is an obtuse angle.
B. If $\angle A$ is obtuse, then $m\angle A = 110°$.
C. If $\angle A$ is not obtuse, then $m\angle A \neq 110°$.
D. If $m\angle A \neq 110°$, then $\angle A$ is not obtuse.

49. Casey, our pet iguana, whacked his tail
5.12.5 again into a pile of marbles. One of the marbles went under the couch, never to be seen again. There were 6 red marbles, 11 orange marbles, 4 blue marbles, and 7 multicolored marbles. What is the probability the one missing is either orange or blue?

A. $\frac{1}{28}$
B. $\frac{11}{28}$
C. $\frac{14}{28}$
D. $\frac{15}{28}$

50. Boyle's Law is stated by the formula:
2.12.2 $PV = pv$

Find V in Boyle's Law

A. $V = pvP$
B. $V = pv + P$
C. $V = \frac{pv}{P}$
D. $V = pv - P$

51. Which property should be used to
1.8.2 simplify the expression: $4(3x + 5)$?

A. Commutative Property of Multiplication
B. Associative Property of Addition
C. Distributive Property
D. Transitive property of Equality

52. What is the length of line segment $\overline{WY}$?
4.12.2

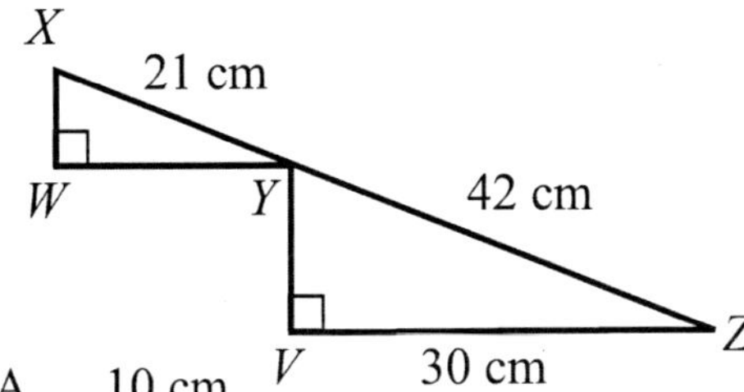

A. 10 cm
B. 15 cm
C. 18 cm
D. 30 cm

53. In the triangle below, what is the
4.12.2 length of side x?

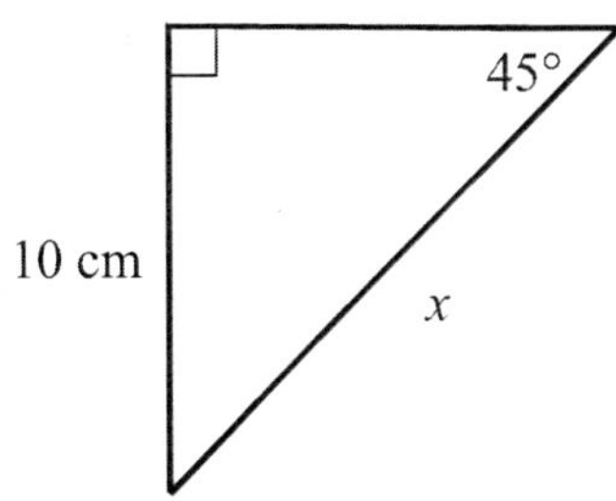

A. 12 cm
B. $10\sqrt{2}$ cm
C. 14 cm
D. $2\sqrt{10}$ cm

54. What is the slope of a line
4.12.5 perpendicular to a line having a slope $\frac{3}{4}$?

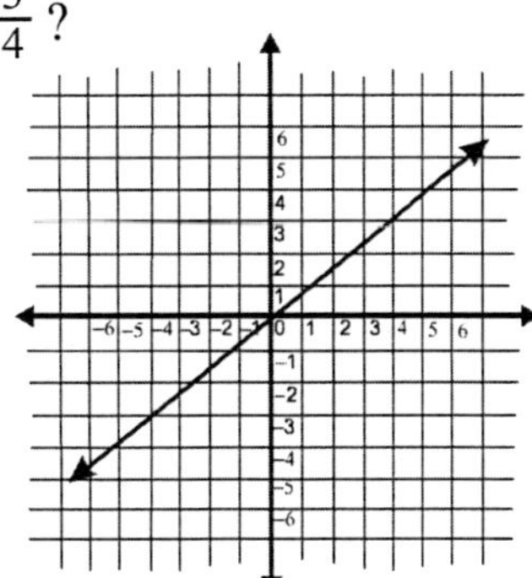

A. $-\frac{4}{3}$

B. $-\frac{3}{4}$

C. $\frac{3}{4}$

D. $\frac{4}{3}$

55. For the following question, use this
3.12.5 diagram:

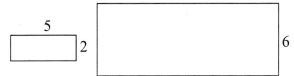

These are similar rectangles. How long is the long side of the larger rectangle?

A. 5
B. 8
C. 15
D. 30

56. Every square has four sides. The shape
4.12.9 I drew is a square. This square that I drew has four sides.
This is an example of

A. inductive reasoning.
B. deductive reasoning.
C. analytical reasoning.
D. none of the above.

57.
3.12.4

Rubberized Radial Tire Sale		
Item	**Warranty**	**Price**
15"	2 year	$39.00
	5 year	$46.00
16"	2 year	$45.00
	5 year	$52.50
17"	2 year	$54.00
	5 year	$60.50
18"	2 year	$63.00
	5 year	$70.00

From the table above, find the price for buying one 16 inch, one 17 inch, and one 18 inch tire each with 2 year warranties.

A. $162.00
B. $229.00
C. $63.00
D. $201.00

58. Alice has 3 hats, 5 blouses, and 4
5.12.4 skirts. How many different outfits can she make?

A. 12
B. 24
C. 60
D. 120

59. Solve:
2.12.2

$$\frac{3x+6}{-2} > -12$$

A. $x < 24$
B. $x > 0$
C. $x > 6$
D. $x < 6$

60. Which of the following tables of values
2.12.6 does not represent a linear function?

A.

MONTHLY RENT

Bedrooms	Rent
1	$550
2	$625
3	$700

B.

x	f(x)
−4	−8
0	0
4	8
8	16

C.

Time	Odometer Reading
7 AM	20825
11 AM	20965
3 PM	21115

D.

COST OF REPAIR

Hours	Cost
1	$72.50
2	$95.00
3	$117.50
4	$140.00

Math Practice Exam 2: Part 1

1.
3.12.4

PAXTON FAMILY

Month	Electric Bill
January	$89.15
February	$99.59
March	$78.99
April	$72.47
May	$99.23
June	$124.69

Look at the chart above. What is the median electric bill for the Paxton family from January through June?

A. $ 55.22
B. $ 94.02
C. $ 94.19
D. $ 99.23

2. Darin is playing a dart game at the
5.12.5 county fair. At the booth, there is a spinning board completely filled with different colors of balloons. There are 6 green, 4 burgundy, 5 pink, 3 silver, and 8 white balloons. Darin aims at the board with his dart and pops one balloon. What is the probability that the balloon popped is **not** green?

A. $\frac{1}{13}$

B. $\frac{10}{13}$

C. $\frac{3}{13}$

D. $\frac{2}{13}$

3. The student council surveyed the
5.12.2 student body on favorite lunch items. The frequency chart below shows the results of the survey.

Favorite Lunch Item	Frequency
corndog	140
hotdog	210
hamburger	245
pizza	235
spaghetti	90

Which lunch item indicates the median of the data?

A. pizza
B. hotdog
C. hamburger
D. corndog

4. In a family of 4 children, what is the
5.12.5 probability that all four will be girls?

A. $\frac{1}{4}$

B. $\frac{1}{8}$

C. $\frac{1}{16}$

D. $\frac{1}{24}$

5\. 2.12.2 Solve for x in the following equation.

$$\frac{6x - 40}{2} = 4$$

A. 6
B. $5\frac{1}{3}$
C. 8
D. $7\frac{1}{3}$

6\. 2.12.2 $7(2x + 6) - 4(9x + 6) < -26$

A. $x > -2$
B. $x > 2$
C. $x < -2$
D. $x < -1$

7\. 1.12.6 $\sqrt{77}$ lies between

A. 7 and 8
B. 8 and 9
C. 76 and 78
D. 5 and 6

8\. 4.12.5 Which of the following is an equation of a line that is perpendicular to the line l in the graph?

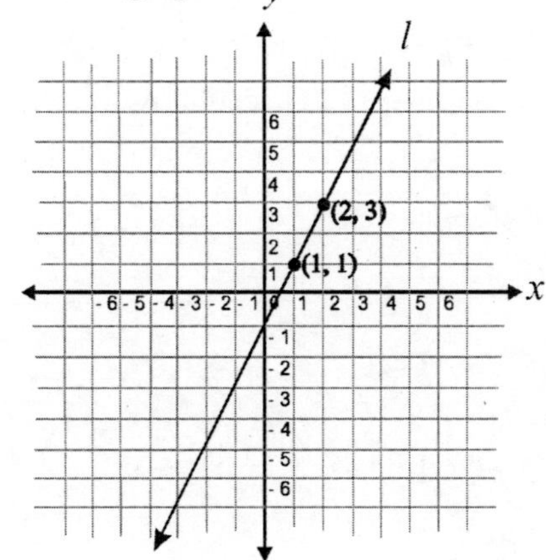

A. $x - 2y = -4$
B. $x - 2y = 4$
C. $x + 2y = 4$
D. $2x + y = 4$

9\. 3.12.5 Cynda wants to buy a daycare center with the measurements below. How many square feet are in the building?

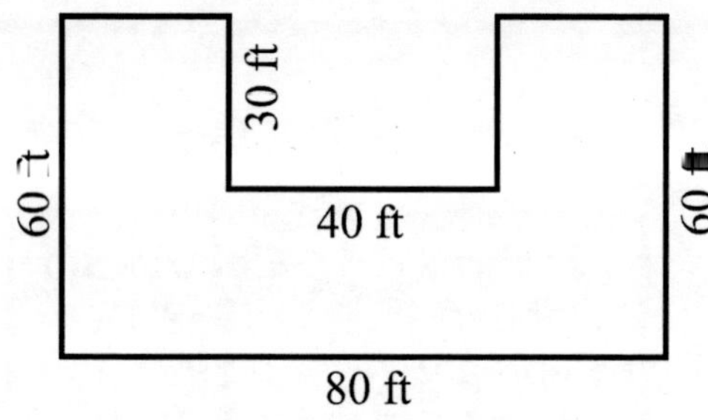

A. 340 square feet
B. 480 square feet
C. 3,600 square feet
D. 4,800 square feet

10\. 2.12.3 Four students have attempted to simplify a mathematical expression. They have four different answers. Which of the answers below is equivalent to the expression?

$2(a + 3b) - 4(3a - b) - (5a + 4b)$

A. $-15a + 6b$
B. $-17a + 9b$
C. $-9a - 2b$
D. $-9a + 9b$

11\. 4.12.6 In the figure below, the measure of $\angle 3$ is 60°. What is the measure of $\angle 1$ and $\angle 4$?

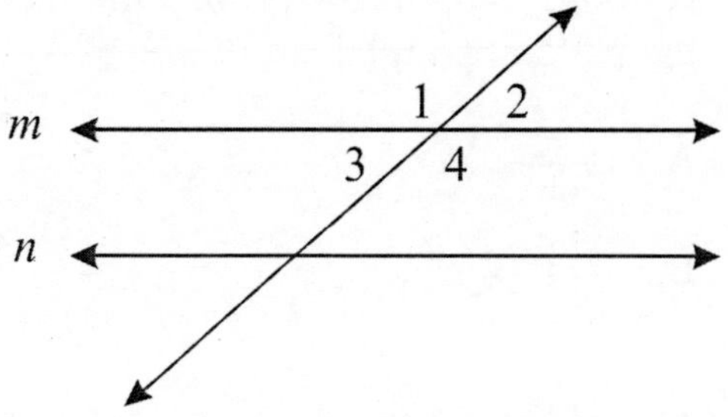

A. 240°
B. 200°
C. 120°
D. 100°

12. Jewel told her friends, "Pick a number
1.12.6 between 0 and 10." They guessed 2, 4, 5, and 8. Jewel said, "The number was $\sqrt{14}$." Which guess was closest?

A. 2
B. 4
C. 5
D. 8

13. If you have a 6 inch cube and decide
3.12.5 when you make the next cube, you are going to double the length of each side, how will the volume be affected?

A. The volume will be 3 times larger.
B. The volume will be twice as large.
C. The volume will be 8 times larger.
D. The volume will be 9 times larger.

14. If the equation below were graphed,
2.12.6 which of the following points would lie on the line?

$$4x + 7y = 56$$

A. (7, 4)
B. (0, 14)
C. (8, 0)
D. (4, 7)

15. In the right triangle below, what is the
4.12.2 value of $\angle x$?

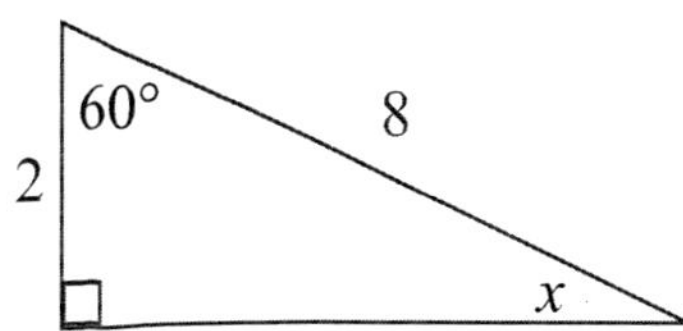

A. 30°
B. 40°
C. 50°
D. 60°

16. Simplify: $\sqrt{32} \times \sqrt{44}$
1.12.7

A. $\sqrt{1408}$
B. 1408
C. $8\sqrt{22}$
D. $2\sqrt{352}$

17. Which of the following is the correct
2.12.4 solution set for the problem below?

$21 + 7|a| \geq -14$

A. {–5, –4, –3, –2, –1, 0, 1, 2, 3, 4, 5}
B. {–4, –3, –2, –1, 0, 1, 2, 3, 4}
C. {1, 2, 3, 4}
D. {–4, –3, –2, –1, 0}

18. The right circular cylinder shown below
3.12.5 has a radius of 5 cm and height of 12 cm.

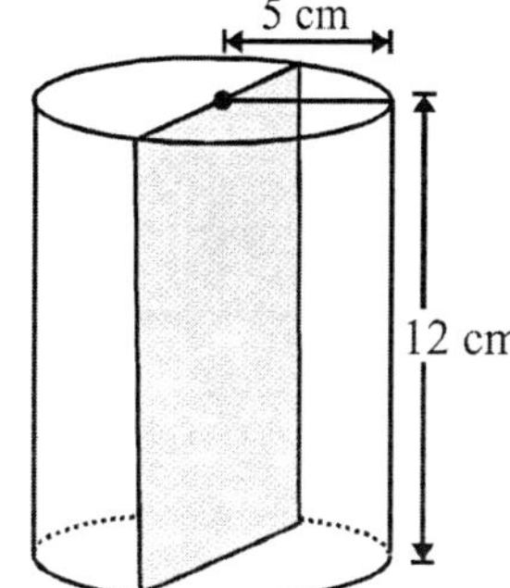

If the cylinder is split in half by the shaded cross section, what is the area of the cross section?

A. 60 cm^2
B. 94.2 cm^2
C. 120 cm^2
D. 942 cm^2

19. In the triangle below, what is the
4.12.7 length of side x?

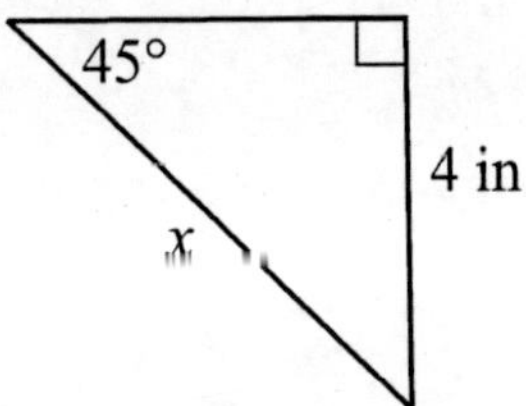

A. $4\sqrt{2}$
B. 5 in
C. $\sqrt{8}$
D. 8 in

20. Solve: $y^2 - 4y - 12 = 0$

2.12.6 A. $\{2, -6\}$
B. $\{-2, 6\}$
C. $\{3, -4\}$
D. $\{-3, 4\}$

21.
4.12.1

Oakwood High School	
Seniors	225
Juniors	225
Sophomores	250
Freshmen	300

If you wanted to display the number of students in each class on a circle graph, what would be the measure of the central angle representing the freshmen?

A. 108°
B. 83°
C. 80°
D. 90°

22. Which ordered pair is a solution for the
2.12.5 following system of equations?

$-3x + 7y = 25$
$3x + 3y = -15$

A. $(-13, -2)$
B. $(-6, 1)$
C. $(-3, -2)$
D. $(-20, -5)$

23. How many subcommittees of 3
5.12.4 members can be formed from a club with 10 members?

A. 30
B. 120
C. 720
D. 3000

24. Jane has 5 different playing cards in
5.12.4 her hand. In how many different orders can the cards be arranged?

A. 5
B. 25
C. 120
D. 240

25. Which of these is the equation that
2.12.1 generalizes the pattern of the data in the table?

x	f(x)
−3	−5
−1	1
2	10
5	19

A. $f(x) = 3x$
B. $f(x) = x + 3$
C. $f(x) = 2x + 6$
D. $f(x) = 3x + 4$

26. Find the range of the following
2.12.4 function for the domain $\{-2, -1, 0, 3\}$.

$$y = \frac{2 + x}{4}$$

A. $\{0, \frac{3}{4}, 1, 1\frac{1}{4}\}$

B. $\{0, \frac{1}{4}, \frac{1}{2}, 1\frac{1}{4}\}$

C. $\{1, -\frac{1}{4}, \frac{1}{2}, 1\frac{1}{4}\}$

D. $\{\frac{1}{4}, \frac{3}{4}, \frac{1}{2}, 1\frac{1}{4}\}$

27. What are the factors of $x^2 + 48x + 135$?
2.12.3

A. $(x + 3)(x + 45)$

B. $(x + 5)(x + 27)$

C. $(x + 9)(x + 15)$

D. $(x - 3)(x + 45)$

28. $\sqrt{96} =$
1.12.6

A. $4\sqrt{6}$
B. $2^5 \times 3$
C. $\sqrt{90} + \sqrt{6}$
D. $96\frac{1}{2}$

29. The Sweet Shoppe has 6 flavors of ice
5.12.4 cream, 4 toppings, and 3 kinds of sprinkles. How many different sundaes can be made using one flavor of ice cream, one topping, and one kind of sprinkles?

A. 13
B. 19
C. 36
D. 72

30. In the right triangle below, what is the
4.12.7 value of x?

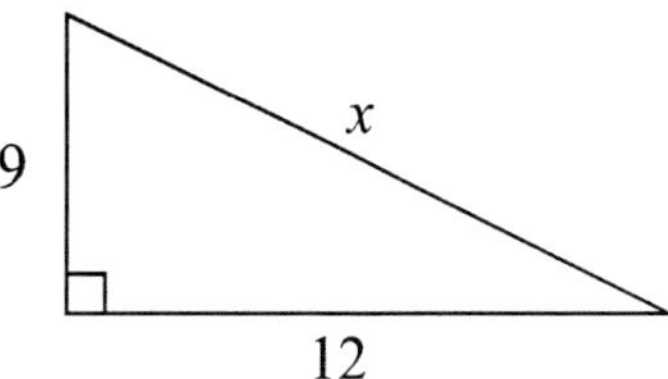

A. 15
B. 225
C. 112.5
D. 36

Math Practice Exam 2: Part 2

31. A box has a volume of 1440 cubic
3.12.5 inches, a height of 10 inches, and a square base. What is the length of one side of the base?

A. 6 inches
B. 12 inches
C. 24 inches
D. 48 inches

32. Simplify: $\sqrt{64}$
1.12.6

A. 4
B. 8
C. 16
D. 32

33. Which of these graphs represents
2.12.4 $-2 \leq x < 3$?

A. −6 −5 −4 −3 −2 −1 0 1 2 3 4 5 6

B. −6 −5 −4 −3 −2 −1 0 1 2 3 4 5 6

C. −6 −5 −4 −3 −2 −1 0 1 2 3 4 5 6

D. −6 −5 −4 −3 −2 −1 0 1 2 3 4 5 6

34. If the measure of $\angle 2$ is 50°, what is the
4.12.6 measure of $\angle 4$?

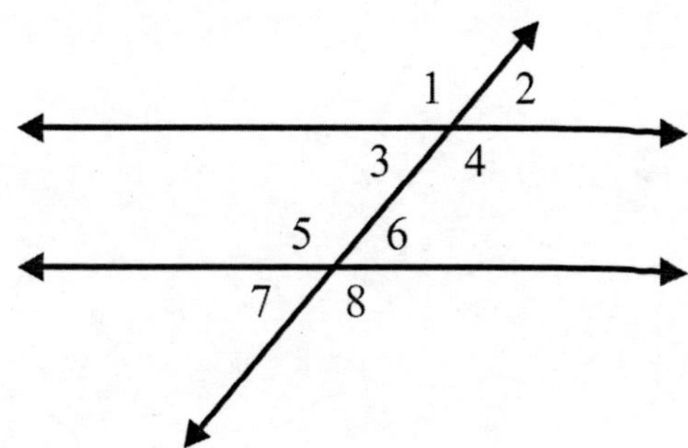

A. 50°

B. 100°

C. 130°

D. Not enough information given.

35. What is the slope of the line in the graph below?
4.12.5

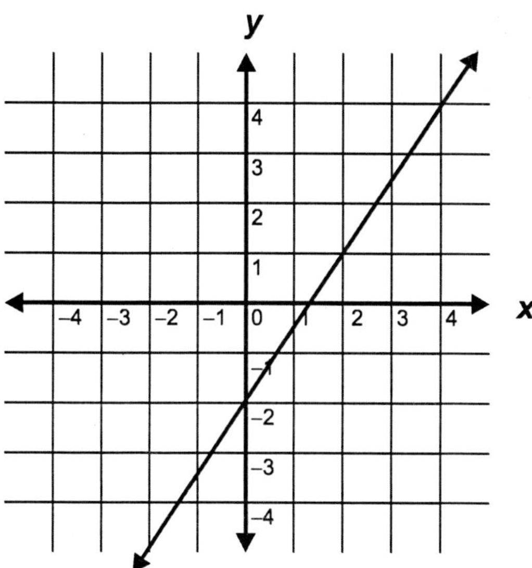

A. 3
B. −3
C. $\frac{2}{3}$
D. $\frac{3}{2}$

36. The box-and-whisker plot below shows the quiz scores of 20 history students.
5.12.2

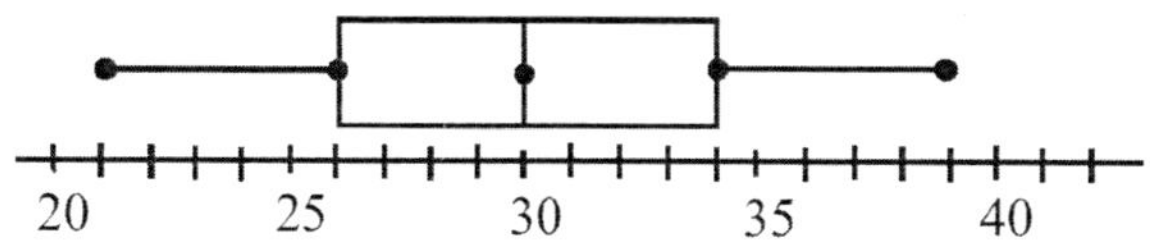

Which of the following statements is NOT a correct interpretation of the box-and-whisker plot?

A. The median quiz score was 30.
B. Approximately 5 students had quiz scores of 34 or above.
C. The lowest quiz score was 21.
D. Approximately half of the students had a quiz score of 30.

37. Simplify $(5x - 4)(x + 2)$
2.12.3

A. $5x^2 - 2$
B. $5x^2 - 8$
C. $5x^2 + 6x - 8$
D. $5x^2 - 18x - 8$

38. The local hardware store has four different style grills for the summer sale. The price of the four grills are listed from least to greatest, and they are represented in the 2 × 2 matrix below.
2.12.5

$$\begin{bmatrix} \$120 & \$155 \\ \$160 & \$230 \end{bmatrix}$$

If the store is having a sale where everything is 20% off, what is the new price of the grill that is the second most expensive?

A. $32
B. $124
C. $184
D. $128

39.
5.12.1

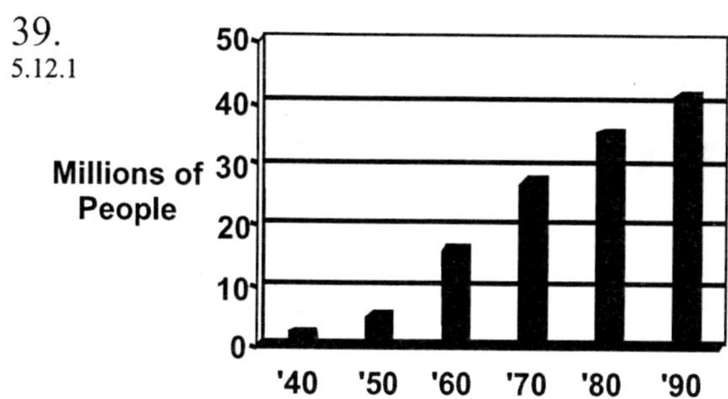

How many people received social security checks in 1975?

A. 32 people
B 32,000 people
C. 320,000 people
D. 32,000,000 people

40. 2.12.5 Which ordered pair is a solution for the following system of equations?

$$5x + 7y = -15$$
$$4x - 6y = 46$$

A. (4, –5)

B. (–3, 0)

C. (5, –5)

D. (–1, –7)

41. 4.12.6 What is the sum of the measures of the interior angles of a pentagon?

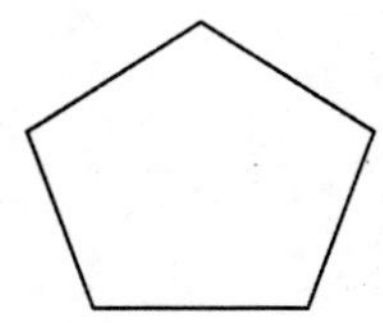

A. 50°

B. 360°

C. 420°

D. 540°

42. 1.12.8 Which property is used in the following equation?

(3 + 12) + 17 = 3 + (12 + 17)

A. distributive
B. commutative
C. identity
D. associative

43. 3.12.5

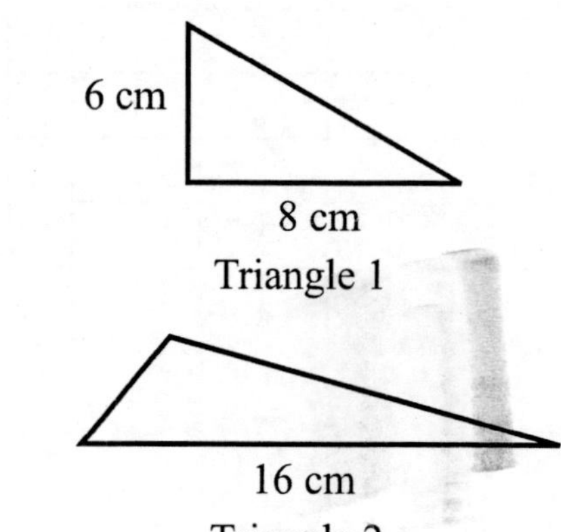

The areas of the triangles pictured above are equal. What is the height of the second triangle? (Figures are not drawn to scale.)

A. 1 cm
B. 3 cm
C. 4 cm
D. 5 cm

SOCKS-A-PLENTY WAREHOUSE		
Item	**3 pair price**	**6 pair price**
Anklets	\$4.50	\$ 8.50
Sport Socks	\$6.50	\$11.50
Support Socks	\$9.00	\$15.75
Dress Socks	\$6.00	\$11.25
Please add 10% for shipping and handling		

44. What would be the total cost of 18 pairs of Sport Socks and 12 pairs of Anklets?
3.12.4

A. \$ 56.65
B. \$ 51.50
C. \$ 57.00
D. \$ 61.60

45. The two trapezoids below are similar.
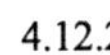
4.12.2

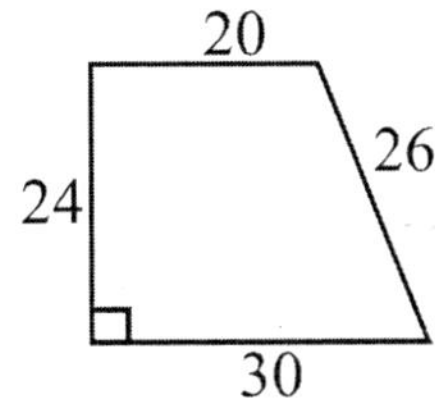

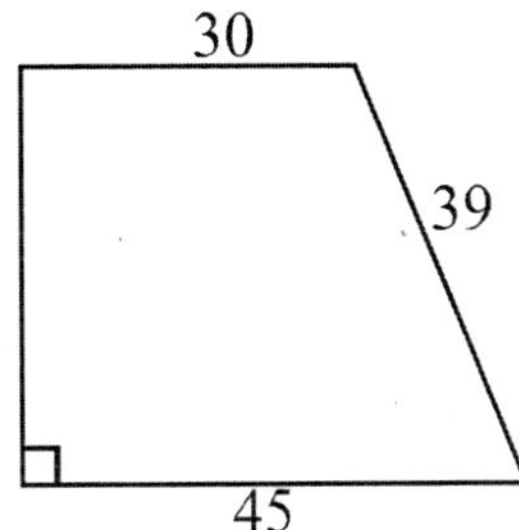

What is the height of the larger trapezoid?

A. 16
B. 30
C. 32
D. 36

46. Solve $ab + cd = 20$ for b.
2.12.2

A. $b = 20 - cda$
B. $b = \dfrac{20 + cd}{a}$
C. $b = \dfrac{20 - cd}{a}$
D. $b = 20cda$

47. For the following pair of equations, find the point of intersection (common solution) using the substitution method.
2.12.5

$$3x + 3y = 9$$
$$9y - 3x = 6$$

A. (1, 2)
B. $(1\frac{3}{4}\ 1\frac{1}{4})$
C. (1, 1)
D. $(\frac{1}{3}, \frac{1}{6})$

48. What speed would a plane average on a trip of 4000 miles for 6 hours?
3.12.3

A. 150.0 miles per hour
B. 666.6 miles per hour
C. 366.6 miles per hour
D. 1500.0 miles per hour

49. Factor: $b^2 - 2b - 8$
2.12.3

A. $(b - 4)(b + 4)$
B. $(b - 2)(b + 4)$
C. $(b + 2)(b - 4)$
D. $(b - 2)(b - 2)$

50. Solve $a^2 + 2a - 8 = 0$.
2.12.6

A. (–2, –4)
B. (4, –2)
C. (–4, 2)
D. (2, 4)

51.
5.12.2
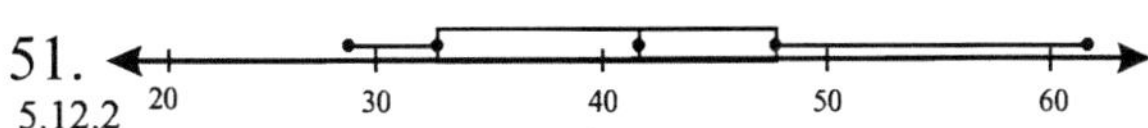

What is the range in the box and whisker graph above?

A. 33
B. 40
C. 42
D 48

52. Solve: $10 + 3(2x - 6) \leq 4(7 - 2x)$
2.12.3

A. $x \leq -1$
B. $x \leq \frac{18}{7}$
C. $x \leq -10$
D. $x \leq 10$

53. What is the slope of a line parallel to a
4.12.5 line represented by the equation
$x = 2y - 3$?

A. -2
B. $-\frac{1}{2}$
C. $\frac{1}{2}$
D. 2

54. What is the range of the function
2.12.4 $y = 2x - 8$ for the domain
$\{10, 11, 12, 13\}$?

A. $\{9, 9\frac{1}{2}, 10, 10\frac{1}{2}\}$
B. $\{5, 6, 7, 8\}$
C. $\{9, 10, 11, 12\}$
D. $\{12, 14, 16, 18\}$

55. Which of the following computations
1.12.8 will result in an irrational number?

A. $8\sqrt{5}$
B. $1\frac{1}{2} \div \frac{2}{3}$
C. $2.4 \div 3.1$
D. $7 + 3.7$

56. Which line is a secant of the circle?
4.12.1

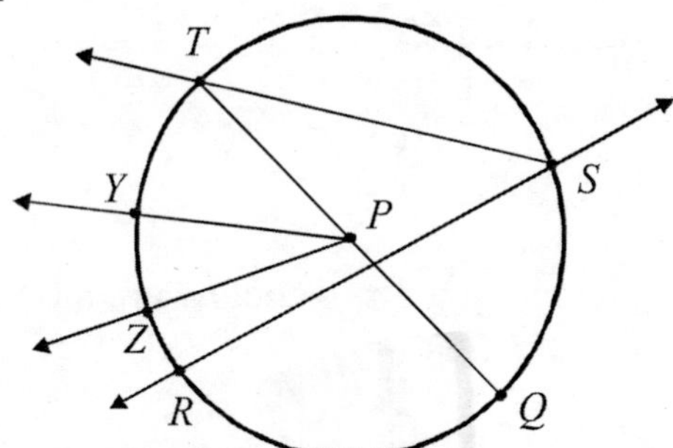

A. $\overline{RS}$
B. $\overrightarrow{PZ}$
C. $\overline{QT}$
D. $\overleftrightarrow{RS}$

57. Adrianna has 4 hats, 8 shirts, and
5.12.4 9 pairs of pants. Choosing one of each, how many different clothes combinations can she make?

A. 288
B. 21
C. 48
D. 96

58. Which of the following is the most
3.12.2 precise measurement?

A. $3\frac{1}{4}$ cups

B. $5\frac{1}{8}$ cups

C. $7\frac{1}{2}$ cups

D. 8 cups

59. Quinton measured the length of a wall
3.12.2 with two different tape measures. Tape 1 measured 15 ft 3 in. Tape 2 measured 15 ft 3.5 in. What is the tolerance between the two measures?

A. ± 0.5 in.
B. ± 1 in.
C. ± 0.25 in.
D. ± 0.05 in.

60. Laura is seen walking from her car to
4.12.9 her home with a brand new baseball bat. Laura's husband Eric loves baseball and tomorrow is his birthday. Therefore, Laura has bought a new baseball bat for Eric.

This is an example of what kind of reasoning?

A. deductive
B. inductive
C. neither
D. not enough information given